Solitude of a Humble Genius - Gregor Johann Mendel: Volume 1

The Unfathomable Mendel

Jan Klein • Norman Klein

Solitude of a Humble Genius - Gregor Johann Mendel: Volume 1

Formative Years

Edited by Paul Klein

Springer

Jan Klein, Ph.D
State College
PA 16801
USA

Norman Klein, M.A. (Master of Arts)
Key Biscayne
FL 33149
USA

Additional material to this book can be downloaded from http://extras.springer.com

ISBN 978-3-642-35253-9 ISBN 978-3-642-35254-6 (eBook)
DOI 10.1007/978-3-642-35254-6
Springer Heidelberg New York Dordrecht London

Library of Congress Control Number: 2013948128

Printed on acid-free paper

Springer is part of Springer Science+Business Media (www.springer.com)

We dedicate this work to Dr. Anna Matalová in recognition of her extraordinary contribution to Mendel scholarship. As a continuator of the Czech tradition of Mendel study founded by Jaroslav Křiženecký and Vítězslav Orel, Dr. Matalová was the head of the renowned Mendelianum of the Moravian Museum at Brno until her retirement. She served as the Editor-in-Chief of the Folia Mendeliana, a journal devoted exclusively to the study of Mendel. She organized several Mendel Forums, meetings that brought together Mendel experts from different parts of the world. She shared graciously her knowledge of Mendel, which is second to none, with many pilgrims to the founding place of genetics. Above all, however, she has made numerous portentous contributions revealing new aspects of Mendel's life and work. We owe Dr. Matalová the stimulus that brought us on the path to this present work. She has infected us with her enthusiasm for Mendel for which we are extremely grateful.

About the Authors

Jan Klein is the director emeritus of the Max Planck Society for the Advancement of Science and Frances R. and Helen M. Pentz Visiting Professor of Science, Department of Biology, Pennsylvania State University. Several happenstances seem to have predestined him to become ultimately Mendel's biographer. He was born in a small Silesian village, a mere cycling distance from Hynčice, Mendel's birthplace. Like Mendel, he also grew up on small farm and studied at the same school as Mendel, the *Gymnasium* in Opava. He earned a PhD in genetics, to which he then devoted his research career. And he is the recipient of two Mendel medals, one awarded by the Moravian Museum at Brno and the other by the Czech Academy of Sciences at Praha.

Norman Klein is a freelance illustrator, who holds a Master of Fine Arts degree in Illustration and Animation with a Minor in Graphic Design and Art History. He lives in Miami, Florida.

Acknowledgments

This book would not have come into existence without the help of many individuals and some institutions. Below we name some of these and specify for what we are particularly thankful to them.

NK

I would like to thank my parents, who have made all this possible, not just in the obvious sense, but through their unwavering support and belief. My mom, who, if there are any saints walking this earth, is surely one of them. My dad, for giving me the opportunity to come on board this unique project and for pushing me to new heights of endeavor.

My brothers, Daniel and Paul, without whose support, encouragement and sense of humor, I would have packed it in long ago. . .Even if one of them is an Arsenal fan!

And Sharokina, for providing welcome opinions and research, over the many years this project has been going on. . .Not to mention putting up with the artistic temperament!

Thank you all, from the bottom of my heart.

JK

I thank:

Paul Klein for his gentle editing and for movie tips as an antidote to an overdose of Mendel. Dr. Dagmar Klein, University of Miami, Miami, FL, for consultations concerning Moravian cuisine. Drs. *Akie Sato* and *Noriyuki Kuroda* of Tsurumi University, Japan, for help in taming my Mac; and *Akie* also for her friendship, discussions, and assistance in the preparation of the manuscript. *Milada Horáková* and Ing. *Zdeněk Horák*, Praha, for their friendship, hospitality, searches for hidden information, and connections. Prof. *Masatoshi Nei*, Penn State University, for never-ending debates and for access to his office facilities. Drs. *Nikolas Nikolaidis* and *Dimitra Chalkia*, the former of the California State University Fullerton, for their friendship and disputations about Greek, Greeks, and Greece, but not only. Mgr. *Antonio Rivas*, Prior of the St. Thomas Abbey, Praha, for his live demonstration of the Augustinian habit, answering many questions, and for a Catalogue listing

Augustinian friars at Staré Brno in Mendel's time. Dr. *Josef Mrowetz*, Stuttgart, for friendship, help in interpreting Hölderlin, Janáček, and other conundrums. *Lynne Yakes*, Tübingen, for providing a somewhat different interpretation of Hölderlin and for editorial assistance in the prehistoric phase of this book's inception. Dr. *Herbert Tichy*, Tübingen, for information about Mendels in southern Germany; *Daniela Čáslavská*, for reminding me how I once was and for help in a search for Alžběta Rejčka's likeness. Dr. *Eva Mrhačová*, Ostravská univerzita, the fourth of the three musketeers, who once studied at Mendel's *Gymnasium* in Opava, for her help in sorting out Mendel's ancestors, and for information about Mendel that was new to me. Prof. *Milan Myška*, Ostravská univerzita, for his works on the history of Opava's *Gymnasium*. Drs. Colm Óh Uigin and *Tomoko Steen*, the former at NCI-Frederick and the latter of the Library of Congress, Washington, DC, for listening to my endless discourses on Aristotle and Mendel, and for helping me to access important material; Colm also estimated the degree of Slavic admixture in Mendel's genome. And *Val Berzins*, for reading and criticizing nearly all the chapters.

My special thanks go to the staff of the *Penn State University Libraries*. Never before have I depended so much on a library as when writing this book. And never before have I encountered a library that responded so perfectly to my needs as this one: well organized, efficient, friendly, and helpful—a pleasure to deal with. I particularly value the service of the Interlibrary Loan Department, which I often challenged with a request that was not easy to fulfill; yet it always acquitted itself of the task with flying colors. Many thanks for the help!

JK, NK, and PK

Retracing Mendel's travels along the Amber Route, we were assisted at the different stations by persons to whom we would like to express our gratitude.

Vražné/Hynčice: Ing. *Vladimír Nippert,* mayor, Obecní úřad, for enabling us to visit the Mendel estate while in the process of reconstruction and for answering many questions about Mendel's native house and the region.

Lipník nad Bečvou: *Božena Musilová*, librarian, Městská knihovna, for taking us on a memorable tour through the city, including braving a climb of a dilapidated church tower; and for her information about the Piarist school in Mendel's time.

Opava: *Jana Indrová*, librarian, Zemský archiv, for letting us examine the parish register recording the births of Johann Mendel and his four sisters. Ing. *Antonín Klein* and *Milena Kleinová* for old newspaper articles on Mendel.

Olomouc: Prof. *Vítězslav Bičík*, Palacký University, for walking us through the city and arranging appointments for us. Dr. *Pavel Urbášek*, head of the University Archive, for access to the *General Katalog der Schüller*, photocopies of the relevant pages, and for answering our questions. Prof. *Jiří Fiala*, Palacký University, for consultation and publications regarding the university's history.

Brno. Dr. *Anna Matalová*, for a tour of the *Mendelianum* when it was still located in the St. Thomas Abbey at Staré Brno, for showing us the Mendel relics, and for supplying us with fundamental Mendelian literature. Dr. *Jiří Sekerák*, for keeping open our channel to the new *Mendelianum* and so to the original sources.

And Ing. *Radan Květ*, for being our guide through the city, for his perduring interest in our writing, and for reading and commenting on our manuscript-*Vienna*: Dr. Robert Stangl, *Universitätsbibliothek, Institut für Botanik und Botanischer Garten der Universität Wien*, for a photograph of the *Botanisches Museum.*

Special thanks are also due to Dr. *Andrea Schlitzberger* of the Springer-Verlag for being so patient and not giving up on us.

Prologue

Si puo? Signore, Signori[1]—as we present this book to the reader, we are reminded of a story about the Moravian composer Leoš Janáček. It is said that one day he came into a bookstore in Brno demanding "that history of the Bohemian brewery." Perplexed and not knowing about the existence of such a book the apprentice consulted the owner of the establishment. The proprietor, familiar with the maestro's sense of humor, thought only for a short while and then pulled out from the stacks a four-volume biography of Bedřich Smetana by a Czech musicologist known for his tendency to cover auxiliary material at great depth. After the maestro left the store, the proprietor explained to the astonished apprentice that Janáček alluded ironically to the fact that the musicologist devoted much of the first volume of his biography to the history of Bohemian beer brewing because several generations of Smetana's ancestors happened to be brewers.

The present book is not about brewing beer—but booksellers beware: If customers ask for treatises on Aristotle, the history of Europe, or the life of Silesian peasants, they might be referring facetiously to our book, for we deal with these topics to an extent that some readers might find disproportionate. We do not apologize for casting our nets so broadly for we intend to catch fishes that might have eluded other biographers. We use the broad approach, for example, to point out that Mendel really starts where Aristotle left the subject some 2,000 years ago and thus give the proper perspective on the achievements of both Aristotle and Mendel. Or, to drive home the message that Mendel was neither of German, Austrian, or Czech nationality, as various writers claim depending on their own nationality, but a Silesian. Suspecting that you might not know much about Silesia, we try to give you a taste of the complexities involved in the formation of present-day European nations. This particular chapter also serves to demonstrate the falsity of the assumption that language-based nationality classifications always mirror the genetic compositions of the nations involved. And the third case of our broad net-casting—the detailed incursion into Mendel's youth and his peasant roots—is meant to dispel

[1] "By your leave, Ladies and Gentlemen." Tonio in Ruggiero Leoncavallo's *I Pagliacci*.

the common perception that his rustic origin has disadvantaged his intellectual development. We argue, on the contrary, that it endowed him with a healthy dose of rationality, which made him immune to the Romantic fever of the century into which he was born. It enabled him to see through the Romantic haze and thus to keep his feet firmly on the ground, when all around him others were engaging in flights of fancy. Alas, the same endowment alienated him intellectually from the establishment with the result of a nearly total lack of reaction to his discovery.

We think that the three words we have chosen for the title of our book—solitude, humbleness, and genius—characterize Mendel best. By "solitude" we do not mean isolation in terms of social interactions for Mendel's behavior showed no such tendency. On the contrary, his colleagues at the Abbey in which he lived and the schools in which he taught, as well as the student whom he taught all perceived him as a congenial and amiable person. It was through his research that he had ended up being alone, without a single person who could understand the direction on which he set out. This solitude, which lasted for the last 30 years of his life, was exacerbated by social isolation in his last decade, and then continued, after his death, until the end of the century. It looked as if the world would never learn about his discovery.

On Mendel's humbleness agree all those who had known him and whose testimony has been recorded. Their characterization of Mendel is not a mere charity to the deceased *à la de motuis nihil nisil bonum* (of the dead nothing but good) for it is supported by all the facts we know about his life. Indeed, the long neglect of his discovery supports Mendel's humbleness best. There are historians who argue that had Mendel been cognizant of discovering the laws of heredity, as they are now commonly attributed to him, he would have said so and would have advertised vigorously his discovery. Since he does not mention any such laws and does not even use the word "heredity" in his main works, he was, according to these historians, apparently unaware of the significance of his results. These modern-day critics, living at times which consider self-promotion a virtue and invasively malignant advertisement a good thing, do not seem to understand that Mendel was brought up in a family in which the dictum "self-praise stinks" was part of their moral code. He therefore must have had an aversion toward dishonest generalization of what his data revealed. It was his humbleness that restricted his interpretations to what his data actually demonstrated.

Finally, by "genius" we do not mean the Romantic vision of a demonic individual bearing his head above the clouds and uttering bits of wisdom to the commoners as if casting pearls before swine. Mendel was not at all of this type; his genius was cryptic, hidden to the extent that none of his contemporaries might have thought of him as being a genius. Indeed, some of the historians mentioned above deny Mendel being a genius and try to present him instead as a lucky fumbler who did not know what he was doing and by chance arrived at results whose meaning he did not fully grasp. This, however is a minority view held by scholars who apparently have only a superficial knowledge of Mendel. By contrast to them, all scientists who have read Mendel's *magnum opus* have been awe-stricken by it and hold it for a work of a genius. We expand on all these points in the text at the appropriate places.

To appreciate fully Mendel's contribution and the greatness of his genius it is necessary to understand what he did, how he did it, and to view his work in the context of what was known and believed in his time. It is for this reason that in this book we place so much emphasis on providing the necessary backgrounds and contexts wherever they might help to understand the issues involved.

A few *technical comments* regarding this book: The text of each chapter is divided into sections and subsections, which will enable readers to choose parts they want to read and others they might want to skip. Nearly all *figures* are hand-drawn by N.K. They are either original or based on old anonymous prints; where the author of the original is known, proper attribution is given. An additional figure (Fig. S1) appears as *supplementary material* online and can be downloaded from http://extras.springer.com/2013/978-3-642-35253-9.

State College, January 2013 Jan Klein
Key Biscayne, January 2013 Norman Klein

Contents

Contents of Volume 2

1 Prelude: Heredity, Sex, and Species: The Greek View

"Τηλέμαχ', οὐδ' ὄπιθεν κακὸς ἔσσεαι οὐδ' ἀνοήμων·
εἰ δή τοι σοῦ πατρὸς ἐνέστακται μένος ἠΰ,
οἷος κεῖνος ἔην τελέσαι ἔργον τε ἔπος τε,
οὔ τοι ἔπειθ' ἁλίη ὁδὸς ἔσσεται οὐδ' ἀτέλεστος.

Homer: *Odyssey*[1]

October has a special significance to the modern scientist, because in this month the Karolinska Institutet in Stockholm announces the year's winners of the Nobel Prizes in three scientific disciplines (as well as in other fields)—medicine and biology, chemistry, and physics. Those scientists who believe that they have made breakthrough discoveries in one of these disciplines await the announcements with hope and trepidation, all others with curiosity. For although there are other awards that recognize the significance of scientific discoveries, none of them carry the prestige that a Nobel Prize does. The accolade is accompanied by great media interest, which then usually lasts until the actual awards ceremony in December. The laureates, however, continue to enjoy a special status among their peers afterward, which often leads to a small avalanche of other awards. They also become adornments to the institutions with which they are affiliated, as well as to their native towns and nations. Outspoken laureates become media gurus, to whom journalists like to turn to for their comments on a variety of political, social, and scientific issues. They remain in the limelight for as long as they are willing to cooperate with the news hunters. For the rest of the laureates, the limelight fades gradually. Nevertheless, they are assured of immortality, even if it may only be restricted to a mentioning of a name followed by a few explanatory lines in a larger encyclopedia. For fame is fickle and the memory of humankind proverbially short—and it is not too difficult to understand why. Nobel Prizes in the three disciplines mentioned have been awarded yearly, with a few exceptions, since their inception at the beginning of the twentieth century. As there are one to three laureates in each discipline each year, in the more than 100 years of award giving, the awardees have grown into a small crowd. Who could remember all their names and accomplishments? Even the practitioners of the three disciplines can at best name fewer laureates than they have fingers on one

J. Klein and N. Klein, *Solitude of a Humble Genius - Gregor Johann Mendel: Volume 1*,
DOI 10.1007/978-3-642-35254-6_1,

hand. And so all we can expect the active memory of humankind to retain are but a few names that stick out far above the Nobel Prize standard. These are the names of scientists, whose discoveries have changed or have led to a change in the way that humanity views the world. They are scientists like Albert Einstein, Max Planck, and Niels Bohr in the twentieth century; Charles Darwin, Alfred Russel Wallace, and Gregor Johann Mendel in the nineteenth century; Isaac Newton in the eighteenth century; Galileo Galilei in the seventeenth century; and Nicolaus Copernicus in the sixteenth century, when modern science began to emerge.

The inclusion of Mendel in this absolutely top class of scientific giants might surprise some readers, who may be used to thinking of him as a good-natured, pious monk, toiling for years in his small garden, crossing pea plants, until he stumbled upon the observation that their characters segregated at specific ratios. We shall argue in this book that this portrayal of him is nothing more than a myth. We shall argue also against the slander that he cheated, as some biographers have declared, and against the variety of postmodernist claims of Mendel not being a Mendelian (carrying his experiments to disprove Darwin, not carrying any experiments at all, and so on and so forth). We shall show all of these claims to be nonsensical, due to those authors' insufficient knowledge of Mendel's work and of the circumstances under which he labored. We shall show Mendel as being aware of the implications of his discovery, which did nothing less than overturn the more than 2000 year long dominance of the Aristotelian view of heredity and replace it with a modern corpuscular view. But before we turn to Mendel, his life, and his work, we must explain what exactly this old view was and why it prevailed until Mendel's time. What follows will not be easy to read, for it will take us to the heart of Aristotle's philosophy. Hopefully, a reader who perseveres through these difficult parts will come out rewarded with an understanding of the background against which Mendel's achievement must be pitted in order to grasp its real significance. But first a cartoonist's view of the central issue.

Heredity Counter Generation

On a sunny Sunday afternoon a young couple strolls through a park with their newborn son in a baby carriage. As they meet a family friend, he leans over the carriage and exclaims: "How cute! He looks just like his father!" (Fig. 1.1). This scene, which must have played itself out time and again in various versions through the ages, epitomizes one of the most profound mysteries of life: the mystery of *generation* or *reproduction*. These two words derive from the Latin verbs *generare* and *producere*, respectively, both of which mean, "to bring forth," "to give rise," "to bring into being," "to beget," "to procreate," or "to give birth." The addition of the prefix *re-* to *producere* emphasizes a second meaning of both words, namely, that besides the act of bringing forth, they also imply a *resemblance* between that which is brought forth and its originator. In the processes of life, the originator is the *parent* and that which originates the *progeny* or *offspring*. The second meaning of generation (reproduction) is most succinctly expressed by the phrase "like begets

Fig. 1.1 Cuckoo's egg or the incorruptibility of heredity

like." The begetting can be either *sexual* (i.e., involving the union of male and female germ cells) or *asexual* (i.e., not involving such a union). The resemblance between the offspring and its parent has two aspects. The one aspect is that the new individual is normally of the same *kind* (*genus* in Latin) or *species* as the parent (the human species in Fig. 1.1). The second aspect is that within a given species, the offspring resembles the biological parent in a particular feature (the bulbar nose in Fig. 1.1) which is absent in many other individuals of that species. Let us call this transmissible feature *character*, and the phenomenon of transmission *heredity* or *inheritance*. We see immediately how the terms sex, species, and heredity tie neatly together in the concept of generation (reproduction). This concept was developed in ancient Greece in the fifth century Before the Current Era (BCE) by Aristotle and then incorporated into the foundation of Western thought. There it persisted, virtually unchallenged, until the nineteenth century. In that century, however, it underwent a radical reinterpretation, when the speculations on which it rested were subjected to experimental verification. The term "generation" was then largely abandoned in its original meaning (though it eventually acquired other meanings). The three components of generation (sex, species, and heredity) developed into separate sciences: reproduction together with developmental biology, evolutionary biology, and genetics, respectively. The man, who single-handedly accomplished this transition from generation to genetics, was Gregor Johann Mendel. If we are to appreciate fully the significance and greatness of his accomplishment, we must try to grasp the circumstances under which the generation concept arose and also go into some detail of the concept itself. The aim of this chapter is to do just that. Here we give a brief introduction to the intellectual climate in which the ancient Greek philosophy emerged, followed by an equally brief description of two of its themes which are relevant to the present discussion, and then devote the rest of the chapter to Aristotle's generation concept.

The Mutiny of Reason

In the seventh century BCE, what later came to be called Greece was a loose conglomerate of independent, competing, and sometimes warring city-states strewn on the coast along the Mediterranean Sea. Only a common language, shared gods, and similar culture united the city-states. Like other peoples of that time, the Greeks used gods to explain phenomena and events they could not explain otherwise. Thus, they attributed thunderstorms to Zeus sailing the thunderclouds and hurling thunderbolts; earthquakes to Poseidon stomping his feet and thrusting his trident into the ground; winds to Boreas, Zephyr, Notus, and Eurus, each blowing his breath in a different direction; and so on. These explanations were so simple that even the dimmest person could grasp them and so make sense of the world. But for some people, they seemed a bit too simple. Toward the end of the seventh century BCE, a group of savants initiated a movement that expressed dissatisfaction with the traditional view of the world and developed a new view, from which gods were largely expelled. Two words then came to differentiate the traditional and the new views: *mythos* and *logos*. Initially the words had a similar meaning, but as they evolved, they acquired diametrically opposite connotations.[2] The Greek word *mythos* originally meant "speech" or "thought" but gradually came to stand for "a traditional story of ostensibly historical events that serves to unfold part of the world view of a people or explain a practice, belief, or natural phenomenon".[3] The term *logos* might have originally meant "word" or anything connected with the use of words, for example, a "narration." In this sense it was used interchangeably with *mythos*.[2] Later, however, it assumed a new meaning. As the Romans began translating Greek texts into Latin, they rendered *logos* as *ratio*, in certain contexts. This Latin noun was derived from the verb *reri*, which originally meant "to calculate," and later also processes mentally resembling calculation, such as "to reckon," "to think," and especially to think in a particular way—"to reason." *Logos* thus came to be translated as *ratio*, in the sense of "reason" and *reasoning*. In this special sense, "reason" became nearly synonymous with "cause," and "reasoning" came to mean the kind of thinking in which thoughts followed each other in a cause and effect combination. Other names that came into use for this form of thinking were *rational* and *logical*. These two terms, however, had originally slightly different meanings. As Greek savants established certain rules of thinking and termed the study of these rules *logike* (logic), *logical* became the kind of thinking that adhered to the principles of logic.[3] The Greeks began thinking logically

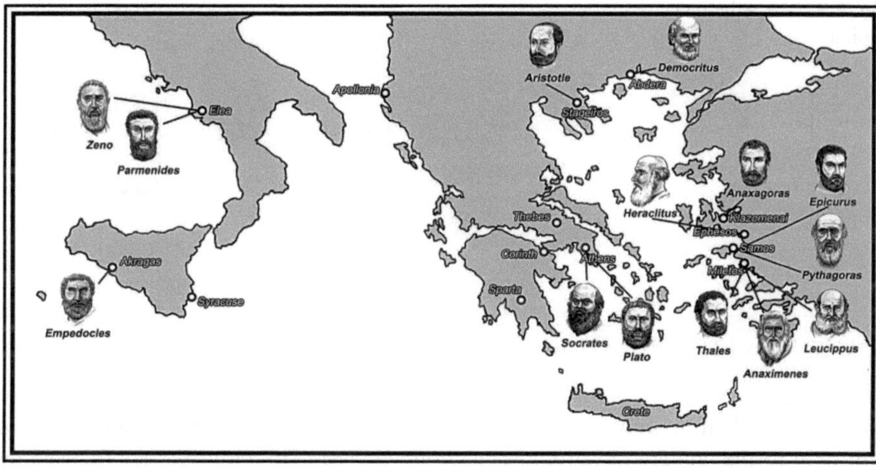

Fig. 1.2 *Frogs about a pond*: the geography of ancient Greek philosophy. The quote is from Plato's *Phaedo*; the "frogs" are the ancient Greeks and the "pond" the Mediterranean Sea, which they colonized

already when they were still in the mythological period of their development.[4] They then conceived stories that were myths by their function (explanation combined with an entertainment) and because they moved back and forth between natural and supernatural but had a tendency toward rationality. Poetic and rational thinking mixed freely in these myths, and it was only a question of time for the rational mode of thinking to prevail over the poetic.

The prosperity of the upper classes, the propensity of the Greeks to use rational thinking, and the absence of organized clergy had led to the rise of a breed of self-supporting freethinkers engaged in an intellectual intercourse that resembled an soccer game. The object of the game was to score points not with a ball but with thoughts challenging the opponents to a response. The game was conducted by verbal exchanges at gatherings or symposia,[5] at schools founded by leading savants, and by means of papyrus scrolls on which the authors recorded their thoughts in writing. One of the first such schools arose in the Greek colony of Miletus, an ancient port on the western coast of Asia Minor (Fig. 1.2). The founder of this *Milesian school*, Thales of Miletus (c. 624–c. 543 BCE), and his followers Anaximander of Miletus (c. 610–c. 546 BCE), Anaximenes of Miletus (c. 585–c. 525 BCE), and others focused their inquiries on nature, which the ancient Greeks called *physis*. Their means of inquiry were observation and rational thinking aimed at explaining the world by the operation of natural (material, physical) agents. Because of their focus on *physis*, they came to be known as *physiologoi* (singular *physiologos*), "those who spoke about nature." They were contrasted with *theologoi* (singular *theologos*), "those who spoke of gods," the thinkers who evoked gods (*theoi*) to explain the operation of the world. Together, the *physiologoi* and *theologoi* began to be spoken of as *philosophoi* (singular *philosophos*),[6] "those who loved wisdom." Since the word

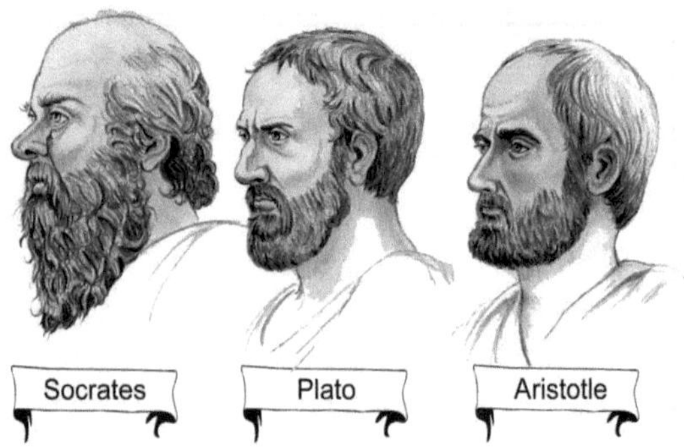

Fig. 1.3 The Athenian Troika

"physiologists" ultimately acquired a more specific meaning, to avoid confusion, we will refer to the ancient physiologists as *philosophers*. The philosophers focused on two fundamental questions regarding the nature of *reality* (i.e., that what is): first, What is? and second, Does it change and if so how? Since the interest in this chapter is the origin of new individuals (genesis, generation), we begin with the second question and come to the first later. Moreover, we restrict our attention to those philosophers, who made the most significant contributions to this subject. They are Heraclitus, Parmenides, and the Athnian Troika – Socrates, Plato, and Aristotle (Fig. 1.3).

Change or No Change?

For the ancient Greeks, "change" was nearly synonymous with "motion." The common view of change was that it represented a transformation of one thing (*A*) into a different thing (*B*). During the transformation, thing *A* ceased to exist and a new thing, *B*, came into being. The process thus involved three states: *Being* (the existence of thing *A*), *Becoming* (the coming into being of thing *B*), and *non-Being* (the cessation of *A's* existence). Early in Greek philosophy emerged two diametrically opposite views of change—that of Heraclitus and that of Parmenides. Heraclitus denied the existence of Being and proclaimed all existence for Becoming. Parmenides, in contrast, denied Becoming and held all existence for Being. Expressed simplistically, Heraclitus claimed that all is change all the time, whereas Parmenides maintained that there is no change in the world at any time. To a commonsense person, these extreme views are both preposterous, for it is apparent that some things change, while others persist. But some ancient Greek philosophers were not commonsense people. Heraclitus of Ephesus (c. 535–c. 475 BCE)[7] argued that things appearing to persist in reality change so slowly that we do not notice it. Rocks crumble, mountains erode, metals corrode, and living things age and die. There is nothing in the terrestrial realm that lasts forever. There is never anything of which we can say that it *is*, because while we think of it as a particular thing A, it has already become something else. There is no Being, there is only

Becoming. If, however, everything flows, as Heraclitus says, then you cannot ever say that something is or is not. What remains constant in the ever-changing world is not the substance or substrate but a process—the process of change. As things change, substances perish, so that there is no single substance or element that is common to all things, not even water or air, which the Milesians held for just such substances. When water changes to air, it "dies" in the process and there is nothing left of it in the air, and the same applies to a change in the reverse direction. Change is so ubiquitous that it itself must be regarded as the real nature of reality.

Like Heraclitus, Parmenides of Elea (flourished in the early fifth century BCE)[8] challenged the commonsense view of change but from a very different position, in which he arrived at the conclusion that change is a logical impossibility. Here is his argument: We start with the statements that *A (Being) is*, whereas *B (non-Being) is not*. The latter statement, however, is nonsense, says Parmenides. Stating that something is not is talking about nothingness, but about nothing there is nothing to be said. As for the former statement, it asserts that *A* is *A* and nothing else. If we then say that *A* is changing into *B*, we must ask: Where does the *B* come from? There are two possibilities. Either it comes from nothing, but this cannot be because we just said that about nothing there is nothing to be said. Or *B* comes from *A*, but this is also impossible because we also said that *A* is *A* and nothing else, otherwise it would not be *A*. If *A* had a trace of *B* in it, then saying that *A* is *A* would not be true, and saying that *A* is *A* and *B* would violate the logical law of contradiction, which asserts that something cannot be two things at the same time. Hence *A* cannot logically change into anything ever. In fact, there is nothing in the world that can change into something else. Not only that, but also there can only be one thing in the world, only One Being, for where would the other Beings come from? Not from nothingness and not from traces of other Beings in the One Being. Furthermore, if the One Being is without a trace of other Beings, it can be said to be homogeneous, exactly alike throughout. Also, since a Being can never change, it remains forever the same, undifferentiated and featureless. Since it cannot come into Being from non-Being and since it cannot turn into non-Being, the One Being is eternal. For the One, time does not exist; the One has no past and no future—it only is. Since it is full everywhere and since it cannot move (remember: motion is a change!), it cannot go anywhere. Parmenides' is a very unappealing vision of the world. Unmoving, unvarying, featureless, uneventful, his is a world without a past and with no future, without evil but also without goodness, without sadness but also without joy. Parmenides' logic seems impeccable, except for the mysteries of where the philosopher himself fits into this picture and how the illusion of many different things arises. Obviously, the senses are deluding us, but in the world of One, there should not be any senses in the first place nor should there be individuals endowed with senses. Is the One dreaming up the world of many? Is it hallucinating? Obviously not, for the same ironclad logic that leads to the One also forbids it to display any activity.

Rather than siding with Heraclitus or Parmenides, most contemporaneous Greek philosophers tried to find a compromise solution between these two extreme views. Generally, the solution had the form of postulating two components of reality, one fixed and the other changing. The function of the fixed component was to provide continuity and so dodge the accusation of an *ex nihilo* generation. The second component served to introduce the actual change on the background of the fixed component. The various proposals varied in the degree of sophistication, some of them being no more than a charade obscuring but not solving the real problem. Others, on the other hand, had to be taken seriously by Parmenides himself. Among the latter were two proposals, which had a long-lasting influence on Western thought—those of Plato and Aristotle.

Plato's Myth

Plato (c. 428/427–347)[9] admitted that the physical world is changing constantly but at the same time postulated the existence of another world characterized by constancy and permanence. Since the other world overstepped or transcended the physical world, it came to be called *transcendental*. Plato's postulate of a nonphysical realm might have been inspired by his teacher Socrates (c. 469/470–399 BCE).[10] The latter was interested in defining ethical concepts, but when he stopped people on the street and asked, for example, "What is courage?" he commonly got answers such as "Courage is when a person saves a child from a burning house" or "Courage is when a soldier risks his life to bring his wounded comrade into safety." These, of course, were not definitions but merely instances of courageous behavior. Socrates realized, nevertheless, that they pointed at something shared by all of them and that this shared feature was the definition of courage. Plato extended these thoughts to physical objects such as tables or chairs and realized that all objects of the same kind shared a common denominator which he called Idea—"tablehood" in the case of tables, "chairhood" in the case of chairs, and so on. There was, however, no tablehood anywhere in the physical world; there only were particular tables, and the same was true for the chairhood and the particular chairs, as well as for all the other kinds of physical objects and their Ideas. Where then were the Ideas? Since they were not physical, they had to be immaterial and had to occupy a world of their own, a world without space and time—the transcendental realm. The absence of time made the Ideas timeless and changeless, in contrast to the material objects of the physical world, which were all subject to corruption and death. Being eternal and incorruptible made the Ideas perfect, again in contrast to the physical objects, which had various imperfections in comparison to their corresponding Ideas. Since the physical objects of a given kind resembled, if only imperfectly, their corresponding Idea, there had to be some sort of "communication" between the physical and transcendental realm. Plato suggested that the Ideas "participated" in the generation of each physical object, when it came into being. He did not specify what the participation amounted to, but some of his interpreters compared the process to imprinting a seal onto a blob of warm wax. The seal corresponded to the Idea with an ideally executed original pattern; the imperfectly imprinted pattern corresponded to a particular physical object, and the wax to what Plato called a "receptacle," presumably matter. Indeed, since Plato some philosophers hold matter for something coarse in comparison to purported ideal immaterial substances and the physical world for a degraded version of a transcendental ideal world. And since Descartes (1596–1650 of the Current Era or CE), they try in vain to explain how immaterial substances might communicate with material objects. This conundrum is, however, not the only insurmountable problem with what is sometimes referred to as Plato's doctrine of Ideas; we will mention some of the other problems below when we come to Aristotle. In fact, the doctrine is so full of holes that it is best assigned to mythology. Plato might have been aware of its weaknesses, and this might have been the reason why he had not described it as a whole anywhere in his 26 dialogues. What is now referred to as his doctrine of Ideas has in reality been pieced together by his interpreters from fragments and hints scattered throughout his different works. Plato's first and most prominent critic was his student Aristotle. He was skeptical of his teacher's doctrine not only because it was a relapse into mythology and because of the contradictions and inconsistencies it contained, but also for its implications concerning what did and what did not constitute knowledge. To Plato physical objects—because of their materialness, inconstancy, and perishability—were unworthy of study, for such an inquiry could produce at best only unreliable opinions based on conjectures. The only source of genuine knowledge was the realm of perfect, unchanging, and eternal Ideas, accessible through intuitions and recollections of encounters of the immortal human soul on its sojourn to the transcendental world. The sojourn took place after the death of one individual and before the soul's incarnation into the body of a philosopher coming into being. In contrast to Plato, Aristotle, perhaps because of his family background (he was a scion in a lineage of physicians), had high respect for an empirical approach to a knowledge-gathering process. It is therefore Aristotle, and not Plato, who is the true founder of Western philosophy and science, over which he has held sway over a period of nearly 2000 years. Hence, if we want to learn what the West knew or thought of heredity at the time of Mendel, we must turn to Aristotle. It is for this reason that we devote the rest of this chapter largely to him.

Aristotle[11] and His Inventory of Reality

Aristotle (384–322 BCE) was born into a well-off family in Stageiros, an ancient Greek city on the Chalkidike peninsula. His father was the personal physician of Alexander the Great's grandfather, and his mother brought into the marriage substantial dowry. At the age of 17, Aristotle joined Plato's Academy in Athens, where he then remained for 20 years, until his tutor's death in 347. Whatever else one might think of Aristotle, this much is undeniable: He is matchless in developing the most comprehensive, cohesive, coherent, and systematic philosophical system. Although only some 30 of the more than 150 works he may have authored have survived, this corpus is awe-inspiring in its breadth and depth. It develops and formalizes all branches of philosophy now recognized, as well as many areas, which are now classified as sciences. The coverage is systematic in the true sense of the Greek word *systema* in that its individual parts form a unified whole. And since the system has been developed by a procedure of inquiry based on defined rules and principles, its parts are logically tied together.

In his early works, with characteristically Aristotelian comprehensiveness, he starts his inquiry with an inventory of reality. He founds his dissection of reality on two assumptions: first, humans think in words, and second, the structure of a language reflects the organization of reality. On that basis, he seeks a correspondence between principal elements of the Greek language and the components of reality. The principal structural component of language is a sentence that asserts or denies something about something else. The basic units of such a sentence or *proposition* are the *subject* and the *predicate*. The English word "subject" is derived from the Latin *subjectus*, which is the equivalent of the Greek *hypokeimenon*. In all three languages, the equivalent nouns mean the same thing—"that which is lying under." Similarly, the English word "predicate" is derived from the Latin *predicatus*, a past participle of *predicare*, whose Greek equivalent is *kategorein*. In all three languages, the essential meaning of the equivalent verbs is "to affirm." The noun predicate (*kategoria*) means therefore "that which is affirmed (said) of something." According to Aristotle, corresponding to the subject and predicate are two principal components of reality, which he calls *ousia* and *kategoria*. The proper translation of *ousia* is "being," but the term is more commonly rendered into English as "substance," which also fits better in the context in which it will be used here. *Kategoria* is commonly transliterated as "category," but in the present context, it will be rendered into English as "attribute" or "property." Thus, in linguistic terms, *ousia* (substance) is the subject of whom something is affirmed, and *kategoria* (attribute) is that which is affirmed of the *ousia* (the attribute affirmed of a substance). This division is the first clear-cut separation of things' attributes from the things themselves. Earlier philosophers failed to make that distinction. For example, Thales of Miletus held not only water but also wetness for a substance. Even Plato, according to Aristotle, confused the two concepts, and it was this error that led him ultimately to the postulate of the transcendental realm. But if *kategories* are attributes of substances, there is no need for an outwardly realm; it is much simpler to assume that properties are *in* the substances or, as philosophers like to say, they are *immanent* in things. Wetness, as an attribute of water, must be in water itself; redness of a poppy flower must be in the flower; courage must be in the courageous person; and so on. We point out this now obvious fact because, as we shall learn later, it will become one of Mendel's important postulates on which he would base his experiments (see Vol. 2 Chap. 3). This circumstance is one of the reasons why we think it proper to start this book with a brief expose of ancient Greek philosophy, Aristotle in particular.

In his early works, Aristotle also uses another criterion for grouping items of reality. To the linguistic yardstick described above, he adds an ontological[12] criterion to cluster the items into those that exist independently of other items and those that exist *in* other items. By combining these two criteria, Aristotle differentiates items of reality into four groups, of which only three will interest us here. He calls these three groups *protai ousiai* or primary substances, *deuterai ousiai* or secondary substances, and the *kategoriai* or attributes. The primary substances function as subjects but never as predicates, and they exist independently of all other items. The secondary substances can serve as both subjects and predicates, and Aristotle describes their mode of existence as "being said of." We will explain below what he means by this expression. The attributes function as predicates only and exist only in the primary substances. Let us now have a closer look at these three groups.

The *primary substances* are the individual material objects, which are the subjects of predication; that is to say, they are the things of which statements are made. Each of the objects is a separate, distinct entity that can be pointed out as *tode ti*, that is, *this* (particular object), and taken one at a time (*kath' hekaston*). We identify some of these *particulars* by giving them proper names, for example, this man Socrates, this horse Bucephalus, this dog Rin Tin Tin, or this wolf Lobo. Both Plato and Aristotle consider the particulars real, but to a different degree: Plato as least real, mere shadows in comparison to the Ideas, whereas Aristotle holds them for the most real things that exist, and that is why he calls them *protai ousiai*, the first beings.

According to Aristotle, none of the *secondary substances* can be pointed out as *tode ti*, as this particular thing or person. Instead, they can be "said of" objects. To explain, we need to diverge slightly. It is a truism that, on the one hand, no two primary substances are identical and that on the other, some objects are clearly similar, their resemblance varying in degree. One can arrange objects into a series of inclusive groups or *classes* according to the declining similarity between the groups. In this arrangement or *classification*, at the lowest level of the hierarchy, individual objects are clustered into groups such that the objects within a given group resemble one another more than they do individuals in any of the other groups. At the next higher level, the groups are placed into fewer groups, each of which encompasses those of the lower level that share many characteristics. This process is then repeated level after level until the group with broadest shared characteristics is reached at the apex of a hierarchical pyramid. Aristotle calls the groups at any level of the sequence *eide* (singular *eidos*) and those in the nearest higher-level *gene* (singular *genos*). Hence, at any level the same group is *eidos* relative to the nearest higher-level items and *genos* relative to the nearest lower-level items. In modern classifications of living organisms, each of the different levels bears a distinct designation and the names *eidos* and *genos* are reserved for the two lowest levels. The English translations of these two terms are *species* and *genus*, respectively. So, by the expression "said of," Aristotle means that an object *x1* has been assigned to the species *X*. For example, in the proposition "Socrates is a human being," the predicate "human being" (a species) is said of the subject Socrates (the individual object). In this classification, therefore, the primary substances are the physical objects, whereas the secondary substances are the "kinds" or species.

One part of their Greek name seems to suggest that the secondary substances are *ousiai*, but the second part, the adjective *deuterai*, mitigates their substantiality, suggesting that it is not genuine. The secondary substances cannot exist without the existence of the primary substances since if there were no objects that could be classified into kinds, there would not be any kinds. Secondary substances come into being only secondarily, on the basis of what is given primarily—the physical objects.[13] In a stark contrast to Plato, Aristotle does not hypostatize the "kinds," that is, he does not attribute to them a separate existence. He admits that they are real, but not real in the same sense as physical objects are because their realness depends on the existence of these objects. He draws the conclusion that the "kinds" are embodied or immanent in the objects, rather than existing separately from them. This deduction may seem to clash with one of the two criteria he uses in his classification of reality. Namely, the positing of the "present in" against the "said of" could be interpreted as implying that the secondary substances are *not* present in objects, whereas the above deductions imply the opposite. Aristotle resolves this seeming contradiction by suggesting that attributes and "kinds" are present in the primary substances in different ways: The attributes are incidental to the objects (some scholars compare them to parasites in the body of a host), whereas the "kinds" are essential components of the body, without which the object would not be what it is. The difference between these two modes will become more apparent later when we learn about Aristotle's inquiries into the nature of the "kinds."

"Kinds" and attributes have, however, more in common than their residence in the primary substances: Both are in the same class which Plato might have been the first to describe and which Aristotle calls *to kathalon*, "that which pertains to all." The medieval Latin-speaking followers of Aristotle translate this expression as *universalis*, meaning "belonging to the whole collection of items." Thus, the *universal*, as the term is rendered in English, is that which a group of objects has in common. The term is used both as a noun and an adjective and in either case it is contrasted with a unit—an individual or a thing—called the *particular*. It can be said that the particular is an instance of a universal or that the universal has been *instantiated* in a particular. It follows from the foregoing that universals are of two types: species type and attribute type. An example of the species-universal is the human species—the collection of individuals encompassing Socrates, Plato, Aristotle, and billions of others. An example of an attribute-type universal is "redness," instantiated in a red poppy flower, red balloon, red flag, and a countless number of other particulars. These examples illustrate the fundamental difference between the two types of universal. Red poppy flower, red balloon, and red flag have little else than the red color in common. They are unrelated items of reality, which accidentally share an attribute—the red color. We now know that the redness of these items is the result of the reflection of light of specific wavelengths by substances, which are not the same in the three items. By contrast, the similarity among Socrates, Plato, Aristotle, and all the other persons subsumed under the universal "the human species" is not accidental but the result of their *genesis*. This word, translated into English as *generation*, signifies the development of an individual. In Aristotle's view, which we will expound upon later in the chapter, the male semen transmits the essence of the father to the maternal material homolog and from the combination arises a descendant bearing the same essence as the parents. It is this transmission of the same essence that is responsible for a given "kind" begetting the same "kind" and so for the existence of particular groups—the universals. We will return to the essence momentarily but first we must complete the account of Aristotle's inventory of reality by mentioning the third group of items—the attributes.

Actually, the features that define the essence and so also the species are in reality attributes, which one might call *essential*. But these are different from the accidental attributes that constitute the *accidental* attributes, which may vary in form among the individuals of a given species. Aristotle organizes the accidental attributes into a logical system, which falls back on linguistic analysis of the way people speak when they describe the world. Aristotle asks: What are the most general questions one must ask in order to characterize a thing? And he then proceeds to identify ten such questions and suggests that the answers to nine of them specify groups into which all predicates can be divided. The first of the ten questions identifies the substance. Because the Greek term for "predicate" is *kategoria*, the ten groups came to be known as the *categories*. (Traditionally, substance is included in the list of categories, even though it is not a predicate and hence not a category.) Aristotle's ten questions (categories) are these: What is it? (Substance.) How much of it is there? (Quantity.) What is its nature (Quality.) What is its relation to other things? (Relation.) Where is it? (Place.) When is it? (Time.) In what position? (Position.) In what condition is it? (State.) What is it doing? (Action.) And finally: How is it affected? (Passivity.) They can all be illustrated by a single sentence: "There is a man (Substance), alone (Quantity), looking like a doctor (Quality), and wiser than Hippocrates (Relation); he is in the street (Place), now (Time), walking (Position), barefooted (State), toward a surgery, either to treat a patient (Action) or perhaps to be treated" (Passivity).[14] Overall, the nine non-substantial categories can be divided to three groups covering qualities, quantities, and relations.

Now, to come back to the essence, which, according to Aristotle, distinguishes the different "kinds" from one another, we note first of all that the Latin *essentia*, from which the English "essence" derives, is a botched-up attempt to translate the Greek phrase *to ti en einai*. The literary translation of the phrase is "the what it was to be" but the medieval philosophers, not all of whom were exactly eloquent in both Greek and Latin, puzzling over it took the word *einai*, translated it as *esse*, "to be," and made the term *essentia* from it. In a simplified version, the essence of a thing is that which makes the thing what it is. The "thing," in this case, is not a concrete thing but a "kind," which we from now on will refer to as *species*. Taking as an example the species we are best familiar with, our own, let us now ask: What is the essence of the human species? There seem to be two ways of determining what makes the human species that what it is. One could be to compare it to all other species and enumerate all the differences that distinguish the human species from them. Considering that there is a very large but unknown number of species, this approach does not seem quite realizable. Aristotle therefore chooses the second approach that exploits as a shortcut his classification of nature into a hierarchy of species and genera. He determines the position of the chosen species, in this case the human species, in this hierarchy, identifies the genus to which humans belong, and then ascertains what distinguishes the human species from all the other species subsumed by that genus. In his primitive classification, the genus to which the human species belongs is identified as "animals." In the next step, he therefore determines what distinguishes the humans from all other animals and thus automatically from all nonanimal species in nature. He decides that the distinctive feature of the human species is the capability of rational thinking. He thus *defines* the humans as animals capable of rational thinking and proclaims this capability to be the essence of the human species.

The Hylomorphic Doctrine

The division of substances into primary and secondary, followed by the placement of the secondary into the primary substances, led Aristotle to the realization that the *ousia* might not have a single but dual nature. At the same time, as we will explain later, he also realized that the assumption of a substance's dual nature was necessary to explain movement without falling into the trap set up by Parmenides. The doctrine of a substance's dual nature, as expounded in Aristotle's mature works, came later to be called *hylomorphism*. Its principal thesis is that substance (*ousia*), the basic stuff of nature, is never simple but always compounded of two components, *hyle* and *morphe*. In old Greek, the word *hyle* stood for construction material, especially wooden material, but Aristotle elevates it to a technical term, which is commonly translated into English as *matter*. The Greek *morphe*, familiar from terms such as morphology, can be translated as "shape" or more generally as "form." However, Aristotle also uses a second term for the substance's second ingredient, namely, *eidos*, which he earlier used as a specific designation of the secondary substance. Since he uses the terms *morphe* and *eidos* as nearly synonymous, the implication is that the second ingredient of the hylomorphic substance is the former secondary substance, which he previously identified as species (*eidos*), essence (*to ti en einai*), and as one class of the universals. Consequently, all four terms have now become virtually synonymous, and Aristotle uses them interchangeably, choosing one or another of them depending on the particular aspect of the second ingredient he wants to emphasize. Let us now have a closer look at the two ingredients; for clarity we write "Form" with a capital letter.

Matter and Form are so tightly associated that it is impossible to describe or imagine one without the other. In the world of sensible objects, Form never occurs without matter and vice versa. One can think of pure matter (Aristotle calls it *prote hyle*, first or prime matter) by mentally stripping away all attributes from an object. What remains cannot be perceived because it does not have any attributes human senses could register. How, then, do we know that anything is left at all? We know it indirectly from the analysis of change. Since change does take place, we must assume that there is something that endures through the process and receives a new Form in place of the old one, and we call this "something" matter. Other than that, we cannot say anything about the prime matter; it lies outside the grasp of human knowledge. It does not fit into any of the categories of being. Aristotle assumes that in some respects, matter is the opposite of Form. While Form is an actuality, matter is a potentiality for having a Form impressed on itself. Form is activity, whereas matter is passivity. Form is differentiated and determinable, whereas matter is undifferentiated and indeterminable. Yet, Form is not a thing; it is the *way* in which matter is organized into a functional whole—the hylomorphic object. A modern-day person, brought up in the tradition of Christian matter-spirit dualism might be tempted to seek parallels with the Aristotelian matter-Form dualism. Such temptation must, however, be resisted. In the *prote hyle*, there is no sign of materialness, however one defines it, for in this formless and propertyless "pure matter," there is no sign of anything. Similarly, although the pure Form is Aristotle's Unmoved Mover, which with its helpers runs the universe, it has little in common with the Holy Ghost of the Christian triune God. Also, as we explain later in the chapter, the Aristotelian concept of the "soul" is very different from the "soul" of Christian mythology. In a material object of Aristotle's reality, it is through the action of the Form that the object's materialness reveals itself in the manner of weight or extension in space. Thus Aristotle's matter and Form do not fit into mold of the traditional matter-spirit dualism. The distinctiveness between the matter and the Form is primarily in the opposite's potential and actual or passive and active, and not in materialness versus spirituality. For Aristotle the borderline between the matter and Form domains is fuzzy anyway. Thus, strictly speaking, a block of marble is a formed matter (i.e., a hylomorphic compound), but Aristotle has no qualms about calling it "matter" waiting to receive its Form under the hands of a sculptor. In this sense the terms "matter" and "Form" are relative, and the relativity arises because the imposition of a Form on matter is a complex process passing through many stages in which matter can be informed to different degrees.

Aristotle uses the term "Form" in different senses, some of which are apparent from the equation *deuter ousia* = *eidos* = *to ti en einai* = *to katholon*, since the four terms of this equation can be held for aliases of not only each other but of Form as well. In addition, since Form relates to the appearance of an object, it has obviously much to do with attributes or properties, which can be classified in three groups—essential, accidental, and individual. Above, we described the first two of these groups. The essential attributes are those that determine the essence of a species—they are species specific and hence they define the *substantial Form*. In the second group are properties for which Aristotle uses the term *symbebekota* (singular *symbebekos*), which is translated as *accidents* (from Latin *accidere*, to happen, as by chance) and which we called earlier "accidental." But, perhaps, better would be denoting them as *variable attributes*, for they are responsible for variability within a species. Aristotle is rather noncommittal regarding the origin of this variability but seems to be inclined to attribute it to a potential of the species Form for variation in certain properties under the influence of environmental factors on matter. In the third group are properties, which make the individual object unique—the *individual-specific attributes*. Aristotle argues that *individual Forms* do not exist, but if they do exist, they are of little significance. The latter claim is rooted in Aristotle's contention that scientific knowledge is the knowledge of the universals and that studying particular individuals makes little sense for reasons of their transiency. Some scholars insist, however, that the postulate of the individual Forms is necessary in order to save Aristotle's late works from internal logical inconsistencies.

There is yet another term that Aristotle uses for the Form and that must be mentioned here: *psyche*. Before Aristotle, the term was used with a variety of meanings. In Homer's time, it stood for "breath" and breath was associated with life, just as it is to this day. When Homeric heroes died, the *psyche* residing in their head escaped through the mouth to live a ghostly life in the underworld.[13] What was left behind was a motionless body, and this observation was taken as an indication that *psyche* was somehow responsible for motion. The early Greek philosophers took over this notion and attributed *psyche* to all natural objects, including stones. After all, didn't some stones, specifically magnetite, cause motion of other objects? Subsequently, however, some philosophers, for example, Heraclitus and Pythagoras, made psyche responsible for all non-corporal ("psychic") activities, separated it from the body, and attributed to it a divine origin together with immortality. This trend culminated in Plato, whose concept of *psyche* later became the inspiration for the notion of the *anima* (Latin) and *soul* (English) of the Church Fathers of Christianity. Aristotle, in opposition to Plato, returned to the early concept of *psyche* as the power behind the activities associated with change in matter. His *psyche* is intimately associated with matter, non-divine, and incapable of separate existence from matter; in other words, it is one additional alias of the Form.

Aristotle's dualistic conception of physical objects reopens the question concerning the nature of the *ousia*—the substance or being, whichever translation of the term you prefer. One consequence of the hylomorphic doctrine is that Aristotle is now forced to abandon the distinction between the primary and secondary substances and renew his search for the true substance. In this search he goes through the whole gamut of concepts he has accumulated in the meantime and confronts them one by one with the criteria a substance, in his view, must meet. We don't need to follow him in these deliberations since, if we accept the synonymy equation above, we can limit the number of candidates to three: compound of matter and Form, matter alone, or Form alone. Furthermore, since the first criterion of a substance is simplicity, we can eliminate the compound precisely because of its compoundness. We are then left with two candidates only—matter and Form. The former, the *hyle*, is the substrate (*hypokeimenon*) of being, the determined element of the compound, whereas the latter, the *eidos*, the determining element, is that by which matter is made into an object we perceive through our senses. One might perhaps expect Aristotle to choose matter as the substance, if for no other reason than its materialness underlying the existence of the natural world, but in fact, he gives the title to the Form. The next two sections should help us to understand why.

Aristotle's Theory of Change

Like most post-Parmenidean thinkers, Aristotle seeks to resolve the puzzle of change by the postulate of two components, one lasting through the alteration and the other coming into and going out of existence. He finds these components in matter and Form of the hylomorphic compound. Before going into details of his explanation of change, let us first clarify what he understands under this term since his conception of change is different from modern. He distinguishes between change (*metabole*) and motion (*kinesis*) but often uses these two terms interchangeably, since for him they are manifestations of the same principle. In correspondence with his division of reality into substantial and non-substantial, Aristotle differentiates between substantial and non-substantial (accidental) changes. In the latter he distinguishes three kinds, corresponding to three of the nine non-substantial categories: quality, quantity, and place. The substantial change differs fundamentally from the three non-substantial changes in that it entails extinction of one substance and emergence of another. It is thus a replacement of one substance by another or *generation* (birth, coming into being, Greek *genesis*) and *decay* (death, passing, Greek *phthora*) of substances. In nature, an acorn vanishes when an oak tree grows from it, and an egg ceases to exist upon hatching of a chick. In the *non-substantial* or *accidental changes,* the substances persist through the change and only nonessential attributes (accidents) come and go. There is no need here to go any further into Aristotle's classifications and characterizations of different kinds of change, so we now return to the gauntlet thrown down by Parmenides.

The gist of Parmenides' arguments for the impossibility of change, let us remind ourselves, is in three assertions or principles. First, nothing can come from nothing. Second, everything either is or is not; there is nothing between existence and nonexistence. And third, nothing happens without an explanation or a sufficient reason. Aristotle accepts the three principles but with qualifications, which then enable him to refute the conclusions that Parmenides draws from them. Aristotle argues ingeniously that although indeed nothing comes from nothing, something can come from what is not. What leads him to this claim is his distinction between actuality and potentiality. An object, he says, actually possesses certain attributes and lacks certain others, which it potentially *can* possess. He calls the state of potential possession *steresis* (literally "lack of something"), which is commonly translated as *privation. Steresis* is not nothingness; it is a state in which something is not actually but can be potentially. When the potential is realized and the actual attribute arises, it does not arise from nothing but from something that, though lacking, had the potential to be. To give an example, Socrates is pale and the paleness of his skin is thus his actual attribute. But he has in him the potential for another attribute, which, if actualized, would substitute the paleness of his skin for tanness. All he has to do to actualize the potentiality is to expose his skin to the sun for a few hours. What the change that he thus undergoes amounts to is not a dissolution of the paleness into nothingness and the emergence of tanness from nothingness, for that would indeed violate the first principle of Parmenides. Rather, it is the actualization of a potential tanness (which existed in the state of privation), accompanied by the transition of the actual paleness into a potential paleness (into a state of privation). This process is clearly a *change*, which is possible without the violation of the first principle of Parmenides because in it a thing comes to be not from nothingness but from privation and another disappears not into nothingness but into privation.

As for the second principle, Aristotle points out an error in Parmenides' logic, which, when corrected, refutes the implication that change is impossible. The error is in that Parmenides confounds existence with predication. Parmenides interprets the second principle as implying that existence is all or nothing and that therefore everything existing is complete. By completeness he means that what exists has all the possible attributes that can be predicated of it. Aristotle disagrees. In his interpretation of change, an existing thing is incomplete in that it lacks many attributes it can potentially have and that change is precisely this: the actualization (coming into being) of the potential attributes. Parmenides' mistake is that he takes the "is" in a statement of existence ("something is," meaning it exists) for being the same as the "is" in a statement of predication ("something is an attribute of a subject).

Parmenides uses the third principle to buttress the interpretation of the second. He argues that if something is, there is no sufficient reason for something to become what it already is. Aristotle's reinterpretation of the third principle follows from the qualification he attaches to the second principle. Since, when something is, it can be incomplete in its existence, there is sufficient reason for it to actualize attributes that it possesses potentially—and so to change.

Aristotle does not by any means assert that all the attributes that there are exist as potentialities in an individual object such as Socrates. On the contrary, a single object disposes with only a limited number of "dormant" attributes, which it could potentially "awake" to replace some of the attributes it actually expresses. Thus, Socrates, for example, cannot grow horns like a bull or a beak like a bird because he simply does not have these attributes in his dormant repertoire. Furthermore, during a change, the disappearing attribute cannot be replaced by just any attribute randomly chosen from the repertoire. There seem to be a certain link between the disappearing and emerging attributes in that the two attributes commonly form a pair of *opposites*. Thus, in the above example of non-substantial change, the two attributes, paleness and tanness, are generally thought of as opposites. The concept of opposites existed in Greek philosophy long before Aristotle and several of the Pre-Socratics operated with it in their interpretations of the world. Notably, Pythagoras (sixth century BCE) held opposites for reflections of the fundamental duality in the makeup of the universe. And Heraclitus espoused the view that the tension between the opposites was responsible for the constant flow of things. It was not, however, until Plato and Aristotle that the notion of the opposites had received a systematic makeover. Plato tried, largely unsuccessfully, to integrate the concept into his doctrine of transcendental Forms. The two concepts seemed incompatible and all his effort ended in a hopeless entanglement of speculations. It was then left to Aristotle to develop the first comprehensive theory of opposites, but it would take us too far astray to go into the details of his doctrine. Suffices to say that, in his view, a change always comes from an opposite. He calls this conclusion the *first principle of change,* and since, in his view, change is the foundation of natural science, it is, at the same time, also one of the first principles of this scientific discipline. It fulfills the criterion that Aristotle sets for the first principles of science: It is not derived from any other principle, while everything else is derived from it.

Thus, Aristotle dissects the change into three parts:[15] the initial object, that is, the thing that undergoes the change; the resultant object, that is, the thing that results from the change; and the persisting object, the thing that underlies the change—the *hypokeimenon*, the subject of the change. There are, therefore, three ingredients of change—the two opposites and the subject of the change, the *hypokeimenon*. The two opposites are important, but neither one nor the other persists through the change; this part is reserved for the subject. It is the persistence of the subject that protects Aristotle's doctrine from the corrosive effects of Parmenides' principles. The opposites are attributes, which cannot exist on their own. They inhere in the subject of whom they are predicates, and the subject is a substance. But how does the change take place? It seems that Aristotle favors the possibility that in some cases, one opposite transforms directly into the other but that generally the change is indirect in the sense that it takes places via privations. Either way, the comings and goings of the attributes cannot transpire without the substrate, on which the very existence of the attributes depends. The substrate (the subject) remains through the change. The comings and goings of the attributes of course change somewhat the subject, but since what changes are accidental properties, essentially the subject remains one and the same.

What we have described thus far, however, has been only one of the two major types of change—the non-substantial change. What, then, of the other type, the *substantial change*? Aristotle's answer is that, in principle, it takes place in the same way, provided that we make one all-important modification in the concept of the subject. In early works Aristotle supposes that the primary substance is ontologically simple, which means that it is not further divisible into even simpler, more fundamental components. But this supposition now becomes untenable because it clashes with the requirement that something in it must persist through the change in order to neutralize the challenges of Parmenides' principles. In the non-substantial change, the traffic of attributes alters the subject, but inessentially, so that no substantial change takes place. In the substantial change, however, one substance goes and another comes in its place, so the subject cannot be the persisting ingredient of the change, unless the supposition of the substance's simplicity is incorrect. If the substance were complex, composed of two components, as the hylomorphic doctrine postulates, then one of the components could persist, while the other is changing. Since the change affects the substance in this case and since the component responsible for the difference between two substances is the *morphe*, *Form*, or *eidos*, it must be the Form that changes. Hence the component that stays the same through the change must be the *matter* (*hyle*), which is responsible for the materialness of the substance. And so, in his later works, Aristotle takes the radical step to revise the original concept of substance. He abandons the distinction between the primary and secondary substances and from here on speaks only of *substance* (*ousia*) and means by it that what had previously been the primary substance, but with the big difference that the primary substance was ontologically simple, whereas the new substance becomes complex. It is now a compound consisting of two components, Form (*morphe*, *eidos*) and matter (*hyle*). This assumption solves the problem of the substantial change, in that the Form is now free to come and go, while the matter stays put.

In a substantial change, either a new substance comes into being (in a process termed *generation*) or an existing substance passes out of existence (a process termed variously as *degeneration*, degradation, decay, or perishing). In the living world, the two processes are tightly interconnected, resulting in birth and death cycles, repeated over and over. The death in each cycle is the death of the individual, while the rolling of the cycles assures the continuation of the species. In each cycle, generation and degeneration operate based on the three-ingredient doctrine. In generation/degeneration, the two opposites are the Form and the privation of the Form; the persisting underlying ingredient is matter. The generation is an actualization of the potential Form existing in the state of privation. The degeneration is the reverse of the generation: the passage of the Form from actuality to potentiality, that is, into the state of privation. Both processes involve matter as the component in which the changes take place but which itself remains one and the same. One of Aristotle's examples of a substantial change is the growth of an acorn into an oak tree. The acorn comes with the Form of an oak tree in a potential (privation) state. The growth of the oak tree from the acorn is the actualization of this potentiality. The mature tree produces new acorns, in which the Form of the tree is in the state of privation. The old tree dies ultimately and with it vanishes one source capable of recycling the oak tree Form. Only its matter will survive but it too will be recycled ultimately. In the meantime, however, the acorns the tree has produced will undergo new cycles of generation securing the persistence of the oak tree species Form.

Aristotle argues that to understand a thing fully, we must know what the thing is made of, the means by which it has been made, what it is, and what it is for. Each of these different aspects of a thing has a different *aition*, a different explanatory reason or *cause*. He thus recognizes four different causes: *material cause*, that from which a thing comes; *efficient cause*, the agent imposing the shape and structure upon a thing; *formal cause*, the thing's shape and structure; and *final cause*, the purpose for which a thing exists. In the example of an acorn growing into an oak tree, the material cause is the stuff of the acorn; the efficient cause is the tree that produced the acorn and that gives it the potential of developing into a tree of the same kind; the formal cause is the Form actualizing the tree; and the final cause is the fully developed oak tree capable of producing new acorns and thus propagating the kind. Some of the causes, however, seem to fuse. So, in the case of the acorn, the Form is both the formal and the final cause. Furthermore, both these causes seem to coalesce with the efficient cause, which concerns the Form as well. Here, then the four causes reduce into two, one involving matter and the other the Form. The three Form-involving causes, however, must not be thought of as being identical, for each concerns a different aspect of the Form. The efficient cause can be viewed as representing the Form as the initiator of the development, the formal cause the process of the Form's actualization, and the final cause the goal of the process, which is the perpetuation of the Form and so the perpetuation of the species.

Aristotle's Concept of Animal Reproduction[16,17]

Up to this point, we have dealt with Aristotle's concept of *genesis* (generation) in general, philosophical terms, describing how something arises from something else. Aristotle, however, also offers a more restricted view of *genesis*, limiting the "something" to living things and even more specifically to animals. Why this partiality? One likely reason is nepotism, for he considers himself a member of the animal clan. But the more important reason is that in animals the cycles of *genesis* and decay are much more obvious and regular than in other things. So obvious, in fact, that later-day biologists would call the cycling reproduction—the bringing into existence again of other things of the same kind. Reproduction, however, is not synonymous with *genesis*, for the latter term is broader than the former. Viewed from a modern-day perspective, Aristotle's *genesis* encompasses two different processes, which are now subject matters of distinct scientific disciplines: the generative process covered by reproduction biology and embryology and the transmission of resemblance from parents to offspring covered by genetics. In Aristotle's time the transmission of resemblance was held to be incidental to the generative process concerned with all aspects of the construction of a new individual. We will return to this important difference between ancient and modern views in the next section. Here we restrict the coverage to the generative process and divide it to subsections according to the important issues that arise from it. As a preamble we provide a brief introduction to Aristotle's classification of animals.

The Kinds of Animal. In his classification of living things, Aristotle distinguishes two major groups, animals and plants, and uses generation to set them apart. Animals propagate sexually and plants propagate by other means. Although both animals and plants are divisible into species, only in animals do individuals of the same species fall into two types—males and females—often distinguished by their appearances. That the two types belong to the same species becomes apparent when they procreate: Only a male and a female of the same species can mate and when they do, their offspring are of the same species as the parents, though again differentiated into males and females. In many species the procreation act begins with copulation (coition), during which the male introduces semiviscous fluid, the semen (*sperma*, the seed), in the genitals of the female. There are many modes of animal propagation among the species, the chief ones being viviparity and oviparity. In viviparous animals, the female delivers living young, whereas in oviparous animals she lays eggs, from which hatch the young after an incubation period.

Generation in Aristotle's vocabulary is the coming into being. It is yoked with its antithesis, the passing into nonbeing—the decay, degeneration, or corruption. Aristotle, expounding on Plato, interprets generation as an individual's striving to leave behind, after its own demise, a being of the same species and as close to its own identity as nature's laws allow it. It is an individual's way of participating in eternity. Because it cannot escape mortality, an individual makes every effort to ensure that at least some part of it endures. The meaning of procreation, according to Aristotle, is a continuous renewal of existence and life, for to be is better than not to be, living is better than not living, and being ensouled is better than not being ensouled.[18] Generation covers the entire process of nonbeing to becoming a being, from the conception to the birth or the hatching of an animal. Before Aristotle, other philosophers had also speculated on the biological nature of generation,[18,19,20] but of these theories only fragments survive, and most of them Aristotle has taken into account in his description of the process. In what follows, we limit our account of generation largely to that provided by Aristotle.

The Origin of the Semen.[21,22] Ancient Greeks knew that there was a causal connection between mating and procreation and that underlying it was the male semen or *sperma*. However, the origin of the *sperma* remained controversial, the three main contenders for the site of the *sperma* formation being the brain/spinal cord, the whole body, and blood. Aristotle was a strong supporter of the hemic hypothesis and provided a detailed description of how semen originated from blood. According to him, the foodstuff that animals consume comprises plant and animal substances made up of the four elements (water, air, fire, and earth) at different ratios and of different degrees of complexity. From these an animal extracts nutrients for use in its various physiological functions. The extraction consists in physical and chemical degradation, first in the stomach and then in the liver and spleen. Heat (*thermon*) is the agent effecting the degradation called *pepsis*, which in common Greek is used for material changes caused during the cooking of food in a pot over fire. Correspondingly, *pepsis* is often translated as "cooking," but also as "concoction" in the sense of "boiling together various ingredients" or simply "boiling." According to Aristotle, *pepsis* has two effects: It softens the food and separates it (breaks it down) into its components. The body's internal (vital, innate, connate) heat subjects the foodstuff to three or four rounds of *pepsis*.[23] In each of the rounds except the last, the heat separates the material into fluid nourishment (*trophe*) and a residue (*peritoma*). Also, in each of the rounds except the last, the nourishment is forwarded to the next station for the next round of *pepsis*, while the residue is separated into useful and useless fractions. The useful residue is collected for the final round of *pepsis*, whereas the useless fraction is channeled for excretion, the solid material through the intestine and the liquid stuff through the kidney. The first round of *pepsis* takes place in the stomach, the second in the liver or spleen, and the third in the heart. The forwarding from one cooking station to the next takes place via the blood vessels, of which, however, Aristotle has inadequate knowledge and so postulates some interorgan connections that in reality do not exist.

Aristotle makes the heart the center of nutrition, but also of sensations, emotions, and even intellect. He claims the heart to be the body's hottest place and makes it therefore the central organ for the generation and distribution of the inner heat. He says that the heart is the body's citadel (*akropolis*) or a hearth holding the kindling fire of an animal. It is the seat of the fire of life, the *vital heat*. Life begins with the kindling of the fire and ends with its quenching. The pulsation of the heart resembles boiling caused by the vital heat expanding the blood. The temperature of the vital heat must be regulated lest it burst into flames consuming the body. Refrigeration is accomplished by cold air brought into the heart from the lungs via the pulmonary vessels. This false view of the heart and of blood's nature and movements stood fast until the seventeenth century CE.

Upon their delivery to the tissues, the blood nutrients convert into the body's constituents in a process now called *assimilation*. Aristotle explains the process as an actualization of a potential that blood acquires in the series of concoctions. The key factor in this explanation is the vital heat. What exactly is the vital heat? Before we answer this question, we must introduce two other concepts with which he operates in this context: *psyche* and *pneuma*. The first of the two is what we already know as one of the aliases of the Form. To what we have learned already about the psyche or soul, we need to add here that Aristotle distinguishes three kinds of soul, which he calls nutritive, sensitive, and rational. All living creatures have the *nutritive soul* concerned with nutrition, growth, and generation—the lowest and most basic grades in the expression of life. All animals have, in addition to the nutritive soul, also the *sensitive soul* concerned with sensitivity/perception, desires, and locomotion. Only humans have in addition to the nutritive and sensitive souls also a *rational soul* concerned with rational thinking and all that comes with it. According to the types of soul they possess, living beings fall into three hierarchical categories in the order plants, animals with the exception of humans, and humans. The three kinds of soul reflect, according to Aristotle, three major groups of activities through which life manifests itself. These activities are giveaway signs of actuality (as opposed to potentiality), and since, in Aristotle's metaphysics, actuality is the Form shaping (informing) the matter, *psyche* is, as we already know, another aspect of the Form. It is an aspect that emphasizes the active side rather than the structural (*morphe*) features of the Form. Although all objects, both animate and inanimate, are composed of matter and Form, only the animate objects have a Form characterized by the special activities that go under the appellation of *psyche*. Present-day biology too distinguishes two aspects of living things: the organization of their bodies, which corresponds to Aristotle's *morphe*, and the activities, which Aristotle calls *psyche* but which modern science refers to as the *functions* of the body. Corresponding to these two aspects is the distinction between two branches of biological sciences: morphology/anatomy and physiology. In Aristotle's biology, the two aspects of the Form pertain not to individuals but to groups of individuals that go under the name of *eidos* or species. The three terms (*psyche*, *morphe*, *eidos*) thus come together as three different aspects of the Form. In Aristotle's metaphysics, Form is actuality and matter potentiality. He therefore defines *psyche* as the Form (of a natural body) that has the potentiality of life. To drive this point home, he distinguishes two senses of actuality, often referred to as the first and the second. When a body is capable of exercising an activity but is not manifesting it, for example, in the case of a man asleep, it is said to be in the state of *first actuality*. Whereas, when a body not only has the capability of certain activity but also exercises it, as in the example of a man awake and going about his business, it is in the state of *second actuality*. Taking into account this distinction, Aristotle defines *psyche* as the first actuality of the body that potentially possesses life and he calls such a body *organic*. Hence to say that something is *ensouled* (i.e., that it possesses soul) is to say that it is alive. Obviously Form and soul are not synonyms since soul is only one kind of Form. At the same time, however, Aristotle assigns to living substances a special position in the universe of things in that he makes them paradigmatic of all substances, as if in the strictest and most proper sense, living things were substances exclusively.

Since *psyche* is idiosyncratic with life and life is in every part of a living body, then so must also be the soul. Nevertheless, Aristotle allocates the central seat of all three kinds of soul to the heart. In the light of what we said about *pepsis*, the allocation of the nutritive soul to the heart makes sense. But why does he assign the other two kinds of soul to the heart as well? In the case of the sensitive soul, the simple reason might seem to be his belief that it is blood that transmits stimuli from sense organs to the heart, which then acts as a coordinator of responses to them. This explanation, however, has two hitches. First, the function of blood is the delivery of nutrients to the different parts of the body, and according to his theory, each tissue can only have one function. And second, Aristotle specifically denies the possibility that blood is the carrier of sensory stimuli. He does not exclude, however, the possibility that something else present in blood is the carrier. It has therefore been suggested that the carrier is the *pneuma*.[24,25] Because of its involvement with perception, *pneuma* is also thought to be behind Aristotle's placement of the rational soul in the heart. But what is *pneuma*?

According to Aristotle *pneuma* is air, but not just any air. Not only is it hot, but also the heat is of a special kind, which he variously refers to as "soul heat" (*thermoteta psychiken*) or "vital heat" (*thermoteta zotiken*) and describes it at one point as analogous to the element of the stars. This sentence calls for a brief excursion into Aristotle's cosmology. He views the universe as a set of concentric celestial spheres enveloping the earth in the center. The closest sphere to the earth is that of the moon. This lunar sphere divides the universe into the terrestrial or sublunary realm and the heavenly or the celestial realm. In the terrestrial realm, things are made of the four elements, are subject to change, and move naturally in straight lines, up and down. All celestial bodies are made up of a fifth element (*quinta essentia* in Latin) or *ether*. In Greek *aither* meant originally the pure, fresh air breathed by the gods, and so the word had from the beginning the connotation of divinity. Aristotle's ether is unaging and unchanging, even though it is a material element. In contrast to the terrestrial elements, ether has circular natural motion, which is perpetual. It pervades the entire heavenly realm, but is absent in the sublunary realm. It lacks qualities analogous to hot, cold, dry, and wet, characteristic of the terrestrial elements. *Pneuma* is not ether but it contains something that behaves ether-like in certain biological and psychological situations. Opinions differ on what exactly Aristotle's reference to ether in the context of the vital heat means. The two opposites in the range of views are a straightforward nonmystical interpretation on the one hand and mystical on the other. The latter extends the divinity of the ether in the heavens to the vital heat down on earth. The nonmystical interpretation[24-26] attributes to Aristotle the view that *pneuma* is like ether insofar as they both have a similar effect on certain physiological processes, especially generation. In this interpretation the vital heat in the *pneuma*, though not etheric in its constitution, is credited of being able to endow the male *sperma* with similar power in sexual generation. But if we reject the divinity of the vital heat, what alternatives remain for Aristotle to explain its presumed effects? Here we must remind ourselves that the ancient Greeks commonly identified life with heat and equated digestion with a kind of cooking. These two beliefs go a long way toward explaining some of the connections Aristotle makes in his physiology: fire—heat—life—birth—production of new substances—digestion—generation.[23] The question is: Is the heat produced by fire the same as the vital heat produced by the body? Seemingly not, because the ordinary fire burns things to ashes rather than bringing anything to life. On the other hand, one could argue that fire and vital heat have different effects, but sun and vital heat have similar effects and that it is this latter similarity which Aristotle has in mind when he draws the analogy between ether and vital heat. Indeed, he says that heat can power different processes, much like a baker who uses heat to bake different products such as bread, cookies, or cakes. Taking a present-day stance, we might say that the vital heat is nothing more than energy that powers metabolic processes in the body.[26] It was in fact Aristotle who coined the word "energy" (*energeia*), along with another term—entelechy (*entelecheia*). He uses these two words nearly interchangeably, but they do have subtly different meanings. He applies both to the actualization of potentialities but uses *energeia* for powering of the process and *entelecheia* for its completion.

To sum up, *psyche* as a Form associated with matter is a substance, as is *pneuma* (warm air), whereas vital heat is a power or energy. Aristotle uses all these terms in explications of what we would today call *metabolism* (from Greek *metabole*, change), the set of chemical changes, which provides energy for the construction and degradation of living matter and for the execution of its functions. Aristotle's description of metabolism may seem naïve and primitive, but in reality it comes amazingly close to capturing the essence of the process. It is now known that blood, heat, and sun energy do indeed play critical parts in metabolism; that blood carries oxygen, which constitutes 21 % of the air; that air is critically involved in reactions charging certain molecules with energy; that heat is intimately associated with energy, as indicated by the fact that food energy is expressed in calories, which are units of heat; and that the sun is directly or indirectly the source of energy for most of life on earth. Moreover, the production of energy by a process involving oxygen is often called "burning;" metabolism does indeed start with the processing of food and the production of nutrients; and metabolism does underlie processes such as growth, repair, locomotion, and generation. Understandably, however, when it comes to details of metabolism, Aristotle's explanations are often well off the mark.

The Quest for the Female Semen. Beside the origin of the male semen, the second much debated question among the Greek philosophers in connection with animal generation was whether females produced an equivalent semen and if they did, what was its nature? Virtually all of them agreed that females produced something related to generation, but they differed in their views on the nature of this "something." The most common attitude was to identify the so-called female semen with the monthly blood flow or *menses* (from Latin *mensis*, month), *katamenia* in Greek. This was also Aristotle's position. Simplifying somewhat his complex explanation, one might say that *sperma* and *katamenia* result from the same process that produces nutrients and that generation is akin to the assimilation of nutrients by tissues. The difference between the two processes is in that the *sperma* undergoes an additional round of *pepsis*, which ordinary nutrients do not.

Where the additional round of concoction takes place is unclear. Some scholars read Aristotle as suggesting that it occurs in the male reproductive organs, presumably the testes, while others place the event in the heart. Wherever it happens, the final bout of concoction changes not only the appearance of the product from red to white and its consistency from fluid to viscous but also its potential. A similar process produces, according to Aristotle, also milk in females and fat in both sexes, neither of which possesses the potentialities of the *sperma*. The additional round of *pepsis* is not accorded to the *katamenia*, however. The latter therefore retains the red color of the blood, although it too acquires, at some stage of the process, new potentialities, different from those of the *sperma*. Aristotle claims that the essence of assimilation, the conversion of blood (nutrients) into homoiomerous substances (the same in structure) of the individual tissues, is the transfer of the Form from the tissue to the nutrients. As blood reaches particular tissue, the latter communicates to the former the Form it possesses. In other words, the tissue imposes upon the nutrients its own Form by making them to assume the same organization as the tissue itself has. The vital heat presumably effects this whole transaction.

According to some interpreters,[24] Aristotle's view is that generation operates on a similar principle. His fundamental assumption is that both sexes contribute to the generation of a new individual, but unequally. The male, through his *sperma*, contributes the Form (*eidos*), whereas the female, through her *katamenia*, contributes the matter (*hyle*). The Form of the *sperma* imposes itself on the matter of the *katamenia*, when the two substances meet at coition. The imposition is effected by the vital heat, which the *sperma* acquired at the extra round of concoction. This concept of generation raises numerous questions, to which various Aristotelian scholars give diverse answers, since Aristotle's own answers are mostly ambiguous. In what follows, we introduce some of these problems and a selection of solutions offered. Since, according to the hylomorphic doctrine, all objects consist of matter and Form, the first question one might ask is: What happens to its own Form when the *katamenia* receives the Form of the *sperma* and, reciprocally, what happens to the matter of the *sperma* when it loses its Form? Aristotle's answer to the first part of the question is presumably: The same thing that happens to any object when it acquires a new Form; the old Form goes in the privation mode. Aristotle's answer to the second part of the question is unclear: Does it disintegrate as in decay or does it integrate into newly forming individual? We return to this question later. A second question is at once more difficult and more important: Where and when does the *sperma* acquire the Form, which it then communicates to the *katamenia*? If man begets man, as Aristotle says, then the Form of the *sperma* must be that of the human species (*eidos*) rather than a Form of a tissue as in nutritive assimilation. Hence, the only place in which the *sperma* can take on the species Form is the heart, which Aristotle holds for the center of an individual's life and for the seat of the soul, which is also the *eidos*. The problem is that in Aristotle's system, the Form is actuality, which in the case of living beings is the *psyche*, the *psyche* of a living being is life, and life is organization and activity. But the *sperma* does not show an organization in which one could recognize the species. Aristotle explicitly rejects any kinds of *preformation*, which postulates just such kind of a structure and activity in the semen. One possible resolution of this dilemma is to assume that the Form is, in modern vocabulary, like a computer program, which must be opened to start running. The opening of the species program in the *sperma* is the encounter with a matter receptive to just this kind of program—the *katamenia*. This suggestion comes close to saying that the actuality in the *sperma* is really a *potentiality of actuality* or *first actuality*. We come back to this problem in the penultimate section of the chapter.

The Origin of a New Individual. All this brings us to the third and the most important question: How does the development of a new individual begin? According to Aristotle, the primary requirement for the initiation of the development is the encounter between the *sperma* and the *katamenia* in the uterus of the female parent. The *sperma*, which during the coitus has entered the womb, is informed, ensouled, or impressed with the Form (the nutritive and sensitive soul) of the male parent from which it comes. More specifically, it has been endowed with the capability to enfold the structure and movement (activities, functions) of all the different tissues comprising an adult male of the species. Likewise, in the female parent's heart, the *katamenia* has been prepared for the encounter by the imposition on its matter of receptivity (potentiality) to receive the Form (soul) of the male. Later, the encounter—the mixing of the *sperma* and the *katamenia* in the uterus—came to be called *fecundation* (from Latin *fecundus*, fruitful) and later still, after the true nature of the process had been elucidated, *fertilization* (from the Latin *ferre*, to bear fruit). In the act the *sperma* imposes its Form upon the receptive matter of the *katamenia,* and the Form begins to convert the matter's potentiality into actuality. The intermediary between the Form and the matter is the vital heat associated with the *pneuma*. Both of these are provided chiefly by the *sperma* and come originally from the last concoction in the heart. The association with the *pneuma*, combined with the presence of the ensouled (informed) matter, furnishes the vital heat with a creative power, which ordinary fire-produced heat lacks. Other than that there is no principal difference between the vital and the ordinary heat. One way of interpreting the communication between the Form and the matter in the actualization of the latter's potentiality is to imagine that the vital heat, with the help of the *pneuma*, warms up or cools down the *katamenia* to precisely the temperature required for the particular movement (action) prescribed for the specific time and stage of development. To understand this interpretation, it is important to remember that Aristotle operates with only four elements and four qualities; that among the latter the pair of warm-cold opposites is dominant and determines the dryness and wetness of a substance; and that the ratio of the four elements determines the activity of the substance. It is only against this background that the creative power of the vital heat makes good sense and averts the need to invoke the divine power of the fifth element.

Embryonic Development. Aristotle calls the effect that the *sperma* triggers in the *katamenia sunistanai* or *setting*, which he defines as imparting the proper movement to the *katamenia*, that is to say, a movement with which the *sperma* itself is endowed. In the English translation of the Greek term, the word "setting" is used in the sense of causing a fluid substance to become firm or solid, as in the case of setting milk for cheese. Indeed, Aristotle compares the action of the *sperma* on the *katamenia* to that of rennet on milk:[27] It causes the fluid *katamenia* to coagulate into solid curds. As in the case of milk, where the rennet does not become part of the curds, the matter of the *sperma* does not enter the coagulum, but instead dissolves into *pneuma* and evaporates. The male Form is thus an impulse that sets in motion a process, which then sends an impulse to initiate another process, and so on, in a manner of a chain reaction. Aristotle, who does not know about chain reactions, prefers to think about the *fetation* (the formation of *fetus*, from Latin "newly delivered"), in terms of a mechanical toy, in which the movement of one part triggers the movement of another part, and this a movement of a third part, and so forth. All these transfers of movements follow sequentially the first movement. This, according to him, is how fetation (or embryonic development, as we say today) works. In the case of the mechanical toy, it is the player who provides the first impulse. In the case of fetation, it is the *sperma* in the form of the vital heat from the last round of *pepsis*, which initiates the setting. The embryonic development is as orderly as the transfer of movements in the mechanical toy. The reason for the orderliness is different in these two cases, however. In the case of the mechanical toy, it is in the structure, in the way the toy is constructed. In the case of fetation, the reason for orderliness is the Form. The development follows a purpose, which is the actualization of the same *eidos* as that of the parent. By having the beginning in its end, the species achieves immortality by the concatenation of endless cycles of generation and decay.

Aristotle provides a rather detailed description of the way he interprets the development of the vertebrate embryo. It is based in part on his own observations of chicken egg development. Here we give only a few selected highlights of it.[26] In viviparous animals, the *sperma* and the *katamenia* meet in the vagina and the mixture is then drawn into the uterus, where it coagulates. The surface of the liquid surrounding the central coagulum cools into a scum that develops into fetal membranes around the embryo. A root-like structure sprouts out of the coagulum and connects with the uterine wall from which henceforth the embryo draws nutriments. The embryo's potentiality to develop all body parts actualizes stepwise in a fixed order, in accordance with Aristotle's concept of *epigenesis*, that is, the gradual, successive differentiation of individual body parts. The first organ to develop is the heart. From it then originate all the other body parts. Blood vessels sprout from the heart in all directions and branch out from the main trunks. Blood, which is essentially nutriment, seeps through pores in the blood vessels and develops into flesh, skin, and bones. Shortly after the heart and near it emerges the brain, the chief cooling organ. Other organs then follow in a rapid succession.[28]

Generation in Plants. It may seem incomprehensible that sexuality of plants had remained unrecognized through much of human history. After all, flowers display impudently their sexual parts for every human eye to see and some plants, as if in mockery, even mimic human genitals.[29] Every meadow, every coppice, and every orchard bustles and buzzes with their sexual activities and people have suffered for ages from pollen allergies. Yet, it had not been until some 300 years ago that the first convincing evidence for sexuality in plants had emerged (see Vol. 2 Chap. 2). The usual explanation for the delay is human prudery about anything sexual, but the fact is that societies relatively open about sexual matters had not been more enlightened about plant sexuality than the puritanical ones. In ancient Greece, sex had been free of the odium that Christianity and related religions had later imposed on it, yet, Aristotle, as well as his student and successor Theophrastus of Eresos (c. 371–c. 287 BCE), "the father of botany," knew much less about the sex life of plants than an elementary school pupil knows today. The real reason for the delay must have been therefore something else, perhaps the fact that the true similarity between plant and animal sexuality lays at the microscopic level, which had not become accessible until 300 years ago.

Actually, Aristotle did believe in plant sexuality, but in one that did not express itself morphologically in differences between males and females. Rather, he thought that plants were unisexual and that the reason for this was that they were, in contrast to mobile animals, affixed more or less permanently each to one spot. This feature would have made contact between males and females for sexual union, if such had existed, impossible. Aristotle posits that instead of semen and *katamenia*, therefore, plants produce seeds, which are equivalent to animal eggs. An animal egg consists of two parts: germ, from which a new individual develops, and nourishment. Similarly, a seed contains a germ of a plant and nourishment for the shoot and the first root. A seed arises by concoction of nutrients in the pericarp, the part of the fruit enclosing the seeds. The concoction is effected by vital heat, which originates in part in the plant but comes additionally also from the outside. It is unclear, however, how Aristotle attuned this concept of unisexuality (two in one) with his general theory of generation, in which the male provides the Form and the female the matter.

Spontaneous Generation.[30] Most ancient Greeks, common people and philosophers alike, believed that certain animals arose without parents of their own kind, and Aristotle shared this view with them. The notion that life could spring from a source other than already existing life came to be called *generatio spontanea* in Latin and *spontaneous generation* in English, whereby the word "spontaneous" was used in the sense of growing naturally, without being planted, as in the Latin root *sponte*, "of its own" (the Greek equivalent being *automaton*). Aristotle's problem was to explain how the same species could repeatedly arise spontaneously. In sexual generation, the same species arises again and again because of the Form that the parents provide and that guides the development of the new individual toward this very same species. In spontaneous generation, however, an individual arises from matter such as mud, dirt, or putrefying stuff, which has no resemblance to the organism (fish, frog, mouse) that arises from it. In spontaneous generation, there is no Form that could inform the matter how it should change into the shape of, say, a mouse. Aristotle could not have possibly believed that pieces of dirt could by chance assemble, all by themselves, into a living being such as a mouse. Amazingly, however, he did find a way of sidestepping this improbability problem and came up with an explanation, which—though false because it is based on untenable assumptions—would have had some merit if the assumptions were valid. Let us reiterate: In ordinary generation, the Form is passed from parents to offspring like a baton in a relay race. In animals, the act of passing the "baton" involves two individuals, the male and the female. In plants, the male and the female principles exist in the same individual, and the passing of the "baton" takes place within that individual. In both animals and plants, out of the mix (*kymena*) of the male and female principles develops a new individual. In both cases the male principle contributes the Form and the female principle the special matter of the *kymena*. Also, in both cases, the two principles are generated by concoction, the male principle becoming a special vital heat present in the *pneuma*, and the female principle a special matter. The principles are special in that the male vital heat is programmed to actualize the potentiality present in the matter of the female principle. To explain how spontaneous generation can give rise to the same species repeatedly, Aristotle invokes the same two principles: vital heat programmed to actualize a certain Form and matter with the potentiality to assume that Form. The creator of the programmed vital heat (Form) is, according to him, the sun, and the sources of the special matter are primarily two of the four elements—water and earth. This claim calls for a second brief digression into Aristotle's cosmology.

According to Aristotle, the sun itself is not hot, for it is made of ether, which does not have any of the qualities characterizing the terrestrial elements. Rather, the sun-heat is a mere by-product of the sun's motion and the resulting friction. The friction causes the air below to ignite, and it is the heat of this fire above the earth that we perceive as sun-heat. Here on earth, there are certain environmental pockets where, by chance, specific earth-to-water ratios occur and where, also by chance, sun-heat creates specific temperature conditions that allow certain organisms to develop without parents. Aristotle thus manages to integrate the notion of spontaneous generation into his overall concept of generation, but only by stretching the doctrine of the vital heat to its limits. Within the framework of his speculations, he makes it appear possible that in small environmental pockets, conditions might conspire to mix earth and water in such proportion that at the right temperature of the air, a process might "ignite," which then the changing heat conditions stir toward the development of a living being. After all, this seems to be happening when seeds are planted into the soil or tortoise eggs are deposited in the sand and warmed up by the sun—with one important difference: The seeds and the eggs come from parents of the same species as the germinating plants or the animals hatching from the eggs. As biologists now know, only life in the form of seeds, eggs, and other form of germs can beget life. The way Aristotle imagined it as an interplay of chance (the creation of the right conditions) and necessity (the development steered by these conditions), spontaneous generation can no longer occur on earth. Biologists believe, however, that life on earth once started by *abiogenesis* from inanimate matter under the special primal conditions on the planet, and then evolved gradually from very simple organic structures to more complex forms.

Aristotle's Theory of Inheritance[16,31]

We reemphasize that dividing Aristotle's concept of generation into generative and transmission of resemblance parts is against the spirit of his treatment of the subject. Hence, calling the latter part "heredity" or "inheritance" is an anachronism comparable to a soldier lighting a cigarette in a movie about Ancient Rome. The English terms "heredity" and "inheritance" and their equivalents in other languages, including Greek, had not been introduced into biology until the second half of the nineteenth century (see Vol. 2 Chap. 2). Indeed, the Greek equivalents of these two terms do not appear anywhere in the whole text of Aristotle's *Peri zoon geneseos*, his major work on this subject, now better known under the Latin title *De generatione animalium*.[16] You may find these terms in some translations of the work, but such cases must be taken as anachronistic inaccuracies. To give an example, an accurate, word-by-word, translation of the Greek sentence *pithana de kai ta toiayta martyria taytais tais dokses; oy gar monon ta symfyta proseoikotes ginontai tois goneysin oi paides, alla kai ta epikthta* is this: "plausible are these evidence for these opinions; not only by birth (innate) resembling become to the parents the children, but also by acquired," which in the context of a scholarly translation reads *"Children are born which resemble their parents in respect not only of congenital characteristics but also of acquired ones."*[32] However, in an attempt to make the meaning of the sentence better understandable to a modern reader, an English translator may be tempted to choose the word "inherited" instead of "congenital" as in this translation: *"Children are born resembling their parents both in their whole body and in its individual parts.. .Moreover this resemblance is true not only of inherited but also of acquired characters."*[33] The current Greek word for "inheritance" and "heredity" is *klhronomikothta* or *klironomikotita*, in which the *klhros* or *kliros* stands for a piece of wood used in lottery or voting, and the *nemw* (or *nemo*) means "to distribute." Together, the meaning is "transferring a set of characteristics through birth."[34] Apparently an equivalent to "heredity" and "inheritance" did not exist in ancient Greek. This absence is also suggested by the quote heading this chapter. Here, too, we might be tempted to replace Athena's roundabout talk[1] by saying simply: "If you, Telemachos, have inherited your father's great strength ..." These arguments may seem like linguistic haggling over nuances. In reality, however, the absence of a technical term for the transmission of resemblance in ancient Greek is a telling point. It indicates that at that time, the phenomenon was held for, at best, a side issue to the phenomenon of generation or, at worst, for a deviation from the normal process of *genesis*, whose vital goal was the perpetuation of the *eidos*, the species. The seeming unpredictability in the transmission of individual characteristics surely must have meant to the ancient Greeks that chance, rather than rule played the decisive part in the process. And chance, for them, meant deviation from norm, fickleness, and abnormality. With this view in mind, let us now have a look at some of the mental acrobatics Aristotle had to resort to in order to explain how Socrates, in the process of becoming a human being, got his maleness and his snub nose, short stature, together with the rest of his individual characteristics.

Aristotle distinguishes four levels of resemblance between parent and offspring when he says that humans beget humans. At the broadest level, the product of the procreation act, the offspring, has, like the parents, the features common to all animals, which to Aristotle represent the genus, into which the human species belongs. In his rendering, these features include the possession of both the nutritive and sensitive soul, as well as a body composed of flesh and bones. At the next, the species level, the offspring possesses, like the parents, features common to all human beings and different from all other animals: featherlessness, bipedality, and possession of a rational soul. At the third, the gender level, the offspring resembles either the father or the mother in what would today be called primary and secondary sexual characteristics. It is either a male or a female. Finally, at the fourth level, the level of the individual, the offspring resembles the father, the mother or some more distant ancestor in characters other than those shared by all human beings, by all human males or by all human females.

The fourth level of resemblance differs from the other three levels in that it is based on comparisons of individuals within a group rather than on comparisons of groups of individuals. The comparisons reveal that all individuals, with the exception of identical twins, are unique. The uniqueness is patently apparent to us when we compare individuals of our own species; to become aware of it in other species may require meticulous observation. To recognize uniqueness of individuals, it helps to compare not their overall appearances but rather individual features and thus to view an individual as a mosaic of *characters* and search for differences in *character states*. For example, eye color is a character, whereas blueness, grayness, and greenness of eyes are different character states. The uniqueness of an individual is in the specific combination of the character states. Modern science calls the particular combination of character states the *phenotype* (from Greek *phainein*, to show, and *typos*, type). Thus the uniqueness of Socrates is in the combination of character states such as shortness of stature; swaggering gate; wide-set, piercing, and bulging eyes; upturned, broad, snub nose with flaring nostrils; and wide mouth with large fleshy lips. The appearance of some of the character states seems accidental in that they do not reappear in the offspring; these, therefore, do not interest us here. Others "run in families" and these are the ones that Aristotle uses in the description of the four levels of resemblance. In principle, the resemblance at any of the four levels can be dissected into individual characters; the difference is only how widely the characters are shared. This lack of *principal* distinction among the four levels must also be the reason why Aristotle does not see any need for giving the intraspecies characters a special status under a distinct name. From his point of view, characters at all four levels are inherited. If he were to use this term, though, those at the fourth level are inherited least predictably. Indeed, he might have viewed the fourth-level characters as the proverbial monkey wrench thrown into his philosophical system. Here is why:

The concept of resemblance presupposes *variability*. If things are the same, we speak of *identity*. It is only when they differ that we speak of resemblance. Variability, however, presents a problem for Aristotle in that he has no place for it in his hylomorphic doctrine. If Form determines the appearance of things and if it is immutable, where does variability come from? He follows Plato and blames variability, which is a departure from the norm, on matter. When Form thrusts itself upon matter, the latter resists the imposition. Unless it manages to master the imposition fully, differences from the norm represented by the Form arise. In this view, variation is de facto a deviation from the standard, and if one holds the standard for perfect, the deviation becomes a defect. The seriousness of the defect depends on the level of resemblance at which it occurs. If it occurs at the level of the genus or species, it results in what Aristotle calls a *monster*. If it occurs at the gender level, it leads to the opposite of maleness, which is femaleness. It is in this sense that he calls femaleness a "deviation" because the norm is the male Form. It is at the level of characters differentiating the individuals of the same species that he runs into the problem. Actually, the problem lurks already at the gender level, because Aristotle leaves unexplained the source of the femaleness: Where does the femaleness, the opposition to the maleness, come from? If from the female, does it mean that it is part of the female's Form, and if so, that the female contributes part of her Form to the embryo? The same questions arise at the level of characters differentiating the individuals of the same species. Does it mean that the characters are in some manner part of the male and female Forms? If so, it would mean that Aristotle would have to extend the concept of Form from the species to the individual level. Some scholars think that this is indeed what he is doing. But Aristotle nowhere clearly states that this is what he means, and for obvious reason. Admitting the possibility of individual-specific Forms would undermine the whole metaphysical infrastructure of his philosophy and threaten to collapse it. The problem thus remains unresolved from the interpreters' point of view and perhaps even for Aristotle himself.

Undeterred by these problems, Aristotle pushes ahead and offers a physiological interpretation of inheritance.[16,18,35] His explanation of resemblance between offspring and their ancestors has some remarkably modern features. The first of these is that he seeks an explanation applicable to all four levels of resemblance, and he does so despite the fact that it brings him in conflict with parts of his metaphysical doctrine. His unifying approach to generation makes sense since the basic stance "like begets like" applies to all levels of likeness, from the genus down to the individual. It is therefore reasonable to think that the physiological mechanism responsible for this likeness is essentially the same at the different levels. We will argue later (see Vol. 2 Chap. 3) that this is also the stance Mendel assumed when he began his experiments with the hawkweed. Mendel's followers, on the other hand, started from the position that the study of inheritance at the individual level should be separated from that at the other three levels. They distanced themselves from the old concept of generation and banded together under the banner of *genetics*. It would only be much later that further developments would reveal that Aristotle's unifying stance was right after all. The second modern feature of Aristotle's physiological explanation is that it is based on an approach that could be called *atomistic*. When we notice a resemblance between two individuals, generally we perceive it as an overall likeness, without consciously paying attention to specific characters that the individuals share. Aristotle is perhaps the first to view an individual as a mosaic of characters, each of which has to be dealt with separately. Here again, this would much later become the standard method on which genetics would be founded. The third remarkable feature of Aristotle's physiological analysis of inheritance is that he separates the cause from the effect in the mechanism responsible for the likeness between individuals.[20b] He holds the character for an effect caused by an agent responsible for its appearance. In the terminology of modern genetics, the individual characters are part of the *phenotype*, whereas the agents responsible for the development of the characters are the *genes*, which are part of the *genotype*. The implication of this separation is that the appearance of the characters is a process having its beginning in some material substance. Here once more, Aristotle foreshadows Mendel, who in modern times would become the first scientist to make this distinction.

Aristotle calls the agent that initiates the developmental process leading to the appearance of a character *dynamis* and the process itself *kinesis*. In the present context, *dynamis* is the physical substance in the semen capable of initiating a movement leading to the appearance of a particular character in the developing embryo. The *dynamis* is the special vital heat present in the *pneuma*. Depending on the specific qualities of the vital heat, different movements are triggered and different characters emerge. The variation in the quality of the vital heat also gives the female a chance to contribute to the developing embryo more than just matter. The variation and the consequent interplay of the *dynamei* with the potentialities present in the matter on the maternal side enable Aristotle to explain the different modes of character transmission, which we now call inheritance.

The principle of the transmission is the same at all four levels of parent-offspring resemblance. The standard setting of generation reproduces the male parent as accurately as the resistance of the maternal matter allows it. Any departure from the standard, viewed by Aristotle as a deviation, is the result of interference with the actualization of the male program. Some of the interferences are trivial and result in chance deviation; others are more profound and are effected by movements initiated by the maternal side. The latter deviations can be either of a "changeover" or of a "relapse" type. In the *changeover type*, the embryo acquires a character opposite to that which would have been effected by the specific *dynamis* and *kinesis* of the male semen. The most profound example of a changeover is the switch of the program to producing a female instead of a male embryo. In the *relapse type*, the embryo acquires a character state from a more distant ancestor than the parents. Important points to remember are, first, that for each character there is a separate *dynamis* and *kinesis* and that the different *dynameis* act independently of one another. Not only that, but also, and that is the second point, the *dynameis* responsible for the different levels of resemblance assert themselves independently. Using Socrates as a representative human being, we now illustrate the various possible outcomes of generation on the example of his having children with his wife Xanthippe. For simplicity, we consider only the third (gender) and fourth (individual = Socrates/Xanthippe) levels of resemblance and take all the Socrates-specific characters *en bloc*. The standard outcome of Xanthippe's pregnancy would be a boy, who would take after the father—in fact, it would be Socrates' clone. This, according to Aristotle, would happen if the impregnating *sperma* had a very high content of vital heat in all gender- and individual-determining *dynamei* so that it would have been able to master any resistance put up by Xanthippe's *katamenia*. In the opposite situation, in which Socrates semen's vital heat would be very weak in both kinds of *dynamei*, while Xanthippe's *katamenia* would be strong, the outcome would be a clone of Xanthippe, because the semen would fail to overcome any resistance put up by the *katamenia,* and a complete changeover would be the outcome. Now, if Socrates' gender-determining *dynamei* were strong, but all the individual-determining *dynamei* were too weak to overcome the resistance of the *katamenia*, Xanthippe would have brought to the world a boy, who would take after his mother in all characters. In the opposite situation, an unlucky girl would have been born who would look like Socrates. In all these situations, we have assumed en bloc the inheritance of character at each level. In reality, the probability of this happening is very low. Instead, some of the individual-determining *dynamei* of both the father and the mother could be strong and others weak in random combinations, so that the son or the daughter would take in some characters after the father and in others after the mother. If it so happens that some of the individual-specific *dynameis* of either parent are not strong enough to accomplish mastering, the system goes into the relapse mode. It switches to the corresponding *dynamis* of Socrates' father, and if that does not work, then to Xanthippe's mother. If these two options fail, the switching continues to successively more remote ancestors (Socrates' grandfather, Xanthippe's grandmother, and so on). All these *dynameis* lie "in reserve," so to speak, or, in Aristotle's vocabulary, they are present as potentialities that can be actualized when need be. Aristotle seems to assume that the parents bear in potentiality the whole histories of their respective lineages. In this way, Aristotle explains the reappearance of character states that have been silent over several reproductive cycles (what we now call "generations") and then suddenly reappear in one of the offspring. He goes even further and postulates that the chain of relapses can reach beyond the species into the genus level. In the case of the human species, the genus is, in his reckoning, an animal other than human. When this happens, a "monster" is born, a creature displaying characters of different species. This retracing of characters along a genealogical lineage may seem to be implying an evolutionary scenario, but such interpretation is far from Aristotle's mind.

Viewed from a present-day perspective, Aristotle's theory of inheritance is remarkably modern in its generalizations but wanting in specific details. Its modernity is in its integration of the concept of inheritance into the general notion of reproduction and development, in the recognition of the need to dissect inheritance into units, in the distinction between characters (phenotype) and entities that determine them (genotype), in the formulation of regularities in the pattern of inheritance, in the recognition of the principle of dominance (and recessivity) of characters, and in an attempt to provide a physiological interpretation of the mechanism of inheritance. Although incorrect in detail, his is a very clever interpretation of heredity.

Aristotle's Species Concept[36,37]

The existence of words in the ancient Greek dictionary, equivalent to the English words "man," "dog," "horse," or "blackbird," reveals that the ancient Hellenes had some notion of animal species. They generally distinguished species the way common people have done since time immemorial: by their *appearances*. This is, actually, what the word "species" meant originally in both Latin and Greek languages. The Latin word derived from the verb *specio* means "to look" or "to behold," and the Greek verb *eido* stands for to "see" or "to observe." The words *eidos* in Greek and *species* in Latin even had the connotation of an apparition, of something unreally real, for everybody could say he or she had seen a *particular* dog but not a dog species. Everybody also expected that a pregnant woman would deliver a human being, an impregnated bitch a dog, and a gravid mare a horse. Aristotle was, of course, familiar with this common view of species, but as a philosopher he also reflected on species in more sophisticated ways. In his logical and metaphysical works, he used the word *eidos* in the sense of Form, with all its diverse connotations described earlier in this chapter under a variety of names. An application of the philosophical species concept in biology should not have therefore presented a serious problem to him. Theoretically, Aristotle's concept of species should be applicable to any part of reality and hence also to living things. Yet, since to this day neither biologists nor philosophers can agree on a single species concept, it is perhaps not surprising that scholars cannot even agree on what Aristotle's biological species concept is. According to the traditional interpretation, Aristotle managed to combine metaphysics with biology splendidly. The trouble is, however, that now many scholars are seriously questioning this conclusion.[38] Let us therefore have a quick look at the main points of the traditional interpretation.

To reiterate, in Aristotle's usage, the word *eidos* has two meanings. First, it means the appearance of an individual, its *morphe*, or Form, and second, it stands for a species, a group of individuals sharing the same Form. The connection between these two meanings is the act of generation,[38] that is, the passage of the Form from the parents to the offspring and the resulting close resemblance between the begetters and the begotten. It is on the basis of this resemblance (appearance) that the offspring is assigned to the same species as the parents. It is also on the basis of their appearance that individuals sharing the same Form and so belonging to the same species are capable to interbreed and produce fertile progeny. Furthermore, since the Form is the essence and the essence is what defines a species (a universal), the universal is present in some way in each individual (particular) member of a species. This presence makes the individual general enough to be knowable. Finally, since one interpretation of Aristotle's doctrine makes the Form the ultimate substance (*ousia*), it seems that all the important concepts of his metaphysics are applicable to his biology. Some scholars argue, however, that there are, in reality, several inconsistencies in the traditional interpretation. One of them is the fact that nowhere in his works does Aristotle state exactly what his concept of biological species is. Although he is the first to make a clear distinction between *eidos* and *genos*, and consistently holds the former for a lower classification category than the latter, he nevertheless keeps using both terms at different classification levels so that his species is always relative to the genus but at different levels. One can therefore only guess from the context at what level of generalization he actually uses the *eidos*. To make things worse, frequently he seems to use *genos* even at the lowest level, where one would expect him to use *eidos* instead. Uncertainties arise also regarding Aristotle's use of expressions such as "like begets like." It seems that Aristotle is a bit too permissive in allowing exceptions to this rule, citing anecdotal evidence for mating between species, of which we now know that they do not interbreed to produce viable offspring. All these and several other ambiguities make some scholars wary of accepting the validity of the traditional interpretation.

The impression one gets from the critique of the traditional interpretation is that of a considerable fluidity of Aristotle's species concept. The title of an essay "Aristotle: A Zoology without Species" used by one critic of the interpretation[38] in one of his essays is a hyperbole but the exaggeration is only slight. Aristotle has apparent difficulty in delineating the species both in downward and upward directions on his hierarchical scale of generality. This difficulty is, however, very modern: Biologists struggle with it to this day. It is therefore somewhat paradoxical that some twentieth-century biologists and some historians of biology accuse Aristotle of dogmatism hindering the acceptance of Darwin's theory of evolution. Specifically, they charge him with two major offenses that go under the names "essentialism" and "teleology."[39–42] Aristotelian scholars[43] and some other historians of science[44–46] reject these accusations. The first accusation is based on two claims attributed to Aristotle: first, that the Form is unchangeable and eternal, and second, that the Form is equivalent to the essence. We have seen, however, that Aristotle had problems with the unchangingness of biological Forms (species) and that certainly in the case of the human species, he was well aware of its variability. Note also that he defines the human species relative to the genus "animal" and that modern genetics recognizes groups of genes (e.g., the so-called *Hox* genes) that have remained remarkably constant during the entire animal evolution. Aristotle's position on the eternity of the Form is equally ambiguous. On the one hand, he accepts that Forms come and go in the process of generation/degeneration, but on the other, he introduces the concept of privation, which amounts to an admission of Forms persistence in a potential mode. He fails to clarify under what condition and how long the Form can last in the privation mode. He is probably aware of the difficulties into which eternally existing Forms would bring him. At any rate, he does not need the Form to be eternal in this mode, for the generation cycles seem to assure the Form's *virtual* eternity.

As for the accusation of essentialism, we must first clarify, what is meant by it. *Essentialism* is "a theory ascribing ultimate reality to essence embodied in a thing perceptible by the senses."[3] Since Aristotle's hylomorphic doctrine attributes to physical objects a dual nature, having them composed of matter and Form, the latter of which he also holds for essence, he is by this definition an essentialist. For the same dictionary holds an essence for "the ultimate nature of a thing," and for Aristotle the most fundamental nature of a thing is that which it has in common with all the other objects belonging to the same species and in which it differs from other species. But what exactly do Aristotle's critics mean by an "essentialist species concept?" Their answer is: It is a concept that is distinguished by four characteristics.[47] First, the species consists of similar individuals sharing in the same essence; second, each species is separated from all others by a sharp discontinuity; third, each species is constant through time; and fourth, there are severe limitations to the possible variation of any one species. How well then does Aristotle's species concept match this essential species concept? Well, we cannot say because, as we have seen, scholars do not agree on what Aristotle's concept of biological species is, or whether he even has one. The four points of the essentialist species concept match roughly Aristotle's *logical* species concept, but in the absence of his clear delineation of a *biological* species concept, it is futile to try to confront the four criteria with the latter. All one can say is that there are indications in Aristotle's biological works that can be interpreted as contradicting these points, with the possible exception of the first, but this is so because the concept of essence is rather fuzzy. One thing is clear, however: Aristotle is no evolutionist. The fundamental axiom of his metaphysics is that the universe has no beginning and no end. It enables him not only to escape the Parmenidean existence-nonexistence paradox but also the need to explain how things got the way they are. In this sense he does not need to be an evolutionist.

Teleology is the belief that there is purpose in nature. The belief exists in many variants differing in the coverage of natural phenomena and processes, the four main levels of coverage being organs, individuals, communities, and the world as a whole. Of the four, only the first two interrelated levels are relevant to the subject matter under discussion here. In the case of organs, it is the complexity of their structure that invites thoughts about their function and thus their purpose. In the case of individuals, it is the complexity of their development from an undifferentiated material. In both cases, the phenomena and processes give the impression of being goal-driven, an impression that there is a purpose or predetermined end (*telos* in Greek) which they strive to attain. The purpose seems to be an organ best adapted to its function and an individual best fitted for the environment in which it lives. An analogy with a human-made product—a house, say—forces itself to one's mind. The construction of a human dwelling requires building material, a blueprint designed by an architect, and workers to carry out the work. The analogy makes one think: Who is the designer of organs and individuals in the organic world? There are two theoretically possible answers to this question. First, there is no designer behind the construction of an organism and its organs; the organism constructs itself in accordance with natural laws. Most biologists think, however, that this explanation is believable only in combination with the theory of evolution by natural selection. According to this theory, the complex design evolves in small steps, each step being the outcome of a process, in which many random variants are rejected and only those that improve the adaptation of the organism to a particular environment (and an organ to a particular function) are allowed to persist. The second, nonscientific, answer invokes a supernatural forces or agents as being responsible for the design of organisms and their organs. The first answer is the "good" teleology accepted by modern biology. The second answer is a "bad" teleology, which some biologists attribute to Aristotle and his followers. Aristotelian scholars argue, however, that careful reading of Aristotle's works does not support the attribution. As we have tried to explain, he admits that the generation of individual organisms is driven toward a specific end, but he argues that the process is self-propelled, rather than being guided by a supernatural force.

Where Modern Genetics Meets Old Aristotle

At this point we might want to ask two questions: What is Aristotle's Form really? And what does it correspond to among the things that we now, more than 2300 years after Aristotle, recognize as reality? The principal assumption of Aristotle's theory of hylomorphism is that reality has a dual nature, consisting of matter and Form. Matter is the stuff of which an object is composed and Form is the way matter is organized. Modern science, especially biology, assumes essentially the same duality in that it distinguishes between the material structure of things and function determined by this structure. In biology there are sciences devoted to the study of structure (biochemistry, anatomy, morphology) and those investigating function (physiological sciences). Both Aristotle and modern science hold the stuff for material and the activity it displays for immaterial, but not in the sense many religions view immaterialness, that is, in the sense of spiritualness. By assuming immaterialness as a part of reality, neither Aristotle nor modern sciences violate the basic tenet of reality's physicalness. For both the manifestation of physicalness is the intimate association of the Form and function with matter (stuff). In this respect, one other parallel between Aristotle and modern science is the abundance of aliases for the immaterial component. Perhaps the closest that modern science comes to Aristotle's Form is the term *information* (note the obvious etymological relatedness of these two terms), especially in regard to the phenomenon which Aristotle calls "generation" and which modern science splits into heredity and reproduction. First in physics and then also in biology, information came to be a measure of order, applicable to any structure and any system.[48] It quantifies the instructions needed to produce a certain organization. Its counterweight is *entropy*, which is a measure of disorder.

In modern genetics one now speaks commonly of heredity as of genetic information. The physical carriers of genetic information are the DNA and their twins, the RNA molecules. The two kinds of nucleic acid are strings (two in the case of DNA and one in RNA molecules) of repeated subunits called *nucleotides*. There are four types of nucleotides, whose names are abbreviated to single letters A, C, G, and T (or U). It is the order of these "letters" that stores the information in the DNA molecules analogously to the letter order and grouping in forming words, sentences, and chapters in a book. All the information necessary for building, for example, a human being and keeping it alive is contained in 23 pairs of very long DNA molecules (chromosomes). The two strands of a DNA molecule enable it to duplicate itself and then pass a copy from each of the two parents to the offspring in the act of reproduction. It has been suggested[49] that Aristotle's concept of the Form prefigures the existence of DNA molecules. Aristotle, of course, knew nothing about molecules and the concept of information was alien to him, but he correctly grasped the essence of reproduction by postulating that something immaterial (the Form, the information) stored in a material substance (matter) was all that was required to give rise to a new individual in the process of generation. His grave mistake was that he reduced the female's contribution to a default situation. Only when the transmission from the male failed in part or as a whole, did the female get a chance to have things her way. In this he followed an ancient Greek belief that reality arose out of the interplay of two primal principles, the formative-male and receptive-female principles.[50]

The parent-to-offspring transmission of information is one aspect of generation, in which Aristotle prefigures modern genetics; another is the development of an embryo into a new adult individual. Modern genetics still struggles to get the whole picture of this very convoluted process. The essence of the process as it is now known is this. When the DNA molecules of the male and the female find themselves together in the same cell, the fertilized egg, the deciphering of the information encrypted in them begins. In a complex biochemical process, the first messages are transcribed into RNA from specific sites of the DNA molecules, and the messenger RNAs are translated into proteins, which fall into two categories. In the one category are proteins necessary for the growth and differentiation of the arising embryo. In the second are RNA and protein molecules that target new sites on the chromosomes and activate them to produce a second-generation RNA and protein molecules. These fall again into two categories—the effector and regulator molecules. The former join the teams that build the developing embryo, whereas the latter return to the chromosomes and find new sites to activate. And so the process continues step by step, in each step some of the previously activated sites being shut off, while new sites are activated, until all the informative sites along the chromosomes have been visited and used at the right moment when their contributions toward the development were called for.

The question then arises: What determines the order in which the different sites are activated? A popular answer was: a genetic program.[47] It assumed that the DNA contained a program specifying the order of activation and that the order was the result of the program's unraveling, similarly to the unfolding of a computer program. This explanation collapsed, however, when the complete DNA sequences of several different species were determined and no such program was found in them.[51] At this point, some researchers (those who had a healthy respect for history) remembered Aristotle. You may recall that he faced the same problem in his theory of generation. He solved it by postulating a progression similar to the operation of mechanical toys of his time. We do not know what the toys looked like, but presumably they were cleverly constructed automata operating on the principle of a chain reaction. In it, the first stimulus triggered a reaction, which then triggered another reaction, and so on, resulting in the puppet's performance of a small act. This is, in principle, how now some geneticists imagine the molecular control of embryonic development. In each step of the cascade described above, the regulatory molecules are specific for those sites of the DNA molecules that need to be produced to trigger the next step of the progression. But how could have such an enormously complex chain reaction arisen? Aristotle did not have to answer this question because in his eternally existing universe, things just happened to be the way they were. Modern science, however, does believe in the beginning of the universe and in the evolution of the organic world. And it has a powerful theory of evolution by natural selection, which accounts for the complex phenomenon of embryonic development by postulating a stepwise accumulation of complexity.

How Does Mendel Fit into All This?

Modern genetics, a discipline in which 5-year-old articles are considered too ancient to be cited, has rediscovered Aristotle. A clear sign of the rediscovery are articles with titles like "Aristotle and modern genetics"[52] or "Genomic metaphysics."[53] Behind the resurrection is not only the realization that Aristotle's views are relevant to some of the ethical issues the discipline raises but also in relation to some of the new genetic findings. What is now happening in genetics is nothing less than a conceptual (the Kuhnians might say paradigmatic) shift. After a century of focus on the study of single or simple character inheritance, genetics is turning to the examination of an organism as a whole and to the development of individual organisms. This was the focus that characterized ancient Greek philosophers and physicians, with Aristotle at the forefront. Their interest was in the generation, the rise of an individual animal from what then appeared to be amorphous emissions of sexual organs. Their interest in the transmission of individual characters, such as the color of the eyes, the shape of the nose, or the texture of the hair, was not strong enough to compel them to coin a separate name for the phenomenon.

When later, in the Middle Ages, the West European civilization assimilated—through the Romans, Arabs, and Jews—much of the Greek culture, it took over also the Aristotelian concept of generation in its entirety, including the views of sexual reproduction, species, and heredity. This tradition then persisted without any significant modifications until the nineteenth century, when the three parts of this package split and became the subjects of separate studies. Mendel, whose studies on heredity founded the genetics of the twentieth century, therefore started essentially from a counter-Aristotelian platform, although, ironically, it was Aristotle who introduced the *symbebekota*, the accidental qualities or characters, as a distinct component of reality. It was Mendel, however, who liberated the characters from the generation package and gave them a life of their own. By doing so, he effected the first great paradigm shift in the study of heredity since Aristotle. A second shift appears to be under way now, and although it sometimes gives the appearance of being counter-Mendelian and pro-Aristotelian, when it is completed, it will probably be a synthesis of the two. It is our aim in this book to unravel how the first shift came about, and to do this, we needed to describe the base from which the new paradigm issued. May we be forgiven if we have been a bit overzealous in our enthusiasm for Aristotle!

References and Notes

[1] *"Never, Telemachos, you will be feeble or thoughtless hereafter. If indeed is instilled in you the great force of your father—Such was the man he was to achieve both action and speeches—Then your journey will not be fruitless or lacking fulfillment."* Homer: *The Odyssey*, translated by Rodney Merrill. The University of Michigan Press, Ann Arbor, MI 2005. Book 2, verses 270–273

[2] Peters FE (1967) Greek philosophical terms. A historical Lexicon. New York University Press, New York, NY

[3] Merriam-Webster's Collegiate Dictionary (1996) 10th edn. Merriam-Webster, Springfield, MA

[4] Frede M, Striker G (eds) (1996) Rationality in Greek thought. Oxford

[5] The Greek word *symposion* means drinking together (*syn-*, together, *posis*, drink). It is also the title of one of Plato's dialogues describing a long drinking party, in which the participants debate the nature of love. In general, *symposion* was a social gathering of intellectuals, in which a discussion upon a chosen subject was aimed at collecting the opinions of different authorities

[6] The first definition of a philosopher can be found in Plato's *Symposium*, but the term was introduced much earlier. It was apparently used by Heraclitus and by the historian Herodotus in the sense *philosophoi anders*, men who love wisdom

[7] Osborne C (1997) Heraclitus, Chap. 3. In: Taylor CCW (ed) Routledge history of philosophy, Vol I. From the beginning to Plato. Routledge, London

[8] Palmer J (2010) Parmenides and presocratic philosophy. Oxford University Press, Oxford, UK

[9] Kraut R (1992) Cambridge companion to Plato. Cambridge University Press, Cambridge, UK

[10] Abbel-Rappe S (2009) Socrates: a guide for the perplexed. Continuum, London

[11] (a) Vella JA (2008) Aristotle: A guide for the perplexed. Continuum, London. (b) Grene M (1967) A portrait of Aristotle. The University of Chicago Press, Chicago. (c) Barnes J (ed.) (1995) The Cambridge companion to Aristotle. Cambridge University Press, Cambridge. (d) Schields C (2007) Aristotle. Routledge, London

[12] Ontology is the knowledge of being (from Greek *ontos*, "being," and *logos*, here "knowledge." The term was introduced in the form of *Ontologia* by the German philosopher Jacob Lorhard (1561–1609)

[13] Studtmann P. Aristotle's categories. http://plato.stanford.edu/entries/aristotle-categories/

[14] Adorno TW (2000) Metaphysics: concept and problems. Stanford University Press, Stanford, CA

[15] Cohen SM. http://faculty.washington.edu/smcohen/320/archange.htm

[16] Aristotle (1943) Generation of animals. English translation by Peck AL. Harvard University Press Cambridge, MA

[17] Aristotle (2001) On the parts of animals. Translated with a commentary by Lennox JG. Clarendon Press, New York, NY

[18] Meyer H (1918) Das Vererbungsproblem bei Aristoteles. Philologus. Zeitschrift für das klassiches Altertum und sein Nachleben 75:323–363

[19] (a) Stiebitz F (1937) Biologické základy antických názorů o dědičnosti. A. Píša, Brno. (b) Lenz HO (1856) Zoologie der alten Griechen und Römer. Becker'sche Buchhandlung, Gotha. (c) Lesky E (1959) Die Zeugungs- und Vererbungslehren der Antike und ihr Nachwirken. Verlag der Akademie der Wissenschaften und der Literatur in Mainz in Komission bei Franz Steiner Verlag, Wiesbaden

[20] (a) Johannsen W (1917) Die Vererbungslehre bei Aristoteles und Hippokrates im Lichte heutiger Forschung. Die Naturwissenchaften 5(24): 389–397. (b) Henry D (2006) Aristotle on the mechanism of inheritance. J History Biol 39: 425–455. (c) Meyer H (1918) Das Vererbungsproblem bei Aristoteles. Philologus. Zeitschrift für klassische Altertum und sein Nachleben 75: 323–363. (d) Grmek MD (1991) Ideas on heredity in Greek and t: Altertum Roman antiquity. Physis 28: 11–34

[21] Longrigg J (1985) A seminal "debate" in the fifth century B.C.? In: Gotthelf A (ed) Aristotle on nature and living things. Philosophical and historical studies presented to David M. Balme on his seventieth birthday. Mathesis Publications, Pittsburgh, PA, pp 277–287

[22]Coles A (1994) Biomedical models of reproduction in the fifth century BC and Aristotle's Generation of animals. Phronesis 40:48–88
[23]Boylan M (1982) The digestive and "circulatory" systems in Aristotle's biology. J Hist Biol 15:89–118
[24]Freudenthal G (1995) Aristotle's theory of material substance. Heat, pneuma, form and soul. Clarendon Press, Oxford, UK
[25]Bos AP (2003) The soul and its instrumental body. A Reinterpretation of Aristotle's philosophy of living nature. Koninklijke Brill NV, Leiden, The Netherlands
[26]Solmsen F (1957) The vital heat, the inborn pneuma and the aether. J Hellenic Stud 77:119–123
[27]Rennet, in case you have forgotten, is a substance secreted by the membranes of the fourth compartment of the calf stomach. It contains the enzyme rennin (chymotrypsin), which digests proteins in milk and so causes its coagulation or curdling. Rennin is used for curdling milk in the production of cheese. Enzymes contained in certain plant juices, for example fig juice, can also curdle milk
[28]Cardwell JC (1994) The development of animal physiology. Med Library Hist J 2(4); 225–242; 3(1); 50–77, 1905; 3(2): 141–154, 1905; 3(3): 189–198, 1905; 3(4): 275–282, 1905; 4(1): 101–107, 1906; 4(2): 206–210, 1906
[29]The orchid *Clitoria ternatea* and the mushroom *Phallus impudicus*. See the frontispiece of Alec Bristow's *The Sex Life of Plants*. Holt, Rinehart, and Winston, New York, NY 1978
[30](a) Lennox JG (2005) Teleology, chance, and Aristotle's theory of spontaneous generation. Aristotle's Philosophy of Biology. Studies in the Origins of Life Science. pp. 229–249. Cambridge University Press, Cambridge, UK. (b) Balme DM (1962) Development of biology in Aristotle and Theophrastus: Theory of spontaneous generation. Phronesis 7: 91–104
[31]Henry D (2006) Aristotle on the mechanism of inheritance. J Hist Biol 39(425–455):2006
[32]See Reference[17], pp. 52–53
[33]Stubbe H (1972) History of genetics from prehistoric times to the rediscovery of Mendel's laws. Translated by Waters TRW. The MIT Press, Cambridge, MA. (The quotation from Aristotle is on p.31 of Stubbe's book and the sentence in the Greek version of Aristotle's *Peri zoon geneseos* appears in section I 117.721b 29.)
[34]Drs. Dimitra Chalkia and Nikolas Nikolaidis, Personal communication
[35]Witt C (1985) Form, reproduction, and inherited characteristics in Aristotle's generation of animals. Phronesis 30:46–57
[36](a) Lloyd AC (1965) Genus, species and ordered series in Aristotle. *Phronesis* 7: 67–90. (b) Balme DM (1965) ΓΕΝΟΣ and ΕΙΔΟΣ in Aristotle's biology. Classical Quarterly 56: 91–104
[37]Zirkle C (1959) Species before Darwin. Proc Am Phil Soc 103:636–644
[38]Pellegrin P (1985) Aristotle: A zoology without species. In: Gotthelf A (ed) Aristotle on nature of living things. Philosophical and historical studies Presented to David. M. Balme on his Seventieth Birthday. Mathesis Publications, Pittsburgh, PA, pp 95–115
[39]Cain AJ (1959) Taxonomic concepts. Ibis 101:302–318
[40]Simpson GG (1961) Principles of animal taxonomy. Columbia University Press, New York, NY
[41]Hull DL (1965) The effect of essentialism on taxonomy—two thousand years of stasis. Br J Phil Sci 15:314–366
[42]Mayr E (1963) Animal species and evolution. Harvard University Press, Cambridge, MA
[43]Balme DM (1980) Aristotle's biology was not essentialist. Archiv für Geschichte der Philosophie 62:1–12
[44]Winsor MP (2001) Cain on Linnaeus: the scientist-historian as unanalyzed entity. Stud Hist Phil Biol Biomed Sci 32:239–254
[45]Wilkins JS (2009) Species. A History of the Idea. University of California Press, Berkley, CA
[46]Stamos DN (2003) The species problem. Biological species, ontology, and the metaphysics of biology. Lexington Books, Lanham, MD
[47]Mayr E (1982) Growth of biological thought Diversity, evolution, and inheritance. Belknap Press of the Harvard University Press, Cambridge, MA

[48]Loewenstein WR (1999) The touchstone of life. Molecular information, cell communication, and the foundation of life. Oxford University Press, New York, NY

[49]Delbrück M (1971) Aristotle-totle-totle. In: Monod J, Borek E (eds) Microbes and life. Columbia University Press, New York, NY, pp 50–55

[50]Gutmann J (ed) (1963) Philosophy A to Z. Grosset & Dunlap, New York, NY

[51]Kupiec J-J (2009) The origin of individuals. Translated from the French by Margaret and John Hutchings. World Scientific, Hackensack, NJ

[52]Vinci T, Robert JS (2005) Aristotle and modern genetics. J Hist Ideas 66:201–221

[53]Mauron A (2002) A genomic metaphysics. J Mol Biol 319:957–962

Silesia and Moravia: Mendel's Tortured Homeland

2

Václav Hanka[1]

A change of scenery—another place, another time. Nearly 2,000 kilometers northwest of Athens and some 2,000 years after Aristotle, a lonc, bespectacled figure, clad in a monk's habit, stands on the Veselský Hill, the most eastward promontory of the Jeseníky Mountains in Silesia (Fig. 2.1). He is Gregor Johann Mendel, the abbot of the St. Thomas Abbey in Staré Brno, on one of his rare visits to the place of his birth, the village of Hynčice down in the valley. As his gaze shifts from the village to the opposite side of the valley, he rests his eyes for a long while on the panorama of another mountain range in the blue haze, the Beskydy Mountains. His expression brightens, as if a pleasant memory had crossed his mind.

We are in Gregor Mendel's homeland. His ancestors came to this hill sometime in the sixteenth century, if not earlier, to settle down in the village of Veselí, where they barely eked out a living by farming the rock-strewn fields. Later, they managed to acquire farmsteads down in the valley, where the fields were more fertile. Nevertheless, they still had to toil hard on them. Johann Mendel, the only son of Anton Mendel, through hardship to himself and to his parents, escaped the slavery of the *robota* (menial labor) and all the other forms of exploitation forced on the peasants by the landlords. Luck was on his side, when he was admitted to the St. Thomas Abbey and, then again, when he was elected to the highest office he could hope to attain—that of the abbot. Now he himself had become a landlord entrusted with the large estates in Brno's environs. It was a long journey from his humble childhood on a farm in a small village to this position as a respected prelate of an institution in a big city; a long journey from Hynčice in Silesia to Brno in Moravia. In the next four chapters, we will describe the important events of this odyssey. But before we do, we must explain what Silesia and Moravia are and briefly describe their

J. Klein and N. Klein, *Solitude of a Humble Genius - Gregor Johann Mendel: Volume 1*, 45
DOI 10.1007/978-3-642-35254-6_2, © Springer-Verlag Berlin Heidelberg 2013

Fig. 2.1 Mendel strolling on the Veselský Hill, the place of origin of his ancestors. His birthplace—Hynčice—is in the valley behind him, the Moravian Gate, and the mountains in the background are the Beskydy. The hawkweeds, one of his experimental plants which he might have first encountered there, still grow on the hill

history; for without this knowledge, some of the events in Mendel's life would make little sense.

Geographia Mendeliana

Paste a map of Europe on a cardboard, cut the board along the continent's coastline and along the ridge of the Ural Mountains, down to the Caspian Sea, then along the ridge of the Caucasus Mountains, and across the Black Sea toward the Aegean Sea, and when you have done all that, center the cutout on the tip of your index finger. Silesia and Moravia are where the finger supports the map—at the heart of Europe.[2] On a physical map, the identifying feature of central Europe is a mountain system that resembles the fossilized remains of a giant lizard embedded in a slab (Fig. 2.2). The lozenge-shaped head of the monster is the region of Čechy (Bohemia). The four sides of the lozenge comprise mountain ranges, which enclose a plateau of rolling hills, drained by two main river systems: Vltava in the south and Labe (Elbe) in the north. The ranges are the Krkonoše (the Giant Mountains) in the northeastern part of the Sudetes chain, to which also belong the Jeseníky Mountains of Silesia; the Krušné hory (the Ore Mountains) in the northwest; Český les and Šumava (the Bohemian Forest) in the southwest; and the Českomoravská vrchovina

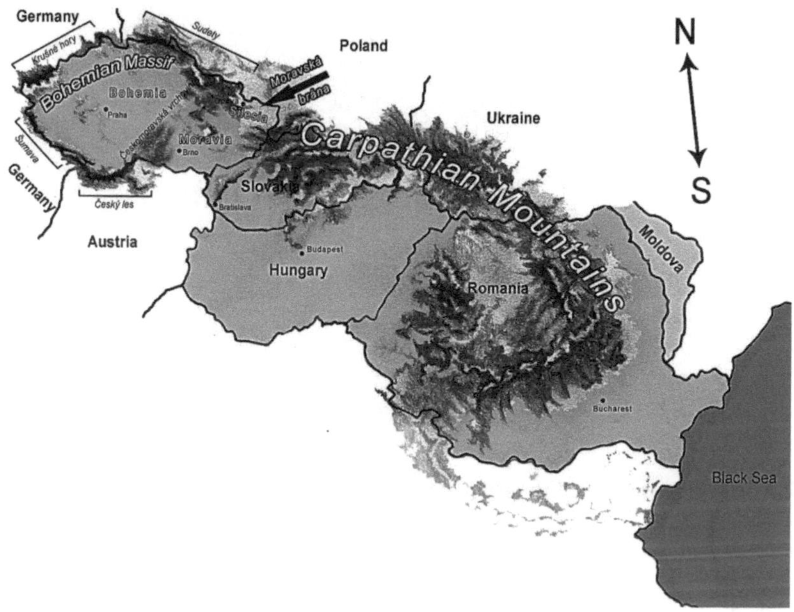

Fig. 2.2 The Bohemian Massif and the Carpathian Mountain systems

(the Czech–Moravian Highlands) in the east. Geologically, this plateau, together with the surrounding mountain ranges, is known as Český masiv, the Bohemian Massif.[3]

The neck of the giant lizard is interrupted by a gap, a wide valley, the Moravská brána or Moravian Gate (Fig. 2.3). The valley separates the Jeseníky of the Sudetes Mountain system from the Beskydy of the Carpathian Mountain system. The latter then extends eastward through the countries of Slovakia, Poland, Ukraine, and Romania. The Moravian Gate widens at its south end into broad, rich lowlands, from which rise, on their western side, the Czech–Moravian Highlands and on its eastern side the White Carpathians. The stretch of land encompassing the Moravian Gate, parts of the Sudetes Mountains on the gate's westerly side, and parts of the Carpathians on the easterly side is *Slezsko* or Silesia. The lowlands opening to the south of it, together with parts of the flanking mountains, down to the border with Austria constitute *Morava* or Moravia (Fig. 2.4). Silesia is drained by the Opava and Odra Rivers; the latter then takes its waters northward through Poland to the Baltic Sea. Most of Moravia is drained by the Morava River or March, which flows southward to join the Danube taking its waters in the easterly direction into the Black Sea. The three regions—Bohemia, Moravia, and Silesia in the sense we described them—are now parts of the modern day Czech Republic, however, not as official administrative units; the name "Moravia" now officially includes Silesia. We use here the tripartite division because it has an important historical significance. The present-day Czech Silesia is only a fragment of the original historical Silesia, which occupied also a large part of modern Poland.

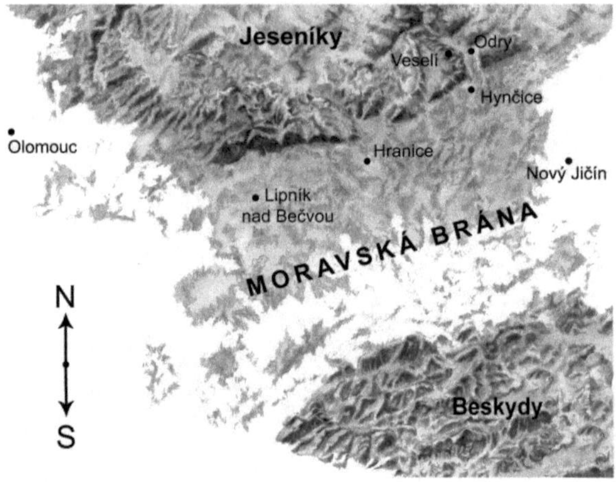

Fig. 2.3 The Moravian Gate (Moravská brána)

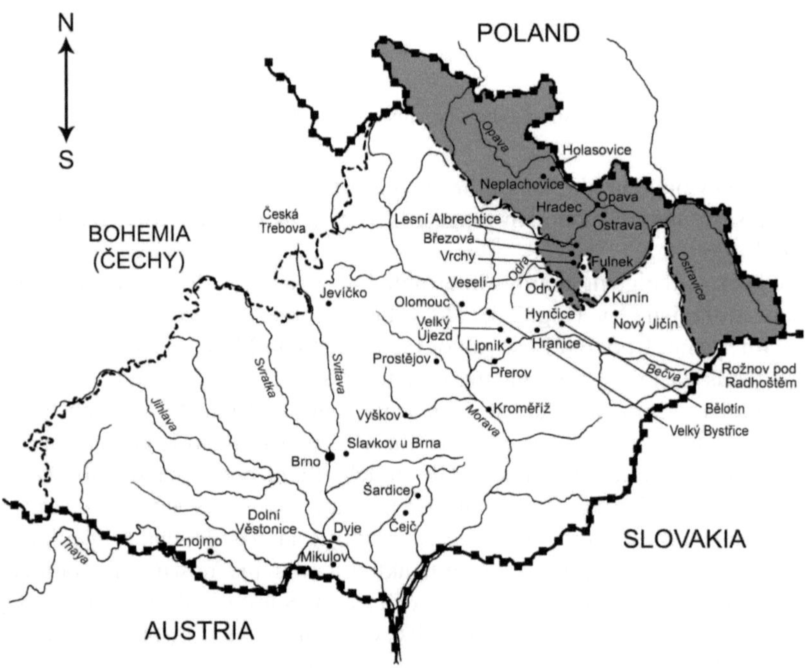

Fig. 2.4 Map of Moravia and Silesia showing main rivers and some of the places mentioned in the text

The Birth of the Land[3]

Human life is so short and the history of the earth so long that the thought of lands like Bohemia, Silesia, and Moravia not always having been there is alien to us. Yet a long time ago not just these three lands but Europe as a whole did not exist. When exactly Europe came into existence is difficult to ascertain because its birth had been a protracted affair extending over hundreds of millions of years. Geologists believe that continents come together periodically to form supercontinents, which then break up into pieces, which then disperse, only to come together again in new combinations in a new cycle. The latest supercontinent, the Pangea, existed between 290 and 250 million years ago and then began to break up. Europe was assembled from the pieces of this fragmentation. First northern and eastern Europe were welded from at least two pieces, then two or more pieces—now constituting central Europe—were added to the block, and finally the amalgamation of additional mini-continents completed the assembly by creating western and southern Europe. The assembly involved collisions between the fragments and the wrinkling of the mini-continents' margins into mountain chains. The four mountain chains at the margins of the Bohemian Massif arose as a consequence of the Massif's collision with the partially assembled northeastern Europe during the Hercynian (Variscan) mountain-building process between 350 and 280 million years ago. The Carpathians were created when central and southern European mini-continents were pushed by the African continent against this assembly. During this Alpine mountain-building process, which began approximately 65 million years ago and continues to this day, the eastern margin of the Bohemian Massif slid under the western rim of the emerging Carpathians, and the suturing of these two edges produced Moravia/Silesia. The suture runs deep below the surface, approximately along the line connecting the cities Ostrava, Vyškov, Přerov, Brno, and Znojmo (Fig. 2.4). Therefore, the area of Silesia/Moravia west of the suture is derived from the Bohemian Massif, whereas the part east of it comes from the Carpathian block. Prior to their amalgamation, parts of or the entire mini-continents were underwater over extended periods of time accumulating marine sediments. The region that became Silesia/Moravia was mostly underwater until about 210 million years ago, but even after this date, it was claimed by the sea more than once and remained underwater over long periods of time. Since approximately 25 million years ago, though, Silesia/Moravia has remained a dry land.

In the last two million years, the physiography of Silesia/Moravia has been forged by the cycling of climatic condition, a period called the Ice Age or Pleistocene. Global cooling of the earth began already some three million years ago, but one million years later, a pronounced fluctuation of temperature set in, alternating cold phases, the *glacials*, with warmer ones, the interglacials. During the cold phases, which lasted on average about 100,000 years, the snow in the northerly latitudes of the Northern Hemisphere failed to melt, even in the summer. Instead, the snow compacted and turned into ice. Large masses of this ice formed on land and these glaciers then grew into even larger ice sheets. Under the stress of their own weight, these ice sheets crept outward in all directions. In the interglacials, lasting about 10,000 years each, the movement of the ice sheets was reversed—advancement was replaced by retreat. In Europe, the Finno-Scandinavian ice sheet was centered on northern Scandinavia, but it covered much of northern Germany and Poland as well. At the time of its greatest advance, a long narrow extension of the sheet, an ice tongue, projected all the way to Silesia and through the Moravian Gate into northern Moravia. The enormous weight of the ice sheets crushed boulders, pulverized smaller rocks, and abraded the earth's surface. The leftover rubble, generated and exposed upon the retreat of the ice sheet, was transported once more over long distances by the front and the sides of the advancing ice mass of the next glacial phase, to be deposited eventually as moraines. Dust storms similarly transported the fine particles of the uncovered, loosened soil over even longer distances to create deposits of the high-quality loess over large regions of Silesia/Moravia. Running water brought the larger particles into rivers, which then laid them down as alluvial deposits along their course. Thus, the glaciers and ice sheets remodeled the landscape, broadening some valleys and blocking others, changing the course of the waterways, and enveloping much of Silesia/Moravia into a new cloak.

Prehistoric Moravia/Silesia[4]

Humans are a restless species. They travel to peddle their goods, gods, and ideas. Individuals, families, tribes, and entire nations wander in search of better conditions. Marauding hordes and whole armies overrun lands on false pretenses or without any pretext at all. Over the millennia, every corner of this planet has been trampled by adventurers, merchants, missionaries, conquistadors, and tourists, some more than others. A corner like Silesia/Moravia, which lies in the heart of an old continent, has sustained particularly heavy traffic during its long history. The mountain chains of central Europe had been a hindrance to human migration, especially in the north-south direction. But Silesia/Moravia, whose chain is interrupted by the Moravian Gate, has offered a convenient crossing, connecting the Polish, Ukrainian, and Russian plains in the northeast with the Hungarian Plain and the Vienna basin in the south. Both beasts and humans have used this passage since time immemorial. During the Pleistocene, herds of large grazing mammals—elephants, mammoths, horses, wooly rhinoceroses, and reindeers—used this passage to reach the Silesian/Moravian tundra, which was abundant with grass, moss, lichen, heath, and sedge. Early humans, scavengers and hunters alike, followed the herds.

When exactly humans first reached Silesia/Moravia is uncertain, but there is credible evidence for their presence about 700,000 years ago. Who they were is unknown, for no bones of the first inhabitants have thus far been found. But judging from the type of stone tools that they left behind, they may have belonged to the species of upright humans, *Homo erectus*. The upright people probably lived in Silesia/Moravia only in the warmer phases and only in small numbers, retreating to the more amicable climatic regions farther south whenever the conditions in Silesia/Moravia took a turn for the worse. They may have courted Silesia/Moravia in this manner for about half a million years, until they were replaced by a new human species about 250,000 years ago. Experts cannot agree on what this species should be called, tentatively referring to it as archaic human, ante-Neanderthal, or pre-Neanderthal.[5] The presence of archaic humans in Moravia is again inferred from the type of stone tools found in the younger layers and found at places outside Moravia in association with the fossilized remains of this species. Some 100,000 years ago, the archaic humans were replaced by Neanderthals, *H. neanderthalensis*, whose presence in Moravia is documented not only by a particular type of stone artifacts but also by fossilized bones found at several sites. The disappearance of Neanderthals from Moravia some 40,000–30,000 years ago coincides with the appearance of modern humans, *H. sapiens*, the "mammoth hunters." Numerous archeological sites scattered throughout Silesia/Moravia attest to their presence. World-famous are Předmostí near Přerov in central Moravia and Dolní Věstonice on the slopes of the Pavlovské vrchy in southern Moravia. One of the most celebrated finds at the former site is the remarkable stylized female figure engraved on a tusk. The latter site is renowned for the black Venus statuette and the enigmatic, strikingly modern carving of a woman's face, both dated to a period between 27,000 and 25,000 years ago. As the last glaciation drew to a close and the ice sheets retreated to northern Scandinavia some 10,500 years ago, the tundra gave way to deciduous and mixed forests and the cold-adapted fauna was replaced by deer, moose, wild pig, and aurochs.

Archeologists call the period of human presence in Silesia/Moravia from c.700,000 to c.10,000 years ago the Old Stone Age or Paleolithic. Within this period, they distinguish a succession of cultures characterized by material objects the bearers of these cultures left behind. As the identity of the bearers is not known, archeologists name the cultures either by the sites at which they are well represented or by the prevailing characteristic implements found at these sites. In Moravia, archeologists recognize at least 14 distinct Paleolithic cultures. They distinguish them primarily by the types of stone tools used. The Paleolithic was followed by the New Stone Age or Neolithic, marked by the use of finely polished stone tools with holes drilled into them, pots fashioned from wet clay and hardened by firing, and the domestication of plants (wheat, barley, and others) and animals (e.g., sheep and goat). These innovations transformed humans from nomadic hunter-gatherers into settled farmers. The Neolithic period of Silesia/Moravia lasted from c. 6000 to c. 2000 years BCE, and its successive periods are distinguished primarily by the type and ornamentation of the pottery made. At least ten different cultures succeeding one another during the last 2,000 years of the Neolithic period alone have been documented. Metals came to full use in the succeeding two periods, the Bronze Age extending from 2000 to 1500 BCE, followed by the Iron Age, which lasted until the historical times. The occupants of Moravia in the Iron Age are the first peoples to emerge from the general anonymity of the prehistoric ethnic groups, the first Moravians with a name. They are the Celts, and they are the bridge between European prehistory and history. Since written information about the ancient Celts has reached us via the writings of the classical authors of Greece and Rome, one can speak here of a protohistorical period of Moravia. The Greeks gave the names *Keltoi* or *Galatai* to the Celts, whereas the Romans, who fought them and ultimately conquered them, referred to them as *Galli*, or Gauls, in the region called *Gallia* or Gaul. It was a contrived, all-encompassing name, for the Celts were not a unified nation but rather a conglomeration of tribes. Although they had names for the different tribes, they apparently had no designation for the whole conglomerate. The people who settled down in Bohemia and Moravia called themselves Boii, and it is from this designation that Bohemia derives its name. The different tribes spoke different dialects of a common Celtic language, and some of the dialects evolved into the different languages of the Celtic descendants living in present-day Ireland, Highland Scotland, The Isle of Man, Wales, and Brittany.

The Language Tree of Europe[6,7]

In books and articles on Mendel, you can read that he was born in Heinzendorf, studied at Troppau, and entered a monastery in Brünn. But if you look for these places on present-day maps, you will not find any of them. At most, the books may state something like "Brünn, now Brno in the Czech Republic," implying that in the post-Mendelian time some other nation usurped the place. For this reason we provide the background necessary for the understanding of Moravian/Silesian history that will follow in the next section.

Modern Europe is divided into roughly 50 countries, each inhabited by at least one ethnic group, that is, by a group of people with a distinct culture. The cultural differences between the groups include distinct economies, customs, intellectual achievements, arts, values, beliefs, institutions, and language. Linguists believe that the European languages and some Asian tongues have developed from a single common Indo-European language. By quantitative analysis of the similarities and differences between the extant and extinct Indo-European languages, they have come to the conclusion that the diversification of the European tongues occurred in manner resembling branching of a tree. Indeed, linguists depict the diversification of languages by diagrams that resemble an abstract tree (Fig. 2.5). Since the linguists know how rapidly languages evolve on average, they can calculate the time of divergence of the individual branches. And since it is people who speak the languages, linguists surmise that the branching points correspond roughly to the divergences of groups of people, for instance, by their migrations to separated geographical regions. All these speculations might have been dismissed as fantasies were it not for the fact that they are supported by archeological data on the one hand and genetic data on the other. Archeologists like to say "stones speak" and metaphorically speaking it is true. Indeed, from "stones," or more broadly, "material remains" of human societies, much can be deduced about the lives and culture of the people who left them behind. As for the geneticists, they know that genes change with time at a more or less constant rate. Consequently, whenever groups of people separate geographically, they diverge in their genes. By comparing genes of different populations, geneticists can estimate how long ago the populations diverged and draw trees similar to those that linguists draw. The remarkable thing is that there is a reasonably good match between the language and the gene trees,

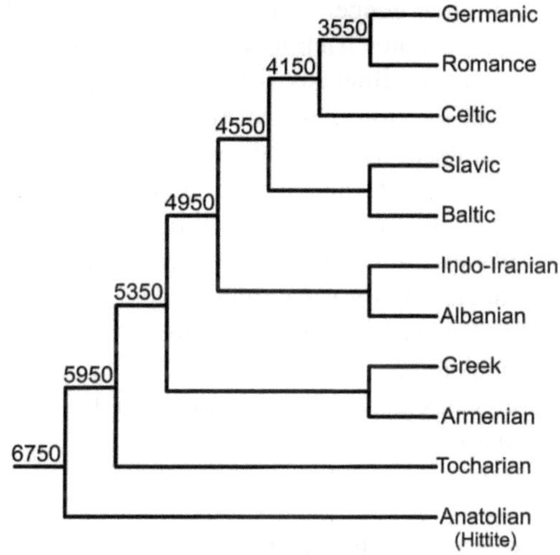

Fig. 2.5 Tree diagram depicting the divergences of Indo-European languages. The nodes mark the divergence time in BCE of two branches. The tree is based on the data of R. D. Gray and Q.D. Atkinson

and both trees can be brought into agreement with the stories the stones "tell" the archeologists.

Focusing on the Indo-Europeans (not all European inhabitants are of Indo-European origin) and grossly simplifying a great amount of data, we obtain the tree in Fig. 2.5 summarizing the origin of modern Europeans. It is uncertain where the people speaking the ancestral Indo-European language lived. There is, however, a general agreement that they did not live in central and western Europe. Presumably, they inhabited either eastern Europe or Asia's Near East. Groups of people left this homeland at different times to colonize southern, central, and northern Europe. The first to move out of the ancestral region were the Greeks. They settled down on the southern parts of the Balkan Peninsula and from there colonized the shores and islands of the Mediterranean Sea like "frogs around a pond."[8] They separated from the Indo-European stem around 5350 BCE and in their new homeland developed their form of philosophy, science, arts, and politics, which became the foundations of Western culture. The Greek civilization then spread through military conquest by Alexander the Great and adoption by the Romans. In Greece it persisted through the Byzantine Empire.

After the Greek, the second major language branch to separate from the ancestral Indo-European stem was that which led to the Slavic and Baltic languages. After their separation some 4550 BCE, the speakers of the Slavic/Baltic languages remained at an as yet unidentified place, presumably somewhere in eastern Europe. We shall return to them shortly. Next, the third branch, which diverged about 4150 BCE, brought the Celts to Europe (Fig. 2.4). The presence of the Celts is associated with the European Iron Age, whose onset is conventionally dated to the beginning of the first millennium BCE. The discrepancy between these two dates, if not caused by a gross overestimation based on the language-divergence data, suggests that the Celts might have been in Europe (or somewhere else) much longer than their association with the Iron Age suggests.

The divergence of the last two major Indo-European branches, Romance and Germanic, dates back to approximately 3550 BCE. The oldest Romance language known to the linguists is the Old Latin spoken by the Romans. The conventional date of the founding Rome is 753 BCE. The earliest traces of settlements archeologists have found under the modern city of Rome date back to the ninth century BCE. Hence, there is again a discordance in dates. Romans, however, had not been the only Latin-speaking people in central Italy at that time, and this fact could explain why the emergence of Latin seems to predate the founding of Rome. The ancient Latin-speaking people had presumably come from the homeland of the Indo-Europeans, wherever it might have been. Their ancestors and the ancestors of the ancient Germanic people might have been the last to depart from it, with the Latin-speakers heading to southern Europe and the Germanic speakers toward northern Europe.

The Path to Modern Europe[9,10]

Rome started as a small city-state akin to the Greek city-states sprinkled around the Mediterranean (Fig. 2.6). The resemblance was not just superficial, for the Romans copied many things from the ancient Greeks; they adopted the Hellenic culture wholesale. Their initial mode of expansion was not maritime, however; unlike the Greeks, they built up and trained an army based primarily on foot soldiers. Their infantry then marched through the Italian Peninsula into western and central Europe. Along the way, it subjugated the conquered people and began to "civilize" them, which meant founding cities, constructing roads connecting them, establishing efficient local administrations, developing an economy, and promoting literacy based on Latin as the language of communication within the whole Roman realm. In this part of Europe, the Roman expansion stopped at the rivers Rhine and Danube. The Romans accepted the fact that the lands eastward of the Rhine and northward of Danube were occupied by what they called "barbarians," whose subjugation and civilization was not worth the effort. While it had been growing, the Roman state turned from a kingdom to a republic, and then from a republic to an empire. At the peak of its expansion around 100 CE, the Roman Empire consisted of 43 provinces covering most of Britain, all of western and southern Europe, whole Asia Minor, and much of northern Africa (see Fig. 2.6). The Latin spoken in the different parts of the empire eventually developed, with the contribution of the local languages, into the modern Romance languages which include Italian, French, Spanish, Portuguese, and Romanian. Starting from about 180 CE, though, the long period of the empire's decline began. In 395 CE it was divided into western and eastern empires, and in 476 CE the Western Roman Empire collapsed. The Eastern Roman Empire survived until 1453 CE in a form commonly referred to as the Byzantine Empire, with Constantinople as its capital.

The collapse of the Western Roman Empire had both internal and external causes. The latter were the "barbarian," primarily Germanic, tribes, which had been "nibbling" at the empire's borders from the time they became neighbors.

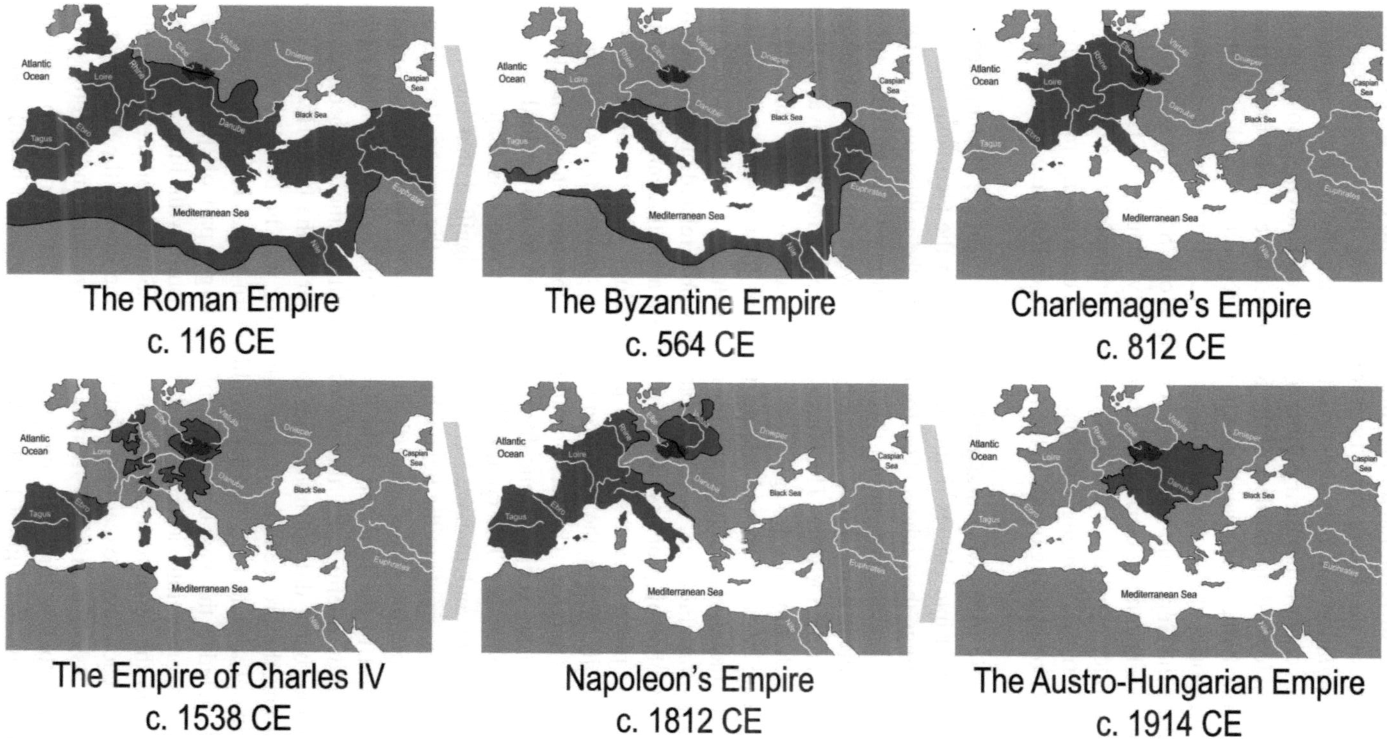

Fig. 2.6 The lands of the Czech crown (Bohemia, Moravia, and Silesia) in the history of Europe. The six panels are six snapshots of Europe at intervals of roughly 500 years. The Czech Republic, comprising these three lands, is the highlighted small blob in which the rivers Labe/Elbe and Wisla/Vistula originate in the heart of the European continent. It is anachronically shown in all six panels in relation to the six major European empires (highlighted areas) culminating around the indicated dates. The message of the figure is that the empire-makers held the inclusion of the heart into their realms for vital

The Germanic peoples differentiated in northern Europe into many tribes, which since the beginning of the fourth century were becoming increasingly restless because of overpopulation and pressures exerted on them by nomadic migrants from Asia. The restiveness led to their pulling up stakes and moving, mostly in the southward direction toward the borders of the Roman Empire. Under the increasing pressure of the barbarians, the Romans began reducing the size of their empire. Some territories they vacated voluntarily (e.g., Britain or the province of Dacia), in others the Germanic warriors forced them out. Step by step, formerly Roman regions were becoming Germanic kingdoms. By about 525 CE, nearly all the territories that once constituted the Western Roman Empire were occupied by different Germanic tribes, warring with one another, each trying to grab more land. By this time two opposing trends in the portioning of Europe were becoming apparent.

Although Germanic tribes were organized differently than the Roman society, they too were stratified into classes. Some individuals in the Germanic societies enjoyed higher standings than others. The former were the nobles or aristocrats, the latter the commoners. The nobles were concentrated around the tribal chieftains, most of them being brave warriors. They were tied to the chieftain by a bond based on an honor code. The warriors pledged to remain loyal to their leader and fight with him for his glory even to death. In return, the leader pledged to take care of his men and reward them for their service with riches and accolades. By the eighth century, the rewards included a *fief*, for which the Latin term was *feudum*. Fief, fee, or *feudum* are all words related etymologically to "cattle" as a symbol of property. In the court of a ninth-century Germanic king, a fief meant a piece of land—with all the buildings standing on it and the peasants inhabiting it—which the ruler ceremonially granted (*invested*) to his man in return for his service. The act of granting established a new relationship between the lord and his subject, henceforth considered to be the lord's *vassal* (from Latin *vassus*, servant). The contract between the two was sealed by two ceremonies—investiture and homage. The *investiture* signified that the vassal was invested (given the rights to the fief), whereas during the *homage* the vassal-to-be swore to become the lord's man (from Latin *homo*, human being, man), that is, to be faithful to the lord and fight for him. This oath was called *fealty* (from Latin, *fidelitas*, fidelity). By the act of investiture, the fief became the vassal's possession, but it did not make him the owner of the land. He was free to use the land and the proceeds from it in whichever way he desired, but if he were to die without a male heir, the fief would return to the lord. The right of using the land included the vassal's option of granting a part of the fief to his own man and so enter into a lord-vassal relationship with him. Since the subvassal could do the same to a sub-subvassal and so on, a complex hierarchy arose, in which at each level, except the highest and the lowest, the noble was at the same time a vassal to his lord immediately above him and a lord to his vassal immediately below him. The hierarchy meant that there were different kinds of nobility, from the highest to the lowest, distinguished by titles such as prince, duke, count, and baron, ruling principalities, duchies, and counties, respectively. This whole system of lord-vassal relationships based of fiefs, loyalty, and military service came to be known as *feudalism.* The political consequence of this system was very different from that of the Western Roman Empire. The latter was based on a strong central administration in Rome and a series of strong local administrations in the provinces controlled firmly from Rome. The feudal system of the Germanic tribes favored independence of the principalities, duchies, and counties. The system held together because of the bonds between the lords and their vassals at each level of the hierarchy and so pulled Europe out of the chaos into which it was sinking after the disintegration of the Western Roman Empire. At the same time, however, the system resisted any efforts to introduce a centralized control into it. Formally, a king or an emperor was at the apex of the hierarchical pyramid, but if he himself was not the most landed person of the kingdom and did not possess the strongest army, he had little chance to force his will onto the lords. By the twelfth century, feudalism had spread from France to England, Spain, and then through most of Europe, reaching its heights in the thirteenth century and beginning to decline in the fifteenth. The consequence of feudalism was the partition of Europe into small kingdoms, each of which was then splintered further into duchies and other smaller units. This was the first trend in the political development of Europe: the fragmentation of the continent into a large number of political modules loosely assembled into kingdoms. The modules, even though they were parts of a larger assembly, were, to a large extent, independent.

The second trend was the opposite of the first. The splintering was contravened by lumping the modules into larger agglomerates—kingdoms or empires. The lords of the modules strived to retain the independence of their modules, whereas ambitious kings aspired to rule large realms. Which trend prevailed depended largely on the relative political strength of the monarchs and the lords. When the lords managed to hold together, they generally succeeded to curtail the monarch's expansionist tendencies. When the monarch was strong and sagacious, he often succeeded having his way. It was a high-level political game, in which the modules were the pawns and the moves in the game were military conquest, inheritance, and marriage. In countries like France or England, the monarchs largely succeeded in imposing central control over their realms. In Germany, the monarchs introduced and kept alive for a while an institution they called the Holy Roman Empire, but it was a paper tiger until it was taken over by the Habsburg dynasty and its center shifted from Germany to Austria. As for Germany itself, it remained fragmented into several dozens of mini-states until the nineteenth century. Even today, the former division into the larger states is still more apparent in Germany than anywhere else in Europe.

It was not the empire-building trend but the empire-splintering tendency that led to the emergence of modern Europe. The Holy Roman Empire was not the first experiment in empire building. Before it, there was Charlemagne's Empire, in which the emperor forcibly brought together modules that apparently had no desire to congeal. In 843, some three decades after Charlemagne's death, his quarreling grandsons split the empire into three kingdoms with the Treaty of Verdun.

Ultimately, it was from these three pieces that France, Germany, and other European states arose.

The Ascension of the Slavs

The expansion of the Germanic tribes represented the first phase of the Great European Migration. Its second phase was dominated by the expansion of the Slavs in the fifth century. If, however, the timetable of language divergences is to be believed, the Slavs might have been in Europe since 4500 BCE, that is, since at least the Late Neolithic. The comparative linguistic also reveals that relative to other branches of the Indo-European languages, the tongues of the Slavs differ from each other much less. This observation suggests that the early Slavs, wherever they lived, occupied relatively small territory and that their languages diverged relatively late, presumably concurrently with the expansion in the second phase of the migration period. The question of the Slav homeland is unresolved, but two conclusions are generally accepted. First, the homeland seems to have been in Europe, rather than in, for example, southwestern Asia, as some scholars had previously believed. And second, the homeland was in Eastern Europe, in a region of which the ancient Greek and Roman historians had little knowledge. Some 20 different hypotheses have been proposed about the location of the Slavic homeland in Europe.[11,12] Most of them place it somewhere in the region extending from the Black Sea in the south to the Baltic Sea in the north; with the rivers Wisla and Dniepr delimiting the region in the west-east direction. Somewhere in this great expanse of land must have been a much smaller territory, which the Slavs claimed to be their original home.

Besides the uncertainty about their "hiding" place over a period of some millennia, there are some facts that are difficult to accommodate in this scenario. For example, the common Slavic language shows some affinities to the languages of the Balts, Germans, and Iranians, as well as weaker affinities to those of the Celts, Illyrians, and Thracians. The Balto-Slavic affinities, which are the strongest, can be explained by the assumption that the two peoples split some time after the separation of the Balto-Slavic branch from the common Indo-European branch. Moreover, the Balts and Slavs may have remained neighbors in eastern Europe. But the affinities with the other languages presumably arose during the Slavs' "incognito" existence. Apparently, the insulation of the Slavs had not been as tight as it might have appeared at first. The Slavs must have abandoned a nomadic lifestyle after they had settled down in Europe. But farmers with all their essential possessions, families, and livestock do not travel light and certainly more slowly than nomads do. Yet the Slavs seem to have appeared suddenly on the stage of history, perhaps within decades or at most a century after they had left their homeland, on a stage as big as half of Europe. One possible explanation is that they expanded on the nomads' heels into the vacated territories thus avoiding the resistance of the indigenous population that would have otherwise slowed down their progress. Another, related, problem is the size of the Slav population. If they occupied a small territory, their population might have been relatively small.[11] When they then expanded into a vast area, the consequence must have been a tremendous populational rarefaction. Yet the chronicles of the period do not report anything of that sort. Perhaps, therefore, one of the initial assumptions—either a small population size or rapid dispersions—might be incorrect.

Western Slavs' Search for New Homeland

The 13 main extant Slavic languages fall into three groups: eastern (Ukrainian, Belarusian, Russian), southern (Bulgarian, Serbian, Macedonian, Croatian, Slovenian), and western (Polish, Kashubian, Sorbian, Czech, Slovak).[13] It is therefore commonly assumed that the expansion took place in three chief directions. An interesting question is this: Why didn't the Slavs expand in the northern direction? One possible answer is that the expansion started from the north; another is that north of the Slavs were the Balts—peoples related to the Slavs on whose territory they did not want to infringe. The fact that one group of the Slavs expanded eastward suggests either that the homeland of the Slavs could not have been at the extreme east of Europe or that the eastern group stayed put.

The consensus view is that the divergence of the 13 languages from the common Slavic language started after 500 CE when the groups arrived at the places of their current distribution. This recency of the divergence explains the high similarity among the current Slavic tongues. Of interest to us is the western group and in it primarily the Czechs who settled down along the rivers Morava and Opava in what are today Moravia and Silesia, and in the former land of the Celtic Boii—in *Boihaemum* or Bohemia. The first question that arises regarding the group's arrival in its new homelands is how the group's tribes got there. Since we do not know where the tribes came from, it is difficult to answer with certitude how they got to their new homelands. Since we can surmise that the tribes came from the east, only two routes are principally possible—southerly and northerly, with respect to the Carpathian Mountain range. The southerly route presupposes that the ancestors of the Western Slavs crossed the Carpathians from north to south through one of the low passes and then traveled westward across the Pannonian Plain to enter Moravia. The northerly route, which seems more parsimonious, would have taken the ancestors along the Carpathians across the Polish plains westward toward the territories occupied by the Germanic tribes—the Saxons in the north, the Thuringians in the middle, and the Lombards in the area of modern Switzerland. Along the way two parts of the Western Slavs split off. One of them—the future Poles—colonized the Polish Plain. The second group—the Polabian Slavs—settled down around the upper and middle Labe/Elbe River. The remaining tribes of the Western Slavs turned southward and marched toward the Moravian Gate. Along the way several other tribes decided to stay behind to found the Polish part of Silesia. Upon reaching the Opava River, more tribes followed suit—namely, the founders of the Czech part of Silesia. The final three-way split might have occurred somewhere along the Morava River. One part turned eastward to become the Slovaks;

the second part headed westward to found Bohemia; and the third group stayed to become the Moravians. The advantage of this scenario is that it accounts for the different degrees of similarity between the Western Slav languages and also for a few other observations, but there is no direct evidence to support it.

Another question one might ask is: Were there any indigenous inhabitants in the territories the Western Slavs, specifically the Moravians and the Bohemians, chose for their homelands? Not long after *Homo sapiens* arrived in Europe, the continent began to run out of uninhabited places. Migrants could therefore expect that wherever they went, there might be people who reached it before them. There were, however, areas that were transiently depopulated for whatever reason—plague, war, devastating invasion, or simply because the predecessors had left to search for a better place. Since ancient chronicles had not recorded any major conflicts of the Moravians, Silesians, and Bohemians with any local peoples, we might perhaps presume that they found their new homes only sparsely populated and that they assimilated the natives. Although the presence of German, Polish, Magyar, Jewish, and other minorities in the Czech lands is well documented throughout history, there is no evidence of a continuity of any of these minorities going back to the onset of the Slavic colonization. Apparently, the minorities entered the Czech lands at different times after the establishment of the Czech state.

Lands of the Czech Crown

Lužice (Lusacia). One of the main tasks of the king in the confederation of the German duchies was to protect the kingdom from its neighbors, Magyars in the south and Slavs in the east. To this end the kings began converting territories neighboring the kingdom into protective buffer zones, which were called *Marken* (*marches* in English), meaning border regions or frontiers. To create a march, a region was first subdued militarily and army outposts were set up in it to keep it under control. Ostensibly, their function was to protect the core duchies from direct attacks, but they soon developed into distinct provinces themselves, which were then used as bases for launching further territory-grabbing operations. This process of expansion was especially pronounced on the eastern frontier, where under the slogan *Drang nach Osten* (eastward drive) it became a Leitmotif of German foreign policy. The Polabian Slavs, unfortunately, found themselves in the path of one of these eastward drives. Originally, they consisted of many tribes, some of which aggregated into groups. The three groups bordering directly on the Holy Roman Empire were the Obodrites in the north, close to the base of the Jutland Peninsula; the Veleti (Uilci, Liutizians, Wilzians) in the middle; and the Sorbs or Lusatian Sorbs in the south, bordering on Bohemia. The last of the three groups comprised two subgroups—the upper and the lower Sorbs. In the tenth and eleventh centuries, the entire region was ruthlessly Germanized and annexed as a new province to the German confederation. Only small pockets of the original Slavic culture remained, but later most of these were also either Germanized or eradicated.

Slezsko (Silesia). Historical Silesia consisted of two parts, Upper and Lower, which encompassed the southeastern and northwestern portions of the Odra Valley, respectively. Other than that, Lower Silesia had no natural borders; it was a part of the Great Polish Plain, delineated politically but not geographically. Upper Silesia's border was demarcated by mountain ranges on the western and eastern side and politically on the southern side. Lower Silesia was at least twice the size of Upper Silesia, and the two parts had been known by different names depending on who ruled them momentarily. Upper Silesia was either Prussian (German) or Polish, Lower Silesia either Austrian or Czech. In historical times the sequence of its occupants had been the Germanic tribes (Silingi and others), followed by Slavs, and then by German immigrants at different times. Throughout history, either the whole of Silesia or its part had successively been under the rule of the Moravians, the Bohemians, the Poles, the Austrians, or the Germans (Prussians). The native tongues of its inhabitants had been Czech, Polish, or German, and most Silesians spoke a dialect derived from these three languages, depending on the region, town, or even the village they were from. Even those inhabitants who learned the standard version of the language spoke it with an accent that immediately revealed the speaker's origin.

The Slavs who settled on the territory of Silesia comprised the tribes named Slezané, Opolané, Dědičané, Holasici, and others. These tribes divided the territory among them and so laid the groundwork for the political fragmentation that came to characterize Silesia more than the other lands. The fragmentation, in combination with the multinationality of the whole, hampered attempts to unify Silesia under one central administration. It also complicated efforts to incorporate Silesia into a larger realm. In the ninth century, the rulers of Great Moravia may have been the first to attempt this by adding at least a part of Silesia to their state. If they succeeded, which is uncertain, Silesia must have broken away when the state began falling apart. Once the Bohemian dukes dampened their desire to cut each other's throats and paid more attention to international affairs, they probably became the next power that attempted to grab Silesia. By that time, however, they had to deal with two other powers—Poland and Germany—which had the same intentions. In the ensuing tag of wars, the Polish rulers mostly prevailed. In 1356, however, Silesia passed to Bohemia to become a hereditary land of the Czech crown. It remained so until 1526, when the Kingdom of Bohemia with its crown lands became the possession of the Austrian Habsburgs. When the Habsburgs lost the so-called War of the Austrian Succession from 1740 to 1748, they also lost to Prussia (Germany) the entire Lower and a large part of Upper Silesia. What remained, the so-called Austrian or Czech Silesia, encompassed the Upper Odra Valley south of the Opava River. After World War One, a large portion of Lower Silesia was ceded to Poland, and after World War Two, Silesia was divided among Poland, Czechoslovakia, and East Germany. A large proportion of Silesian Germans, most of whom immigrated to Silesia throughout history, some of them on the invitation of Polish and Bohemian rulers, was sent back to Germany.

Morava (Moravia). Assuming that the Western Slavs left their original homeland around 400 CE, they must have reached Moravia-Silesia no later than 500 CE. The former date is an educated guess, whereas the latter is supported by indirect hints that can be found in the writings of ancient historians and geographers. The last occupiers of Moravia-Silesia before the Slavs seem to have been the Buri in the north and the Quadi in the south, both comprised of Germanic peoples. But since archeological indicators for their presence in this region vanish in the fifth century, one can assume that they left before the arrival of the Slavs. Similarly, the Germanic Lombards, whose presence in the southeastern corner of Moravia is supported by archeological data, had apparently cleared this region by that time as well. On their way to Pannonia (Hungary and Romania) and ultimately Italy, the Lombards apparently passed through Bohemia from north to south at the end of the fifth century, but by 526 they had already crossed the Danube and entered the Pannonian Plain. Hence, at the beginning of the sixth century, Moravia-Silesia should have largely been devoid of human habitation. Very little is known about the Moravians and Silesians of the sixth century. Presumably they had been settling down, forming villages, building houses, tending their fields, fighting raiders, and trading with merchants who occasionally passed through. Since time immemorial one of the major south-northern trading routes, the so-called *amber trail*, had lead through Moravia-Silesia.[14] It connected the Baltic with the Adriatic Sea, and its staple trading article was amber as well as any other merchandise momentarily in demand along the route.

The Moravians took early steps toward a unification of the tribes in response to the Frankish threat and, more acutely, to the raids of the Avars. The latter were confederations of nomadic tribes who swept from inner Asia to the Balkans and to central Europe in the sixth and seventh centuries. One of the centers from which they organized their raiding expeditions was Pannonia, which they occupied after the Lombards cleared it, and one of the targets of their raids was Moravia. They developed a sort of a parasitic relationship with the Moravian tribes, in that they raided them periodically to rob them of their goods and abduct their women and men. They made the women their concubines and slaves and forced the men to fight their wars on the front lines. In c. 620 the Moravians successfully rose against the Avars, having been led, according to at least one chronicler, by a Frankish merchant named Samo. He then helped the victorious Moravians organize themselves into a semblance of a state sometimes referred to as *Samo's Empire*. It seems to have covered an area that included the whole of Moravia, Silesia, and Bohemia, as well as parts of Slovakia and Hungary. The empire lasted until Samo's death in 658 and then disintegrated. A true Moravian state emerged, however, some 170 years later. This state of *Great Moravia* lasted from c. 830 to 906. It encompassed, in addition to Moravia and Silesia, part of Bohemia, the southern part of modern Poland, the western part of modern Hungary, and Slovakia. Ultimately, however, it was weakened by the quarreling sons of the last king and fell under the onslaught of invaders from without the realm.

Moravians were the first among the Western Slavs to acquire literacy in their own tongue.[15] It happened on the occasion of their conversion to Christianity in the year 863 CE. Although Frankish missionaries had been active in Moravia for some time before this date, they were not very successful. They did not bother to learn the Slavic language, expecting that along with the religion, the Moravians would also accept the German culture and participate in religious rites (liturgy) performed in a language (Latin) they did not understand. Not surprisingly, the Moravians were not anxious to do this. Recognizing the inevitability of the Christianization of his subjects, Rostislav, one of the Moravian rulers, sent messengers to the emperor in Constantinople with a request to send missionaries to Moravia, who could preach in a language the Moravians could understand. The emperor obliged and the choice fell on two Greek priests, the brothers Cyril[16] and Methodius, natives of Thessalonica (modern Salonika), a seaport in northern Greece (Macedonia). On the Balkans of that time, lived several Slavic tribes and the brothers were familiar with their dialects. To prepare themselves for their mission, Cyril and Methodius chose one of these dialects and translated into it several critical Christian texts. Since, however, the transliteration of Slavic words using either the Greek or Latin alphabet presented a problem, the brothers invented a new alphabet based on Greek but enriched with letters reflecting the specialties of Slavic pronunciation. A modified form of this *Cyrillic alphabet* is used to this day in several Slavic languages, including Russian (but not Czech). The mission of Cyril and Methodius was a success, as far as Christianization was concerned, but soon problems arose with Rome. Along with the splitting of the Roman Empire into western and eastern parts, a schism also occurred in the church, bifurcating it into Roman (Catholic) and Byzantine (Orthodox) forms of Christianity, which differed somewhat in their professed creeds. The pope in Rome, already losing jurisdiction over much of Eastern Europe, was not happy to hear complaints from his emissaries that the Christians in Moravia were orienting themselves more and more toward Byzantium. The pope could have avoided this situation had he responded to earlier Rostislav's request for missionaries. Now, however, the pontiff became concerned with the situation at hand and exerted pressure on the Moravian rulers. Ultimately, one of the rulers, Svatopluk, gave in to the pope's demands, suppressed the Slavic form of service, and ordered a reinstatement of the Latin rite throughout his realm.

As the state of Great Moravia began to disintegrate at the end of the ninth century, it had two neighbors: In the south, formerly the home of the Avars, there were now the Franks, and all around the rest of Moravia were Slavs splintered into different tribes. What happened next is not clear, because for more than a century, Moravia disappears from the historical record. The chronicles continue to chatter about the Frankish and Byzantine Empires, and also about various Slavic tribes with which these two power blocks had trouble, but they are mum as far as Moravia is concerned. We may assume, therefore, that during that period of obscurity, the individual provinces of the Moravian conglomerate broke away one after the other until only the original nucleus remained. When the silence of the chronicles finally broke, Moravia reemerged in the company of new neighbors. On the plains south of the Carpathian Mountains was the forerunner of the future Hungary. In the east was a glimmer of the future Slovakia. In the region stretching from the Middle Danube to the Alps was a cluster of provinces that would one day unite to form Austria. In the west sprang up the Bohemian state. In the north, Germans were in the process of engulfing various former Slavic territories. And in the northeast the Kingdom of Poland was gradually coming into focus. In those one hundred years or so, a new political map of central Europe had been drafted.

Čechy (Bohemia). Bohemia has a geographically well-defined border delineated by the wreath of mountain ranges encircling the inner plateau. This circumstance has proved to be both an advantage and a disadvantage. The advantage was that the mountains functioned like ramparts protecting the inner core from invaders. For similar reasons, however, it was also a disadvantage in that it acted like a corset restricting Bohemia's territorial expansion. Bohemia compensated this disadvantage by forming alliances with friendly neighbors speaking the same or similar languages—Moravia, Silesia, and sometimes also others. These alliances assumed different forms, depending on the historical circumstances.

The Slavs who settled in what would become Bohemia, Moravia, and Silesia were fragmented into many tribes, which distributed themselves throughout the lands to occupy separate regions. The tribal units were at first largely independent of one another and lacked any higher type of political organization. Soon, however, the need for just such an organization began to manifest itself. The Slavs found themselves in the vicinity of two power blocks: the empire of the Germanic peoples, and the Avars. In the sixth century, the Germanic peoples formed many kingdoms, principalities, duchies, and smaller political units within the territories previously occupied by the Western Roman Empire. Some of these units assembled themselves into larger aggregates or empires, which showed a tendency to grow by the acquisition of additional units. The first of the new empire-builders were the Franks, whose ancestors migrated from the areas of the Upper and Middle Rhine to the Roman province of the Celtic Gaul. There, one of their branches produced two successive dynasties, which began assembling a Frankish Empire by subjugating other Germanic peoples one after the other. First, from 481 to 751 the Merovingian and then from 751 to 843 the Carolingian dynasties successively put together an empire that at its height extended from the Atlantic coast to the Labe/Elbe River. Shortly afterward the empire broke down into three, and then two pieces, and from these ultimately arose the modern states of France and Germany. The predecessor of the German state was the Holy Roman Empire, which from 918 until 1024 was promulgated by the Ottonian dynasty of Saxony. Later, from 1024 to 1137, the care of the empire had been passed back to a succession of Franconian emperors, followed by a period of Hohenstaufen kings and emperors from 1138 to 1254. After a break from 1254 to 1273, during which the states could not agree on a candidate, came a period from 1273 to 1437 during which the emperors were selected from a variety of houses, not necessarily German. From 1438 until 1806, when the empire was officially abolished, the imperial throne had become the private possession of the Habsburg dynasty of Austria.

From the entangled Czech history, we mention two episodes, which not only represent major turning points in the development of Bohemia but also illustrate the dilemma a small country faces when it happens to have a political behemoth for a neighbor. In this case, the behemoth was the Holy Roman Empire. The first episode happened at the beginning of the thirteenth century, after the consolidation of the Bohemian tribes under the leadership of the Přemyslid dynasty. From a period of internal struggles and attempted takeovers by external powers, Bohemia had emerged as a country capable of and desiring an independent existence in the form of a self-ruled kingdom. It already had two kings in the past, Vratislav II from 1085 to 1092 and Vladislav II from 1158 to 1172. External authorities recognized both, but in both cases the title was restricted to those particular persons. What Bohemia needed at that stage was the recognition of being a kingdom in perpetuity with guaranties of sovereignty.

A Kingdom Granted and a Kingdom Stolen[17]

From the very broad perspective presented in this chapter, two critical developments in Bohemian history are the establishment of the Bohemian Kingdom by Přemysl I in 1198 and the loss of its independence by the election of Ferdinand I of Austria as its king in 1526. The former achievement was due to Přemysl's remarkable diplomatic skills, as well as his ability to change loyalty in a weather vane fashion depending on the direction of the political wind blowing at the time. Before 1198, Bohemia was a principality, which had no guarantee of independence save for its army. In a couple of instances, Holy Roman Emperors elevated two Bohemian princes to kings for services rendered, but without tenure, meaning that after the kings' death, their domain returned to the principality status. Přemysl's achievement was not only that he himself became a king but also that the two then most powerful men in Europe—the Pope Innocent III (r. 1198–1216) and the Holy Roman Emperor Friedrich II (r. 1215–1250)—recognized the kingdom. The Golden Bull of Sicily, which Friedrich II, then the King of Sicily, issued in 1212, was a document that proclaimed to the whole of Europe the independence of the Bohemian Kingdom. In addition it proclaimed, first, that the election of the King of Bohemia was an internal affair of the Bohemians alone, with which no external power, not even the Holy Roman Empire, had any right to interfere. The Empire's function was merely to pass on to the newly elected king the royal insignia as a formal acknowledgment of the election. This confirmation was a mere ritual step based on the notion of the Holy Roman Empire being the secular ruler of all Christianity. Second, it established the sovereignty of the Bohemian state. And third, it confirmed the hereditary nature of the Bohemian royal title. The special position of the Bohemian Kingdom in comparison to the numerous principalities, duchies, and counties of medieval Europe was later reaffirmed by the Emperor of the Holy Roman Empire and King of Bohemia, Charles IV of Luxembourg (r. 1346–1378, emperor from 1355). He also reaffirmed the historical tie of The Bohemian Kingdom with the margravate of Moravia and with the Principality of Silesia, to which he added also the Upper and Lower Lusatia to form the union of the lands of the Czech crown. It was also Charles IV who ordered

Fig. 2.7 Ferdinand I's perjury before the Czech estates in 1526

a special Crown of Saint Wenceslas (*Koruna svatého Václava*) to be made in 1346 (see Fig. 2.7) and to be used in all subsequent coronations of Bohemian kings.

The second critical development in the history of the Czech lands was the subjugation of the Czech lands by the Habsburg dynasty for four centuries. It was a protracted process, which started with the election of Ferdinand I as King of Bohemia, but the preparation for the takeover actually began with the Habsburg entrée into European high politics. The Habsburgs were an old Swabian noble family, whose name is popularly derived from *Habichtsburg* or Hawk Castle, their ancestral seat in the Swiss canton Aargau. Since, however, the name of the castle first appeared in 1108 in the form *Havichsberch*, it might derive from the Middle High German *hab* or *hav*, meaning "river crossing" or "ford." The family had not been poor but also not terribly rich, ancient, but not terribly distinguished until one of the Electors of the Holy Roman Emperors thought of one of its members as being a well-suited candidate for the imperial throne. And so it happened that Rudolf von Habsburg became the Emperor of The Holy Roman Empire. The rival candidate was everything that Rudolf I of Habsburg (r. 127–1291) was not: one of the richest, most powerful, and best known knights in medieval Europe. This, together with the fact that he was a Slav, disqualified Přemysl Otakar II (r.1263–1278) from the candidacy. After his election, Rudolf surprised his Electors by displaying qualities they did not realize he had, qualities that would become characteristic of the Habsburgs in general—greediness and a hunger for power. He accused Otakar of having illegally acquired four duchies—Austria (Österreich), Styria (Steiermark), Carinthia (Kärnten), Carniola (Krain)—and ordered him to pull out from them. Otakar at first ignored the order but later was forced to obey. Rudolf subsequently proclaimed the duchies the property of the Habsburg dynasty and it was from this nucleus that the expansion of the dynasty's realm began. Rudolf's further interference with Bohemia's internal affairs led in 1278 to a battle at Marchfeld (Moravské pole, east of Vienna), where Otakar's knights alone, without the participation of the Czech nobles, faced the combined armies of Rudolf and the Hungarian king. Otakar lost the battle and his life. In retrospect, Rudolf's ascendancy to the imperial throne can be seen as Act One of the Bohemian Tragedy.

Act Two of that tragedy played itself out 28 years later between Otakar's and Rudolf's descendants. Before his death, Rudolf tried to secure the election of his son Albrecht (or Albert) for the imperial throne. He failed, not only because Otakar's son Václav II, now the King of Bohemia and one of the Electors, opposed it, but mainly because the German dukes found Rudolf I too powerful and did not want to see another Habsburg on the throne. Albrecht, however, did not give up and ultimately succeeded, seven years after his father's death. In the meantime, Václav III (r. 1305–1307) succeeded his father, Václav II (r. 1278–1305), but reigned only for two years before he died at the hands of an assassin. The Czechs, fearing that the Habsburgs might try to grab the vacated throne, immediately elected Heinrich (Henry) of Carinthia (Jindřich Korutanský) the King of Bohemia. Their fears were justified: Albrecht, from his title of Roman king, invoked a right that he did not have and named his son, Rudolf II, the King of Bohemia. To support the Habsburg claim to the Bohemian throne, Albrecht had his son marry Elisabeth Richeza (Eliška Rejčka) of Poland, who was from the Piast dynasty, and the widow of Václav II. In 1306, Albrecht occupied Prague and expelled Henry of Carinthia, while Rudolf besieged the fortress of Horažd'ovice, where the nobles supporting Henry sought refuge. All of this came to an end, though, as Rudolf died in Horažd'ovice of dysentery and Albrecht was assassinated in 1308, which allowed the nobles to once again elect Henry as the King of Bohemia.

The third act of the Bohemian tragedy took place some 200 years after the events just outlined. During those two centuries, the Czech lands went through a period of cultural, economic, and political expansion, followed by a period of great upheavals brought upon the country by religious and social wars leading to the splitting of Christianity into a Catholic faction and various versions of Czech pre-Protestantism. The initial instability that arose from the termination of the Přemyslid lineage was overcome by the inauguration of the Luxembourg lineage. The dynasty then ruled the country for about 100 years and pushed the Habsburg menace into the background. Subsequently, the Polish Jagellonians replaced the Luxembourgs on the Czech throne and when the last of their kings prematurely perished in battle, several candidates showed interest in succeeding him. Among them was Ferdinand of the Habsburgs, at that time an administrator of the Austrian lands over which Otakar II clashed with Rudolf I. Since he was brother-in-law of the deceased king, who was also King of Hungary, Ferdinand aspired for both the Bohemian and the Hungarian crown. The election of the Bohemian king was in the hands of the three estates—

the nobility, the clergy, and the townspeople. Of the different candidates, Ferdinand appeared most suited, although the estates were wary of having another Habsburg as their ruler. In the end they agreed to have him under certain conditions. To these he agreed and at a ceremony (Fig. 2.7) he solemnly swore to pay at least one-half of the debt left by his predecessor, to honor all the privileges specified by the Golden Bull of Sicily, including the preservation of the kingdom's independence, the right of the Bohemians to choose their kings, and the inviolability of the kingdom's borders. He promised further to set up his residence in Prague, to honor the established rights and privileges of the estates, to respect the freedom of religion, and not to choose and crown his successor during his lifetime. While he was swearing to do all these things, he must have had plans already in his head of how to break them all. For no sooner was the Crown of Saint Václav placed on his head, than he began issuing directives contrary to the individual articles of his oath. The Czechs protested at first and when this did not help, they rebelled, but he ignored their protests and suppressed the rebellion brutally.

Ferdinand's ascension to the Bohemian throne in 1526 was a sad day for the Czech people. It began an era of Habsburg rule over the Czech lands that was to last for 400 years. On paper, the Bohemian Kingdom continued to exist, but the Austrian emperors continued to decorate themselves with the Crown of Saint Václav, while imposing their culture, religion, and language on the Czechs. In 1619 the Czechs rose against the Habsburg occupation and elected their own king, Frederick V (r. 1619–1620), but were routed in the Battle of the White Mountain in 1620. The victorious Habsburgs then instituted a regime of reprisals, executions, forced emigrations, Germanization, and forced Catholicization. A period of darkness descended on the Czech lands. It took a conflagration in the form of the First World War for the Czechs to finally free themselves from their oppressors and for the international community to return to them the control over the lands that were historically theirs.

Islets of German-Speaking People in the Czech Lands

The Czechs have had three neighbors throughout their history—Germans, Hungarians, and Poles. While they had their issues with each of them from time to time, it had been the Germans with whom they were uneasy with at the best of times and outright hostile at the worst. But to simply call them Czech neighbors is not quite accurate because in addition to the Germans living outside the Czech lands, there were also those that lived in the settlements within. For a want of a better term, we shall call the latter "immigrants." German immigration into the Czech lands had been under way throughout the entire history of Bohemia and Moravia. The earliest immigrants were missionaries, priests, traders, and merchants. Later, however, nearly all occupations and social classes came to be represented by the incomers. There were times when the ingress of German people was a mere trickle and others in which it resembled a veritable stream, if not a flood. They came for various reasons, some of them invited, others on their own initiative, all of them hoping to find in the Czech lands better opportunities than they had in their homeland. They came either from the Alpine districts of Austria or the various parts of present-day Germany. Many Czech rulers encouraged immigration, especially after the depopulation of the country by the ravages of wars or epidemics. The German immigrants tended to settle down in the border regions, which were the least densely populated, and in the towns and cities, where they could stick together more easily than in the rural areas. As a result, some of the settlements became almost entirely German, forming a *Sprachinsel*, an islet in which German was spoken almost exclusively.

The immigrants showed no desire to become assimilated by the adoptive country, linguistically or politically. Very few of them learned Czech, and for those who did, Czech was only a secondary language. They wanted not only to stay German but also for the Czech lands to become German. In fact, many of them believed that they were settling down in a land that *was* German originally and that it was their mission to make it fully German again. Later, they would be supported in their views by chauvinistic historians, who argued that before the Slavs, the territories were occupied by German tribes and hence that the land was rightfully German. While the validity of the first part of the argument cannot be disputed, if one were to accept the second part, then it could be argued that since the lands were originally settled by the Celts before the Germans ever got there, neither the Czechs nor the Germans, but the Irish and the Welsh are the rightful heirs of Bohemia and Moravia—and Germany for that matter. Faced with this objection, the tendentious German historians then came up with the claim that Germans actually never left the Czech lands and that it is the continuity of their presence that makes them their rightful owners. Even though these historians could not deny that the Quadi and the Marcomanni actually moved out of the Czech lands, they still insisted that a small number of settlers stayed behind and persisted through the Slavic occupation through to historical times. The "evidence" for this claim is purportedly provided by the etymology of the geographical names in Bohemia and Moravia, particularly the names of rivers and towns. According to these historians, the names are clearly German and since the Slavs took them over, the takeover could have happened only if some Germans remained continuously in the land. In reality, however, the names of the rivers (the hydronyms) are not of German but of Old Indo-European or Celtic origin, and their transmission apparently does not require a continuity of inhabitants; a short period of overlap suffices. As for the names of the towns, they fall into three categories in terms of their etymology.[18] In the first category are names whose origin can no longer be deciphered with certainty. Examples are provided by the cities Brno, Opava, and Olomouc, three important stations in Mendel's life. Their German names are Brünn, Troppau, and Olmütz, respectively. The origin of these three names is uncertain, except that a German origin is the least likely of the different possibilities. The German word *der Brunn* is a poetic version of *der Brunnen*, which means spring, well, or fountain, so seemingly the etymology of the name is clear. In reality, however, it is not, because the Old Slavic *brn* and *brnie* means "mud" and the original name may have referred to a castle at a muddy place. Furthermore, equally "obvious" etymologies of Brno/Brünn have been proposed based on Hungarian or Celtic words. So, who is right? We do not know, but knowledgeable linguists seem to prefer the "muddy place" interpretation. (Incidentally, the earliest known references to the settlement, dating from the second half of the eleventh century, call it variously Brnen, Brynen, Birnen, Byrno, and Brnno, which of course helps little to resolve the conundrum.) The names Opava/Troppau are obviously derived from the hydronym of the river on which the city lies. (The German Troppau is a reduction from *an der Oppau*, "on the river Oppau.") The hydronym apparently originated from the Old Indo-European word *apa*/*opa* for water, to which the Slavs added the ending—*ava*, commonly used by them to designate rivers. The ending itself is apparently derived from another Old Indo-European word for water, *ahwa*. The oldest known reference to the Opava River, also from the eleventh century, has the form *Vpa* (=*Upa*). Olomouc/Olmütz is derived from the name of the castle around which the city arose and in the eleventh century records appeared under the names Olmuc, Olomuc, Olomuz, or Olomocz. The name consists of two parts, Ol(o)- and *–múc*, of which the first means "beer" in Old Slavic, while the second is related to the Old Slavic *mútit*, meaning "to make noise, roar." Linguists, however, consider this etymological combination very improbable and argue that the first part may have actually represented a person's name. In other words (no pun intended), the derivation of the name is unclear, but there is certainly no indication of German origin.

In the second category are toponyms in which the name is clearly of Slavic (Czech) origin and was later translated or adapted into German. In both cases, the Czech version appears before the German version in historical documents. Examples, related in one way or another to Mendel, are Lipník (Leipnik), Nový Jičín (Neu Titchein), Přerov (Prerau), and Vražné (Petersdorf), the names in the parentheses being the German versions. In the toponym Nový Jičín, the first word means "new" in Czech and the second, in its original version Tyczin or Dičín, initially meant "the property of a person of wild manners" (from Old Slavic *dikyj*, "wild"). Přerov derives its name from the Old Slavic word for a ditch or moat of a castle, which surrounded the city. Vražné, where Mendel was baptized, derives from the Old Slavic word *vorg*, "to throw" or "cast," which gave rise to *vrch*, meaning a hill. Vražné therefore meant "a settlement on the hillside." The German name, Petersdorf, was introduced later in reference to the Church of Saint Peter and Paul that was built in the village (*das Dorf* = the village). In at least one case, the original Czech name underwent such a drastic transformation that even the Czechs no longer recognize it as their own. One such case is the name of the location where Napoleon defeated the combined Russian-Austrian armies in 1805—Austerlitz, east of Brno. The name sounds German, but in fact, the original toponym from 1234, Novosedlice (meaning "new settlement"), was genuinely Czech. The transformation Novosedlice→Nuzedlicz→Neussedlicz→Neusserlicz→Asserlitz→Austerlitz occurred in less than 400 years. When the name started to change, a new Czech toponym emerged, Slavkov, when a person by the name Slávek acquired a castle nearby. And so for the Czechs, Austerlitz is now Slavkov.

In a small third category are toponyms that were originally German and were later changed to Czech. Most of the settlements that belong to this category were located in the borderland (the Sudetenland[19]), were founded relatively late in history, or both. They were founded by German immigrants, and not by the pre-Slavic settlers. To sum up: The etymology of the toponyms does not support the claim of continuity from the Quadi and Marcomanni to the eleventh century, when the first German immigrants began to arrive into the Czech lands. Until then almost exclusively Slavs inhabited these lands. Taken all together, there simply was no basis for the German immigrants' sentiments that they were reclaiming a land that was originally theirs.

Naturally, the Czechs were not at all happy with the Germanization of their lands. Official or not, they ignored the German names and stuck to the Czech versions, which had been passed onto them from their ancestors. And as soon as it became possible, they purged their textbooks and maps of the barbarisms. It is against this historical background that one should view the Czech sensitivity to the perpetuation of German names for all things Czech. We cannot resist mentioning one typical example. Arguably, the piece of Czech classical music best known to the outside world is *Vltava* by Bedřich Smetana. Never heard of it? That's because you know it under its German title *Die Moldau* from the cycle *Mein Vaterland* by Friedrich Zmetana. Yet the Czechs call the river, which originates and ends in Bohemia, *Vltava* and not *Moldau*, and it was this Czech name that Bedřich Smetana, an acutely Czech-conscious composer, used for the title of the second symphonic poem of the cycle *Má vlast* (My Country). It is therefore tactless, if not outright boorish, to remind the Czechs of some of the darkest periods in their history by referring to the most Czech of Czech works by its German translation. And it is similarly tactless to continue using Brünn, Troppau, and Heizendorf instead of Brno, Opava, and Hynčice, no matter how difficult they may be to pronounce. In this book, we therefore avail ourselves to names in the original language whenever possible and only in cases in which it becomes awkward textually shall we resort to the use of English equivalents.

Returning to the German immigrants, it must regretfully be stated that very few if any of them were convinced by etymological or other arguments. They continued to regard themselves as pioneers who either had returned to ancient German lands or were bringing their superior culture into new lands and so acquiring new territories for Germany. At some point, however, they must have asked themselves: What was Germany in their time? The Holy Roman Empire of the German nations may not have been wholly German, but at least it held most of the German-speaking states under one umbrella. When Prussia began to rise, the immigrants could still identify the Austrian Empire with Germany. But once Prussia began to organize German states separately from the Austrian Empire, there were suddenly two Germanys, one under Hohenzollern and the other under Habsburg control. Things got even more confusing, when Prussia declared war on Austria, and subsequently when Prussia succeeded in uniting the German states with the exclusion of Austria. The German-speaking minority in the Czech lands then faced a problem: To whom did they owe their allegiance? Formally, they lived in a country governed by Austria, but the empire to which the country belonged no longer represented the dominion (the German *Reich*) from which their ancestors came and to which they thought they owed their allegiance. Discounting the Austrians, there were no real Germans left in the new Austrian or the Austro-Hungarian Empire. In all but political respects, the empire seemed to have been overwhelmed by the Slavs (the Czechs, Moravians, Silesians, Galicians, Croatians, Slavonians, Slavs living in Bosnia-Herzegovina) and the Magyars. It was only a matter of time until the empire would fall apart. On the other hand, to openly display loyalty to the Hohenzollern German Empire would have amounted to treason, from the viewpoint of the Habsburg government. So the Germans in the Czech lands were in a real quandary: They were unwilling to integrate into the Czech population, they could not count for much longer on the Austrians, and they lost the political and moral backing from the lands of their origin. The situation was compounded by the growth of Czech nationalism and the increasing hostility of the Czech majority to what they considered to be an arrogant German minority in their midst. In a certain sense, the collapse of the Austro-Hungarian Empire resolved the quandary. It absolved the German minority of whatever allegiance it might have owed to Austria, and the new political system, the democratic Czechoslovak Republic, showed more tolerance to dissenting groups than the Austrian Empire did. The German minority thus began to display openly its sympathies for the ideals of the Third Reich propagated by the nationalists in Germany. In 1939, it provided Hitler the pretext for the invasion and occupation of Czechoslovakia. Many Germans living in Czechoslovakia considered Hitler a liberator from the Slavic oppression. After World War II, the Czechs retaliated for the horrors that the German occupation brought on them, by summarily expelling most of the German settlers from the Czechoslovak territory—an act that they cannot be proud of.

Origin of Mendel's Genes

From all we have learned in this chapter we can say that Mendel's genes were drawn from a stock, a *gene pool*, which probably originated in Anatolia and reached central Europe some 6,000 years ago. During the six millennia, the pool changed in three principal ways. First, some of the genes mutated (changed chemically) and some of these new forms either replaced the old ones or remained in the pool along with the old forms. Second, new forms of genes were introduced into the pool from other pools brought into Europe by invaders from Central Asia: Scythians, Huns, Avars, Magyars, Mongols, and Turks. And third, as the pool spread across Europe, it partitioned into sub-pools, which differed in their composition, but mainly in the frequencies of the different gene variants they contained. This differentiation into sub-pools was mainly the consequence of limitations on gene exchange between populations separated by large distances. The genetic differentiation occurred in parallel with the linguistic and ethnical divergences of the European populations. One can therefore speak of Germanic, Slavic, and other sub-pools. Since the time of

Völkerwanderung, a substantial exchange of genes has been occurring between the various sub-pools, particularly in places like Moravia favored by human migrants. The major contributors to the Moravian gene pool had been Slavs and Germans and to a lesser degree other sub-pools.

The mixing of pools and sub-pools might be expected to have had a beneficial effect in that it enlarged the possibilities for creating new gene combinations. Indeed, if the number of outstanding personalities a gene pool has yielded is any measure of its quality, the Moravian pool must be adjudged as being of a high standard. The list of internationally renowned persons born in Moravia/Silesia includes the composers Gustav Mahler (1860–1911), Pavel Křížkovský (1820–1885), Leoš Janáček (1854–1928), Pavel Vranický (1756–1808), his brother Antonín Vranický (1761–1820), František Vincent Kramář alias Franz Krommer (1759–1831), František Xaver Richter (1709–1781), Antonín Emil Titl (1809–1882), František Václav Míča (1694–1744), and Alois Hába (1893–1973); the painters Alphonse Mucha (1860–1939), Otakar Kubín alias Othon Coubine (1883–1969), and Zdeněk Burian (1905–1981), one of the most influential paleo-artists of the modern era; the novelist Milan Kundera (born 1928); the educational reformer and religious leader Jan Amos Komenský alias Johann Amos Comenius (1592–1670); the historian František Palacký (1798–1876); the philosopher, architect of Czech independence, and the first president of the Czechoslovak Republic Tomáš Garrigue Masaryk (1850–1937); the philosopher and mathematician, the founder of phenomenology Edmund Husserl (1859–1938); the psychologist Sigmund Freud (1856–1939); the logician, philosopher, and mathematician Kurt Gödel (1906–1978); and the physicist and physiologist Ernst Mach (1838–1916). To this could be added an even longer list of poets, novelists, and artists, who attained renown at the national level only, but could in fact measure up to international standards if only their works would be made better known abroad. An impressive contribution to the world of culture and science by a country with less than five million inhabitants!

References and Notes

[1]Václav Hanka (1791–1861) was a Czech philologist, professor of Slavic languages at the Charles University, Prague. He discovered the so-called *Manuscripts of Dvůr králové and Zelená hora*, a collection of poems and songs, allegedly from the thirteenth and fourteenth centuries. It is now generally agreed that the manuscripts are a forgery, but the identity of the forger or forgers has never been established firmly. The song "Morava, Morava, my dear little Morava," which Hanka composed, is commonly perceived as a folk song

[2]Leighton R (2000) Tuva or Bust! Richard Feynman's Last Journey. W.W. Norton, New York, NY. In this book, the author describes how he and Feynman used this method to verify the claim that Kyzyl, the capital of Tuva, is the center of the Asian continent, as the obelisk on the city's main square proclaims

[3]Ager DV (1880) The geology of Europe. Wiley, New York, NY

[4]Podborský V et al (1993) Pravěké dějiny Moravy. Muzejní a vlastivědná společnost v Brně, Brno

[5]Tattersall I, Delson E, van Couvering J (1988) Encyclopedia of human evolution and prehistory. Garland Publishing, New York, NY

[6](a) Gray RD, Atkinson QD (2003) Language-tree divergence times support the Anatolian theory of Indo-European origin. Nature 426; 435–439. (b) Bouckaert R et al. (2012) Mapping the origin and expansion of the Indo-European language family. Science 337; 957–960. These authors argue that the homeland of the Indo-Europeans was in Anatolia (present-day Turkey). To reach this conclusion, however, they used statistical methods which some scholars consider unreliable

[7]Renfrew C (1988) Archaelogy and language. The puzzle of Indo-European origins. Cambridge University Press, New York, NY

[8]The whole quote reads: "I believe that the earth is very large and that we who dwell between the pillars of Hercules and the river Phasis live in a small part of it about the sea, *like ants or frogs about a pond*, and that many other people live in many other such regions." In: Phaedo P (1966) Plato in Twelve Volumes, Vol. 1 translated by Harold North Fowler; introduction by W. R. M. Lamb. Harvard University Press, Cambridge, MA

[9]Delouche F (ed) (1992) Illustrated history of Europe. A unique portrait of Europe's common history. Henry Holt, New York, NY

[10]Cussans T (ed) (1994) The Times Atlas of European History. HarperCollins, New York, NY

[11]Barford PM (2001) The early Slavs. Culture and society in early Medieval Eastern Europe. Cornell University Press, Ithaka, NY

[12](a) Váňa Z (1983) The World of the Ancient Slavs. Wayne State University Press, Detroit, MI. (b) Chropovský B (1989) Slované. Historický, politický a kulturní vývoj a význam. Orbis, Praha

[13]Ruhlen M (1987) A guide to the World's Language Volume 1: Classification. Stanford University Press, Stanford, CA

[14]Květ R (2003) Duše krajiny. Staré stezky v proměnách věků. Academia, Praha

[15]Dvornik F (1970) Byzantine missions among the Slavs: SS. Constantine-Cyril and Methodius. Rutgers University Press, New Brunswick, NJ

[16]Actually his original name was Constantine; he took the religious name Cyril shortly before death

[17]Hora-Hořejš P (1995) Toulky českou minulostí Vol 2. Od časů Přemysla Otakara. I do nástupu Habsburků (1197–1526). Baronet, Praha

[18]Lutterer I, Majtán M, Šrámek R (1982) Zeměpisná jména Československa. Slovník vybraných jmen s výkladem jejich původu a historického vývoje. Mladá Fronta, Praha

[19]The words *Sudetes* (in English), *Sudety* (in Czech), and *Sudeten* (in German) are probably of Celtic origin meaning "boar forest" (from *sud*, boar and *–eta*, forest). It first appeared on the map of the Greek–Roman–Egyptian geographer Claudius Ptolemy (c. 90–c. 168 CE) as the *Soudeta ore* designation for the mountain range constituting the northern side of the Bohemian rhombus of mountain ranges enclosing the Bohemian Massif (see Fig. 2.2). It then reemerged in the sixteenth century CE in the works of the mineralogist Georgius Agricola (1494–1555) in the Latinized form *Sudeti montes*. The expression *Sudetenland* appeared only in the twentieth century as a name for the mountainous border parts of Bohemia and Silesia, as well as pockets in northern Moravia, in which the German-speaking inhabitants had attained majority

The First Decade: The Childhood of a Farmer's Son

3

> **Счастливая, счастливая, невозвратимая пора детства! Как не любить, не лелеять воспоминаний о ней? Воспоминания эти освежают, возвышают мою душу и служат для меня источником лучших наслаждений.**
>
> L. N. Tolstoy: *Childhood*[1]

Childhood is the key to one's personality. It is in this phase of life that a person's character is formed and that the norms of the relationship between the inner world of the individual and the outer world are established. A child's experiences shape the psyche of the future adult. Although not quite the clean slate for which an English philosopher held it,[2] a child's mind is certainly far more impressionable than the mind of a grown-up. To use a cliché metaphor of our computer age, in a child the hardware is already in place, but the programs are still being written. Hence the child's unbounded curiosity, the questions without end, the intensity of feelings never to be matched later in life, the limitless capability of daydreaming, the blurring between the real and the imagined world, the yet undulled sense of wonder, and the unquenchable desire to make sense of the world. In Mendel's case, the nature of the hardware was determined one night in October 1821, when two of an uncountable number of possible gene combinations came together randomly in a single cell. The psyche-forming period then began nine months later and continued for some 11 years, which the child Johann Mendel spent in the house of his parents.

Because childhood is such an important interval of one's life, biographers spare no effort to find out anything they possibly can about their subject's formative years. Alas, frequently their effort yields no more than a few non-revealing anecdotes, for generally neither the parents nor anybody else suspects that a child will one day become the subject of biographers' interest. There are exceptions, of course. If a child matches adult virtuosi in keyboard performance (as in the case of Wolfgang Amadeus Mozart) or demonstrates all Euclid's propositions without taking a single geometry lesson (as is claimed of Blaise Pascal), then the parents make sure that posterity learns about it. Similarly, if a 26-year-old publishes a best seller about his childhood (as was the case of Mendel's contemporary, Lev Nikolayevich Tolstoy), he provides his biographers with a treasure trove of information. But what might you

J. Klein and N. Klein, *Solitude of a Humble Genius - Gregor Johann Mendel: Volume 1*,
DOI 10.1007/978-3-642-35254-6_3, © Springer-Verlag Berlin Heidelberg 2013

expect to learn about the childhood of a peasant's son from a Godforsaken corner of Silesia, who became famous only some 20 years after his death? By the time the first biographers showed interest in Mendel, it was too late to interview any of his childhood pals. And so we know next to nothing, not even a single anecdote, from the first decade of his life. Is it then not, you may ask, overly ambitious to devote a whole chapter to Mendel's childhood? Perhaps, but while we cannot relate any specific events of his early life, we can attempt to reconstruct the circumstances under which his childhood unfolded. Surely, not just concrete incidents, but the general pattern of life in a nineteenth-century Silesian village and farmhouse must have molded Mendel's personality. It is this reconstruction of the conditions and atmosphere in which Mendel grew up that is attempted in this chapter. We begin with the description of the patch of land Mendel called home.

Kravařsko

The Czechs call the land of Mendel's childhood *Kravařsko* and the Germans *Kuhländchen*. Both names derive from an equivalent root, but apparently in different ways. The Czech word *kráva* and the German word *Kuh* mean the same thing, "cow" or "cattle" in English. (The German *Ländchen* is a diminutive of *Land*, which has the same meaning in German as it does in English.) It might seem obvious, therefore, that the two names are translations from one language into the other, but from which into which? Czech historians insist that it was from Czech into German, whereas their German colleagues maintain that it was the other way around. In reality, the origin of the designations may be more complicated than either of the two sides present it.

The name *Kravařsko* is linked to an old dynasty of Czech lords, the Benešovids, Beneš (= Benedict) being a common name in Bohemia. They founded Benešov, a settlement on an ancient trading route connecting Praha and České Budějovice.[3] In the second half of the thirteenth century, a Benešovid by the name Drslav obtained, possibly directly from the king's hands, a village in another part of the kingdom, near the Silesian city of Opava, on the way to Ostrava. The village was called *Kravaře* or Kravařov in Czech, which was later transliterated into German as *Krawaren*.[4] That the transliteration took place in this and not in the reverse direction is indicated by the fact that the name means something in Czech (*kravař* means "a man who owns a pair of draught cows"[5]) but nothing in German. To distinguish themselves from the other Benešovids, the new owners of the village began to use the epithet "of Kravař." Vok of Kravař, the son of Drslav, allied himself with King John of Luxembourg and received from him, for services rendered, estates in a different part of Silesia, in the heart of the Moravian Gate. The estates included Jičín, Fulnek, and Bílovec. Vok upgraded his possessions by founding, in 1313, a new settlement on a rivulet called Jičínka, not far from an old fortress. The settlement then became known as Nový Jičín and the old castle as Starý Jičín. He also fortified the village of Bílovec and colonized it with German settlers, who then called the place Wokenstadt (Vok's town), which became

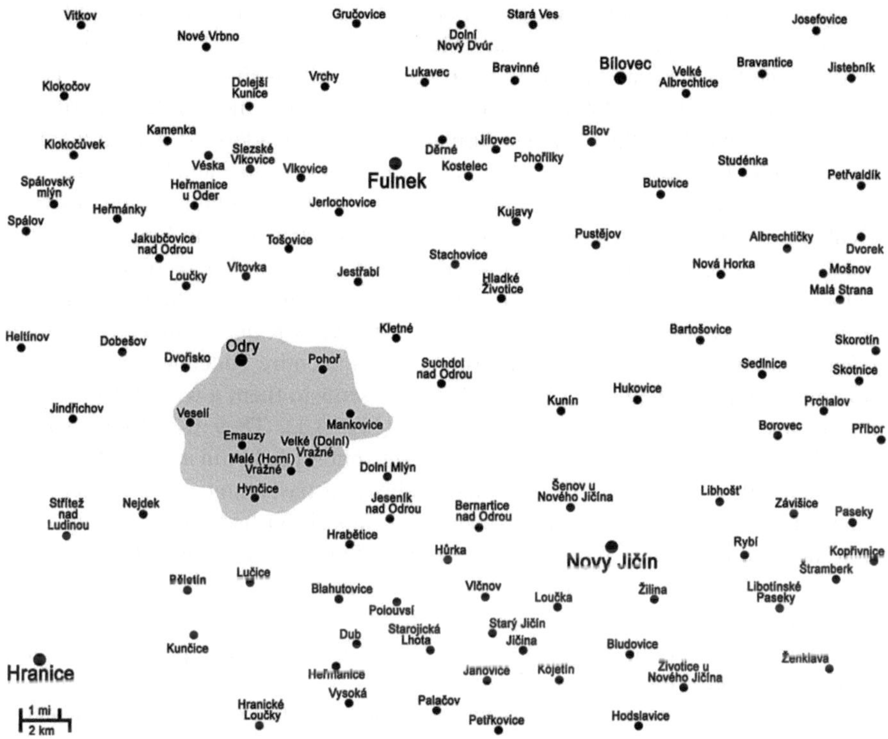

Fig. 3.1 Map of Kravařsko. The *shaded* area is the Odry domain in Mendel's time

Wagstadt after contraction (Fig. 3.1). Vok's descendants later acquired additional estates in the area, so that with time the entire region came to be called Kravařsko after the owners of most of the land.

To understand why some German authors believe in the priority of the term *Kuhländchen*, we need to delve a bit deeper into the history of colonization in this region.[6] Most of the 60 or so settlements in the Kravařsko region had Czech names when first mentioned in documents from the thirteenth and fourteenth century. One can presume therefore that Slavs founded the settlements prior to the twelfth century. Their German names were introduced later either by Germanic immigrants or under the Habsburg rule as part of the Germanization program. The immigrants presumably came from the northwest as an extension of the waves rolling over the Sudetes Mountains. Where these waves of immigrants originated and how and when they reached the region is uncertain. Saxony, Bavaria, and Swabia are most commonly considered as the sources of the immigration, largely based on the distribution of family names, but these deductions are tenuous at best. The first large-scale influx of German immigrants into Moravia-Silesia occurred at the beginning of the thirteenth century under Margrave Vladislav, brother of the Bohemian King Přemysl Otakar I. It then continued to a varying extent under most of the subsequent rulers. Some historians believe that most immigrants entered the region after the depopulations caused by the invasions of Mongolians (1241), Magyars (1252), and Turks (1663/1664); by the Black Death (1348, 1624, 1645, 1680, 1713–1715); and by the Thirty Years War (1618–1648). Others view the immigrations as a more or less continuous process. That whole villages were burned down by invaders or depopulated by diseases cannot be doubted. But whether some villagers managed to escape into the woods and return after the invading hordes had left or whether inhabitants from neighboring settlements resettled the depopulated villages can no longer be determined. Whichever was the case, some of the settlements were recolonized either predominantly or exclusively by Germanic people. It is to these settlements that the German authors apply the term *Kuhländchen*. They view *Kuhländchen* as part of the *Sudetenland*, an all-Germanic area encompassing the entire Sudeten Mountains and some of the adjacent regions. *Kuhländchen* represents to them a wedge driven from the *Sudetenland* into the Moravian territory (Fig. 3.1). They therefore count as belonging to *Kuhländchen* all those villages inhabited predominantly by Germanic immigrants. Since, however, the information about the past ethnic composition of the villages is not always reliable and since the composition had changed with time, different German authors have come up with different delineations of the region.[7] What the different authors have in common, however, is the claim that the *Sudetenland*, including *Kuhländchen*, was part of Germany. This claim is, however, contrary to historical and political reality. Throughout their long pre-World War II history, Bohemia, Moravia, and Silesia had never been part of Germany, which in fact did not come into existence as an independent state until the late nineteenth century. During most of that history, these three lands were part of the Bohemian Kingdom, which had a clearly delineated frontier. The colonization of a part of this frontier by Germanic immigrants did not make these areas part of Germany, just as the colonization of southern California by Mexicans has not made California part of Mexico. Moreover, neither the *Sudetenland* nor Kravařsko was at any time monolithically Germanic. Although some villages were predominantly Germanic and German speakers were indeed in majority in parts of the region, some villages remained predominantly Czech, and others had mixed populations. Intermarriages between German speakers and Czechs were not uncommon, so that, notwithstanding the language differences, the physical, cultural, and behavioral dissimilarities between the two ethnic groups began to diminish. So much so that an early Mendel biographer, who was familiar with the situation in *Kuhländchen*, observed some 90 years ago that "because of their continuous, centuries-long mixing, the German and Slavic inhabitants of the region can no longer be distinguished from each other as races and differ only in their language".[8] And as far as their languages were concerned, it seems that they had been coming together as well. Some German writers were appalled by the dialect the *Kuhländchen* people spoke,[9a] and the biographer, too, complained about its unintelligibility.[8] The regional Czech dialect was similarly incomprehensible to outsiders, but the German and Czech inhabitants seem to have been able to communicate with each other with no great difficulty. One is tempted to think that, had it not been for the politicians, tendentious historians, and agitators, the two groups would have managed to live together just fine.

Fig. 3.2 The Kravař cattle

Coming back to the origin of the term *Kuhländchen*, we note that for centuries the area had indeed been renowned for cattle breeding. On the lush meadows along the Odra River and its many tributaries, the cattle breeders raised a stock valued not only in the territory but also outside of it. Among its ancestors were short-headed Tyrolean cattle, bulls from the Bern region of Switzerland, and central European red cattle.[9b] Generations of inbreeding resulted in the *Kravař cattle* (Fig. 3.2), strong enough to pull a plow or a cart yet a good producer of milk with a high content of fat (then considered to be a desirable character). German writers argue that it was for the cattle breeding that the area came to be called *Kuhländchen*.[7] This may be so, but the fact that long before the development of the breed, the Czechs already knew the area as Kravařsko undoubtedly helped in the acceptance of the German name. While Kravařsko could mean "a land originally owned by the house of Kravař" and similarly *Kuhländchen* "an area in the Moravian Gate occupied by cattle-breeding Germanic settlers," both terms can be applied to an arbitrarily defined region representing the plateau between the Oderské vrchy on the one side and the Beskydy on the other and stretching approximately from Jistebník in the northeast to Bělotín in the southwest (Fig. 3.1).

The average altitude of the plateau was about 300 meters above sea level, with most of the land taken up by arable, fertile fields and luxuriant meadows accompanying the rivers and streams that widened here and there into ponds or marshes. The dominant feature of the plateau was the Upper Odra River, which suddenly changed its course from southeast to northwest near the Jeseník nad Odrou. The primary occupation of the region's inhabitants was crop growing and cattle breeding, supplemented with beekeeping, horticulture, and cheese making. Silver and lead had been mined in the past at Pohoř and the Veselský kopec, the two hills between which lay the city of Odry. The mining was, however, not extensive and was stopped altogether in 1814. At other hills flanking the region, quarries were supplying stones for road and house construction. At the Štramberský kopec east of Nový Jičín were large limestone quarries and kilns for burning the crushed stone, converting it into quicklime. Other, smaller limestone quarries may have been scattered throughout the region. Near clay pits arose brickyards at which the raw material was shaped by hand into bricks and then burned in kilns. The three cities in the Kravařsko region, Nový Jičín, Odry, and Fulnek, had their own breweries and distilleries, and from the eighteenth, but especially in the nineteenth, century they were the sites of industrial expansion, particularly in the various branches of textile manufacture. Factories specializing in the production of wool, cotton, silk, or linen merchandise, as well as dyeing and bleaching shops, were mushrooming. The hatters and milliners of Nový Jičín became famous throughout the Austrian Empire. Before the first railroads were built in the middle of the nineteenth century, all the materials and goods had to be transported and distributed by horse- or oxen-drawn wagons. For some enterprising farmers, this was an opportunity to earn some extra *Kreuzer*.

Each of the three cities, together with all the villages administratively associated with it, comprised a separate district (*okres, Kreis*), and the three districts formed the core of the Kravařsko region. According to the 1890 census, the combined districts encompassed 62 villages, 7687 houses, and nearly 61,000 inhabitants, of which 73 % were of Germanic extraction and only 27 % were Czechs. In the Odry and Fulnek districts, nearly all the inhabitants were Germanic. A marked difference existed between urban and rural populations. In the largest, Nový Jičín district, for example, in the city, 92 % of the inhabitants were Germanic, but in the villages the majority (57 %) was Czech. The census from 1834 indicates that similar proportions of Germanic and Czech inhabitants also existed in Kravařsko in Mendel's time. The difference between the Nový Jičín district and the Odry and Fulnek districts reflects the direction in which the Germanic immigrants penetrated from the Sudetes Mountains into Silesia and Moravia. The difference between the urban and rural areas in the Nový Jičín district betokens a preference of the immigrants for settling down in cities. Since 1946, though, nearly all the inhabitants of the Kravařsko region are Czechs. The city of Fulnek was part of Silesia until 1475, when it was assigned to Moravia. With the city of Odry, it was the other way around. The city of Nový Jičín has always been in Moravia, so that Kravařsko straddled the Moravian-Silesian border.[10]

The Odry Domain

Kravařsko and *Kuhländchen* are names that had no legal, political, or administrative significance. They were folkloric sobriquets given by the people to vaguely defined regions according to what appeared to be their most obvious characteristics. Throughout the Middle Ages and up until the beginning of the twentieth century, the legal and for much of the time also economical unit of land division in central Europe was a *manor* or *domain* (*panství* in Czech, *Herrschaft* in German), owned by a manor lord or a noble house.[11] The various levels of political and administrative organization—the counties, duchies, principalities, kingdoms, and empires—were superimposed on the mosaic of domains. In Mendel's time, the margravate of

Moravia and the duchy of Silesia, the two crown lands of the Bohemian Kingdom in the Austrian Empire, consisted of 287 and 116 domains, respectively, which combined made 403 domains.[12] The division was a legacy of the times when the medieval ruler used to reward his knights for services rendered with a grant of land and thus made them feudal lords or nobles (see Chap. 2). The sizes of the different domains varied considerably, and they also changed with time correspondingly with the fortunes of their owners. In general, however, domains were also units of barter among the lords, changing hands from time to time. The dates of ownership changes punctuate the history of the individual domains.

The first decade of Mendel's life played itself out in the domain of Odry and then in the duchy of Silesia.[13] The Odry domain encompassed some 1,900 hectares (4,695 acres) of land, all of which belonged legally to one lord, with different ones at different times. On the land, there were some 20 settlements—one town (Odry in Czech, Odrau in German) and 19 villages—but the number of villages varied with time as some of them merged and new ones arose.[14] When exactly the town of Odry was founded and by whom is unclear. Its predecessor was a Slavic village Vyhnanov, which was documented for the first time in 1234. It was located on the left bank of the Odra River, at the foot of the Pohořský kopec, the Pohoř Hill, but was abandoned in the second half of the fifteenth century, never to be recolonized again. In its place arose a village around a castle on the right bank of the Odra River. The first documented owners of the castle and of the domain that came with it were the lords of Šternberk, another venerable dynasty of Czech nobles. At different times subsequently, several other noble families claimed the Odry domain as their own, first Czech and after 1620 mostly Germanic and Austrian. In the time of Mendel's parents, the domain was in the hands of Prince Karl Lichnovsky, a patron of Beethoven. The Lichnovsky dynasty was of Polish descent and owned several other domains in Silesia and Moravia.[15] With time the Odry village developed into a town. In the Middle Ages, the difference between a village and a town (city) was not so much in their sizes as in their legal standings. City inhabitants, the burghers, were mostly artisans, merchants, officials, and clerks, who enjoyed privileges and freedoms that the village-inhabiting peasants lacked. The privileges varied from city to city, depending on who founded the city and under which law. A city founded by a king, for example, remained under the direct jurisdiction of the country's sovereign and was thus exempted from the control of the local lord. Likewise, some cities founded by Germanic immigrants were granted the privilege to manage it according to German rather than Moravian or Silesian laws, in some cases even according to the law of a specific Germanic city, Nürnberg, for example.

The Odry castle (manor house)[16] was the lord's residence, from which he ruled the domain. Although the entire domain was legally his possession, it was differentiated into two kinds of land, allodial (dominical), which was at his free disposal, and rustical (urbarial), with which he could dispose less freely.[11] The allodial land included the *demesne* (the estate on which stood the castle or manor house and everything that went with it), the fields that the lord's people farmed for him, and most of the uncultivated land, such as forests, pastures, marches, and ponds. In most domains the allodial fields accounted for about one third of the cultivated land. Some of them were associated with the manor, while others were scattered over the entire domain, for the lord retained for himself the choicest fields, wherever their location might have been. Each cluster of a lord's fields was associated with a group of buildings, which the Czechs called *popluží* and the Germans *Meierhof*, the house of the steward. The buildings included dwelling places of the steward and his helpers, stables, barns, granaries, and all the other accessory structures of a farmstead. The steward managed the whole complex on the lord's behalf. The rustical land, which encompassed all the peasants' holdings, again fell into two kinds, "bought in" (*eingekauft*) and "not bought in" (*uneingekauft*). As the names suggest, in the former case the peasant purchased the land from the lord, whereas in the latter the lord leased the land to the peasant against regular payments of rent. It might come as a surprise, however, that even after the purchase, the peasant did not own the land; the lord did. The peasant had to continue paying dues to the lord as if he were merely renting the land; he could not sell the land or part of it without the lord's permission, and the lord could reclaim it. The only difference between the two types of holding was that the right to cultivate the bought-in land was somewhat more secure than the right to cultivate the leased land. The lord could terminate the lease at any time and evict the tenant from the land; to evict a peasant from a bought-in land was more difficult legally. The lord could, of course, sell the land with the peasant to another lord.

This relationship between the peasant and his lord was based on the feudal principle of bondage established in early medieval times.[17] The principle amounted to the greatest real estate rip-off in the history of Europe. It was conceived by the Romans and copied by the Franks, who then spread it across Europe (see Chap. 2). The system based on it, *feudalism*, flourished from the fifth to the twelfth century and then began to decline gradually in western Europe, when the lords realized that it was an inefficient way of farming. In the Habsburg Empire, however, a sharp reversal of the trend came about in the seventeenth century, especially after the Thirty Years War and particularly in the Slavic lands. This period was marked by an influx of foreign nobility, mainly German and Austrian, into the Czech crown lands, where many domains and estates lost their owners through emigration and confiscation. The foreign nobility resented the Slavic peasants, regarding them as barbarian heretics speaking incomprehensible languages. To a noble, a peasant was an object, a serf, and to a peasant a noble was a scourge sent by God to test his faith. Historians don't seem to be able to agree on what exactly the word "serf" means. Some equate serf with slave because etymologically the word derives from the Latin *servus*, which means just that. If, however, one defines a slave as "a human being who is owned as property by another and is absolutely subject to his will" or as "a bond servant divested of all freedom and personal rights,"[18] then peasants of the Austrian Monarchy did not fall into this category. Since they had, in theory, the right to sue and a few other prerogatives, they were not slaves. But in some corners of the empire, they were not far from that status. More commonly, historians fall back on a definition of serfdom attributed to the Emperor Joseph II. According to the emperor, the Austrian peasants were serfs because they were deprived of the freedom of movement (they could not transfer from one domain to another without the lord's permission, which was seldom granted), the freedom of marriage (they could marry only within their domain, and even for that they again needed the lord's permission, which *was* commonly granted, but for a fee), and the freedom to learn a profession of their choice (with the exception of becoming priests or soldiers).

Severe restrictions of the peasant's personal freedom were, however, only one part of the bondage. The other part was the peasant's string of obligations to the lord, beginning with the dues he had to pay and ending with the services he had to deliver. The dues fell into two categories: monetary, consisting of regular, twice yearly cash payments in the amount determined by the size of the land he farmed, and in kind, which meant he had to provide his lord with chicken, eggs, vegetables, meat, grain, or some other farm produce in amounts specified by tradition. The service had the form of what the French called *corvée*, the Germans *Frondienst*, and the Czechs *robota* (not *robot*, as is commonly claimed). Its most common form was labor in the fields—harvesting and haymaking in the summer, plowing in the fall, sowing in the spring, and all the auxiliary work associated with these activities. Other forms included driving coaches, carting and long-distance hauling, constructing and repairing buildings, hunting, fishing, and in later times also industrial work, such as spinning and weaving for the lord's enterprises. Most of the work was done on the steward farm nearest to the peasant's village, but other jobs, such as the haulage, could take the peasant far away from home. The amount of work was fixed by custom and hence changeable by the lord. Two or three days per week were common, many lords requiring more, especially during harvest time, when the farmer was of course also most needed on his own fields. A "day" meant from sunrise to sunset, with two hours pause for meals and rest, but not counting the time the peasant needed to get to the work place. In all this work, the peasant was required to use his own implements and tools, as well as his own horses and oxen, where such were needed. Indeed, the lord counted on the peasant's use of his own animals and kept at the steward farms cattle for milk and meat, and horses for riding only. The peasants had the possibility of converting the services into cash payments, but most just did not have the money to do this. A peasant could also send a farmhand, if he had one, to work on a lord's field instead of him.

The rent and the *robota* were by no means all the dues the lord believed the peasant owed him. The lord owned the only mill to which the peasant was allowed to bring his grain, and, of course, the milling was not for free. Additionally, the lord owned the local brewery and the distillery, and so the peasant, by drinking beer or liqueur in the local tavern, indirectly paid additional dues. Marrying and dying, too, cost money, which also went to the lord. Yet, this still was not all. When the lord had been given his due, there was still a line of other collectors with outstretched hands—the imperial and local governments, the church, and the local community. The contribution to the government was via a tax on the land the lord owned, which the peasant had the "privilege" to be taxed on. Through the tithe, the peasant supported the church and the local priest, paying one tenth of the produce from his land in cash or in kind. And his contribution to the community was used for upkeep of roads and bridges, for maintenance of the school and the church, for supporting the night watchman and the headman, as well as miscellaneous other

Fig. 3.3 The gallery of Habsburg rulers

services. One does not have to have a degree in economy to realize that after filling all those outstretched hands, the peasant on a small farm never stopped worrying about getting his family through the winter.

There was an enormous injustice in the whole situation. The farmers comprised some 70 % of the population in the Austrian Monarchy. They bore by far the greatest tax burden in the empire. They fed the country. They made the good life of their social betters possible. Without them, none of the technological, scientific, and cultural achievements in the individual lands of the empire would have been possible. They themselves had developed a high level of folk art in their customs, songs, stories, fairy tales, dresses, furniture, and architecture. And from their midst rose geniuses like Gregor Johann Mendel. Yet, during the entire period of the empire's duration, the only person who seemed to know about the peasants' existence was the tax collector. The peasants had no representation in the empire's political life. They had many obligations, but few rights and even those were violated repeatedly. They were reduced to a status bordering on slavery. And when they rebelled, armies were dispatched to bring them under control savagely. It was only when the empire's economy came to the brink of collapse that the rulers began curbing the nobles and made a lukewarm attempt to give the peasants more rights in an effort to achieve an economical recovery. Maria Theresia (r. 1740–1780, see Fig. 3.3 for the portraits of the Habsburg rulers) made half-hearted steps in this direction, but the effects of the maze of directives (patents) she issued in the years 1771–1777 had only been slight, for the nobles chose to ignore most of them. Her son, Joseph II (r. 1780–1790), officially abolished most of the reprehensible features of the bondage, issuing many more patents during the years 1781–1785, but he, too, was unable to enforce the new laws. When he died, his successors Leopold II (r. 1790–1792) and Franz I (r. 1792–1835) reversed most of his directives. And so it happened that Anton Mendel, G. J. Mendel's father, was still obliged to work three days per week for the lord of Odry, as if nothing had changed. It took the revolution of 1848 to abolish serfdom, bondage, and the robota definitively. But this was too late to make any difference for Anton Mendel. Not all the peasants in the Austrian Monarchy were serfs, however. Some peasant families managed to retain their independence all along since ancient times. These *free peasants* or *freeholders* (*svobodníci* or *dědinníci* in Czech and *Freisassen* in German) had no obligations to the lord, only to the king (emperor) and his government. Although quite rare in most places, in some parts of the Monarchy whole communities of freeholders did exist. Furthermore, not all the serfs were equal. Historians recognize several categories of serfs differentiated by their holdings, status, and the type of land they cultivated.

In the lands of the Czech crown, one of the main criteria of the socioeconomical stratification was the size of the land. The basic unit used in this stratification was one *lán*[19] (from the German *Lahn*), which translates into English as a *hide* and amounts to 18.4 hectares (45.5 acres). The villagers were divided into four or five categories, each referred to in the historical records by their Czech, German, or Latin terms, depending on who kept the record and in which period. For most of these terms, there are no precisely corresponding English equivalents. In Czech the three basic categories were *sedlák, chalupník,* and *zahradník/podsedek*; an additional special category was *rychtář*. *Sedlák* (*Bauer* or *Grundbesitzer* in German, *rusticus* in Latin, and roughly *farmer* in English) was someone who owned at least four hectares (ten acres) of land. There were three subcategories of this category: *celoláník* (*Lahner* in German, *one-hide farmer* in English) who held 18 hectares (45 acres) of land or more, *pololáník* (*Halblahner* in German, *half-hide farmer* in English) with at least nine hectares (22 acres) but less than 18 hectares of land, and *čtvrtláník* (*Viertler* in German, *quarter-hide farmer* in English) with at least four hectares (ten acres) but less than nine hectares of land. A farmer also owned at least one team (pair) of horses. *Chalupník* (*Chalupner* in German, *gazarius* or *domunculator* in Latin, *cottager* in English) owned less than a quarter of a hide of land, a few oxen, but no horses. *Zahradník* (*Gärtler* in German, *hortulanius* in Latin, "gardener" in English) and *podsedek* or *podsedník* (*Untersasser*, *Hintersasse*, or *Podsedker* in German) had a small, usually fenced-in plot (hence the name; in Czech "zahrada" means "fenced-in area"). The difference between the two was in that the former "owned" the land (had it directly from the lord), whereas the latter rented it either from a farmer or from the village community. Both grew on their plots potatoes and vegetables for their own consumption and earned their living by working for the farmers for wages or payments in kind. *Rychtář*, *fojt*, or *šoltys* (*Richter*, *Voght*, or *Schultheiss* in German, *sculetus*, *judex*, or *advocatus* in Latin[19] are all terms difficult to translate into English; *mayor*, in the sense of an official village representative, is perhaps closest to them. He was a rich farmer, who could afford buying his title from the lord. If the title was hereditary, he was referred to as *Erbrichter* in German. He owned a big house called *rychta* or *fojtství* in Czech and *(Erb)richterei* in German and had the right to run a pub. He was the head of the village with the right to settle small disputes between villagers. The judicial authorities of the domain dealt with more serious feuds or crimes. The more prosperous farmers had hired hands and servants (*čeleď* in Czech, *Gesinde* in German) working for them.

Fig. 3.4 Hynčice and Vražné: overall topography (redrawn from a photograph)

Hynčice and Veselí

Two places in the Odra domain are of special interest in respect to Mendel's biography—the villages Hynčice and Veselí. The former is the place of Mendel's birth and the latter the domicile of his ancestors. The history of the two villages, like that of most other settlements in the region, goes back to the time when Slavs settled the area. In the surviving documents, Hinczica is first mentioned in 1334 and Wessiele in 1362.[14] The German names of these two villages, Heinzendorf and Wesiedel, appear later as translations or transliterations of the Slavic names.[20] The origin of the Slavic names is unclear, however. "Hynčice" presumably derives from "Hynek" ("Heinz", which is an abbreviation of "Heinrich" in German), which is a personal name, but who this mysterious Hynek (presumably a founder or an owner of the place) was is not known. The etymology of "Veselí" seems obvious, since the word means "merriment" but also "festivity" in Czech ("veselka" being the Czech word for a wedding). It is therefore possible that the founders wished the inhabitants of the new settlement to be jolly or that visitors to the settlement found the people there to be of this disposition. The German name suggests, however, an alternative etymology. The original Slavic name might not have been "Wessiele" but "Ves sídelní,"[21b] meaning "residential settlement," which gave "Veselí" by contraction.

Like many other villages in the region, Hynčice (Fig. 3.4) was founded along a road running beside a stream and so acquired a stretched-out topology. In Mendel's time, it comprised 71 or 72 houses and so represented an average size village for the area.[13] The smallest villages in Kravařsko consisted of some 15 houses, whereas the largest ones had more than 200 houses. The Chapel of Virgin Mary (Fig. 3.5), which one now finds in Hynčice, was not built until 1855 to serve for minor services. The stream that flows through Hynčice went originally under the name Wrasni, is now called Vraženka, but was known to Mendel under the name Rossbach, which meant "pony creek." It originates in the hills near Veselí and is joined by two tributaries before it reaches the village. Although normally idling and

Fig. 3.5 The Chapel of the Virgin Mary in Hynčice

diminutive, after a cloudburst or a long period of torrential rains, it can turn into a raging waterway overflowing the road and even flooding the cellars of some of the houses. In the past, some of its water was diverted into channels, which served as races driving the wheel of a watermill at two different speeds.

On the same road and stream in the northeasterly direction is the village Horní (Malé) Vražné (Klein Petersdorf), followed by Dolní (Velké) Vražné (Gross Petersdorf), the latter extending close to the site, where Vraženka flows into the Odra River. The names Horní (upper) and Dolní (lower) referred to the location of the villages on the creek relative to its source and the designations Malé (lesser) and Velké (greater) to the sizes of the settlements having 46 and 87 houses inhabited by 282 and 511 residents, respectively.[13] Politically, these three villages were separated by a political border, Hynčice and Horní Vražné being in Silesia and belonging to the Odry domain and Dolní Vražné being part of Moravia and belonging to the domain of Nový Jičín. Dolní Vražné had a school, post office, and a Catholic Church consecrated to Saints Peter and Paul (Fig. 3.6), with a rectory and cemetery. Today the three villages are confluent and are administratively joined into one called Vražné. The road along which the three villages are spread out leads in one direction to Mankovice on the left bank of the Odra River and in the opposite direction to the main road connecting Odry with Bělotín (Fig. 3.1). Near the Mendels' house, the road connects with two other roads, one leading from Hynčice to Veselí and the other to Lučice in the opposite direction. Walking from Hynčice to the three

Fig. 3.6 The Church of Saints Peter and Paul in Vražné

nearest cities took respectively one hour to Odry, two hours to Hranice, and three hours to Nový Jičín.

According to Hugo Iltis,[8] one of Mendel's earliest biographers, Hynčicians were *a bit slow, but diligent folk, in both work and life, prone to brooding over God and the world.* Felix Jaschke, a wholesale merchant from Fulnek and a self-appointed chronicler of his beloved *Kuhländchen*, had a somewhat different opinion of them. He traveled extensively over the region, gathering information about its history, lifestyle, customs, and culture. His collected works encompassed several volumes of notes and reports, some of which he published in regional periodicals, while others he left as manuscripts.[22] In one of these manuscripts, Jaschke had this to say about the Hynčicians: *The inhabitants burn lime of excellent quality and then transport it over distances of many miles into villages and towns. Having switched to carting, the farmers no longer give much thought to horse and cattle breeding. And since they deliver merchandise to remote places and quite often stay away from home for weeks and even months, they neglect their fields and farms. In the large cities, they get used to good eating and drinking, and to all the vices prevailing there, and so when they get old at home, they are good for nothing. The proneness to this kind of life affects even peasants who possess only one or two horses, but cart calves to Olomouc. Every week on Thursday, calves from Pustějov, Butovice, Suchdol, Jeseník, Bělotín, and several other places are brought to Horní Vražné and Hynčice, loaded there on wagons, and at two o'clock in the afternoon the wagons leave and reach Olomouc on Friday at six o'clock in the evening at the latest. The wagons then return on Sunday, loaded with vegetables, which they cart to the neighboring towns. On these trips, the carters eat and drink well.* Although Jaschke writes about the contemporaries of Mendel's father, it is rather doubtful that Anton Mendel ever joined the gang of merry carters. Building a new house, tilling his fields, taking care of the orchard, and working three days a week on a lord's land, Anton Mendel could not afford to be away from the farm for even one day. Ultimately, in the middle of the nineteenth century, the extension of the railroad line to the vicinity of Hynčice put an end to the villagers' dabbling in the carting business.

Jaschke's claims notwithstanding, the primary occupation of most Hynčicians must have been land cultivation and livestock production. In Mendel's time, all of them together held 383 hectares (946 acres) of fields and 89 hectares (220 acres) of meadows. They possessed 41 horses and 98 cattle, with none of the families owning more than two horses.[13] From the parish register, it appears that the Hynčice of Mendel's time covered the whole spectrum of socioeconomical strata described earlier, from farmers, through cottagers, to zahradníci and podsedníci. In the farmer category, however, there might not have been more than one or two big, one-hide farmers. Perhaps only the mayor (*Richter*), who lived in a large *Erbeichterei* (Fig. 3.7), might have fallen into this category. All the others were small, half- or quarter-hide farmers, cottagers, zahradníci, or podsedníci. Besides peasants, a blacksmith, baker, butcher, miller, and a skinner (a person who disposed of dead horses, cattle, and other animals) also probably lived in the cluster of the three villages (Hynčice, Horní Vražné, and Dolní Vražné). The cluster had also a general store, a tavern, a post office, a parish church with rectory and cemetery, and two schools. Each village had a herdsman and a night watchman. Hynčice was originally a Czech village but by Mendel's time was inhabited almost exclusively by people of Germanic extraction. When in the history of the village this shift in ethnicity occurred is uncertain. It seems that in the first half of the sixteenth century the village was abandoned for a while and then recolonized in the second half, presumably by Germanic people; it then stayed that way until 1946.

Although in Mendel's time Veselí was of the same size as Hynčice (71 houses, 562 inhabitants in 1834), it differed from the latter in its character.[13] The difference was given by its altitude, location, and ethnic composition. The difference in altitude was only some 270 meters (the altitudes of Veselí and Hynčice being 574 and 300 meters above the sea level, respectively), but this was enough for the distinction between high- and lowlanders in the minds of local people. Veselí was founded on the plateau at the top of the Veselský kopec, a premonitory of the Jeseníky

Fig. 3.7 Hereditary mayor's house (*Erbrichterei, rychtářství*) in Hynčice

Mountains. The village was surrounded by fields, the soil of which was of a considerably lower quality than that of the lowland fields, being poorer in nutrients and richer in stones. Consequently, the farmers of Veselí had to toil very hard and apply generous amounts of natural fertilizers to wrestle out of the fields' yields that would keep them alive. Beyond the fields, on the slopes of the Veselský kopec was a ring of forest, which was the lord's property. Down below the hill, in the valley of the Odra River, was the town Odry, which in the nineteenth century spread out already to the river's left bank, all the way to where the Pohoř Hill begins to rise. The location of Veselí close to the town, the seat of the domain's lord, offered employment opportunities, which the more distant villages lacked. The secluded location on the top of a hill, encircled by a dense forest, at a distance from the main throughway of the Moravian Gate, may have spared the village from visitations by marauding armies and might have been one of the reasons for the continuous presence of a Slavic population in Veselí. Its self-containment might have been a justification for the Veselian's insistence on having their own parish church and their own elementary school.

Ancestors

The common method of studying a person's ancestry is to follow the reproductive descent of individuals with the aim of determining who the person's ancestors were and where they originated. Crucial to this method are the names of the individuals and information about their birth, marriage, and death. In most of Europe in the last 500 years or so, this information has been recorded in parish registers kept by the priests who performed the Christian rites marking these three important events in a person's life. From the register, it is therefore possible to reconstruct who descended from whom through a succession of generations.

Fig. 3.8 Alois Schindler (1859–1930), Mendel's nephew and first biographer

Considering the historical circumstances, Mendel's pedigree is surprisingly well researched, thanks primarily to his nephew, Alois Schindler (Fig. 3.8), the son of Mendel's sister Theresia.[23] According to his own account,[24] Schindler became interested in genealogy already in the fourth grade of the Gymnasium in Brno. There he studied under the loving care and the support of his uncle, Gregor Johann Mendel. Later, after completing his medical studies in Vienna and becoming a practicing physician in Zlaté Hory, he continued to pursue his interest as a hobby. Initially, he concentrated on the ancestry of his own family, but later, when he became aware of Mendel's growing fame, he refocused his research onto his uncle's forbearers.[21,25,26] He drew most of his information from the registers of the different parishes in the Odry domain, mainly from those of Vražné and Veselí, but he extended his search also to other documents in the archives. He published the results of his studies in the form of genealogical tables[21,25,26]—the Schindler's List of the Mendels—as well as in a series of articles presenting his interpretation of the tables[21] and in his correspondence with his contemporaries.[27] All the articles appeared in local, not easily accessible periodicals, but later some of them were reprinted in a book.[28]

Schindler was able to trace Mendel's ancestry backward for nine generations (counting Johann Mendel as the first generation) and identify 494 of the expected 510 ancestors (Fig. 3.10; for a more complete pedigree, see supplementary Fig. S1, which can be downloaded from http://extras.springer.com/2013/978-3-642-35253-9, and Table 3.1). In generations 1 through 5, he identified all the 62 expected ancestors; in generations 6, 7, and 8, he could ascertain 60 out of 64, 86 out of 128, and 64 out of 256 anticipated forbearers, respectively.[26] Most of them lived in Veselí or Hynčice, but some came from other villages in the area—Vražné, Mankovice, Jakubčovice, and Tošovice. The nine generations covered a time span of more than two centuries, stretching from prior to 1613, the year of the earliest record concerning a Mendel, to 1862, the year in which Mendel's mother died. Most of the ancestors eked out an existence by land cultivation as oeasants and cottagers. According to Schindler,[21] to improve their lot, several of them took up gardening as a secondary occupation—a choice he ascribes to their hereditary fondness of flowers. This claim must, however, be treated with caution, for it seems to be based on the mere fact that in the parish register their occupation is given as *Gärtner* (*Gärtler*) or *zahradník*. It would be well to bear in mind that at that time *Gärtner* was an ambiguous term, which could mean either a gardener or a small landholder. It is rather implausible that a village like Hynčice could have sustained a gardener, not to mention several of them. For one thing, it is unlikely that families, in which cash had always been hard to come by, would spend it on luxury items such as flowers; for another, even the smallest landholders managed to set aside a parcel of land on which the mistress of the house would have a bed of asters and a bunch peonies or dahlias. And in the windows, there would always be space for a pot with geraniums or petunias. The strong liking of flowers was by no means a special trait of the Mendel clan. Most central European peasants possessed it, and most still do, as a visitor to even the smallest hamlet in Czech Republic, Austria, or southern Germany can easily be persuaded to believe by the riot of colors in every window. On the other hand, in Veselí, three generations of professional gardeners *did* exist, but these were referred to specifically as *Blumengärtner* (florists), and they were not part- but full-time gardeners, whose customers or employers were the lords of Odry living in the close vicinity of the village. Indeed, it is most unlikely that a farmer holding some five to ten hectares of land could take up gardening or any other job as a secondary occupation; the farm alone kept not only him but his whole family occupied throughout the entire growing season. Only the more prosperous ancestors of Mendel could find the time to function also as headmen.

The oldest record of Mendel's ancestors that Schindler[25] could find in the Kravařsko region was from the years 1611–1627 in the Protestant parish register of Odry. It was a record of a father, Konstantin (Standtke) Mendel, and his two sons, Martin and Blasius, living in Veselí. Of the father's birth, there is no record but Schindler estimated that the date might have been around 1550. The date of the father's death is not known either, but it is known that in 1613, when one of his sons was getting married, he was dead already. If earlier registers existed, they must have been lost in the turmoil of the Husite Wars and their aftermath. The earliest Roman Catholic register of Odry begins with the year 1631, so that there is a gap of four

Table 3.1 Tabular form of the Schindler's List (source, see Chapter 3 notes 21, 25, and 26)

Number	Year of birth	Father's number	Mother's number	Given name	Family name	Inferred ethnicity	Address
1	1822	2	3	Johann	Mendel	–	58
2	1789	4	5	Anton II	Mendel	–	58
3	1794	6	7	Rosina	Schwirtlich	–	13
4	1754	8	9	Valentin	Mendel	–	58
5	1753	10	11	Elisabeth	Blaschke	–	56
6	1751	12	13	Martin	Schwirtlich	–	13
7	1754	14	15	Rosina	Kasper	–	13
8	1725	16	17	Anton	Mendel	–	20
9	1729	18	19	Elisabeth	Weiss	–	26
10	1725	20	21	Andreas	Blaschke	–	58
11	1729	22	23	Elisabeth	Blaschke	–	48
12	1727	24	25	Paul II	Schwirtlich	–	3
13	1724	26	27	Juditha	Anders	–	HV2
14	1717	28	29	Josef	Kasper	–	13
15	1722	30	31	Elisabeth	Münster	–	5
16	1689	32	33	Andreas	Mendel	–	11
17	1699	34	35	Marina	Blaschke	–	41
18	1690	36	37	Martin	Weiss	–	26
19	1694	38	39	Judith	Ertel	–	51
20	1689	40	41	Andreas	Blaschke	–	56
21	1693	42	43	Anna	Grohmann	–	56
22	1692	44	45	Andreas	Blaschke	–	48
23	1701	46	47	Martina	Münster	–	MK17
24	1694	48	49	Johann	Schwirtlich	–	3
25	1692	50	51	Rosina	Kuntschig	–	23
26	1667	52	53	Martin	Anders	–	HV2
27	1684	54	55	Martina	Kunert	–	38
28	1674	56	57	Andreas	Kasper	–	13–19
29	1684	44	45	Justine	Blaschke	–	48
30	1692	60	61	Martin	Münster	–	5
31	1693	62	63	Marianna	Schwirtlich	–	58
32	1656	64	65	Wenzel	Mendel	–	6+VI25
33	1659	66	67	Marina	Wellert	–	KA1
34	1659	68	69	Paul	Blaschke	–	41
35	1670	70	71	Anna	Ertel	–	2
36	1635	72	73	Christoph	Weiss	–	24
37	1639	74	75	Dorothea	Popp	–	45
38	1660	76	77	Johann	Ertel	–	51
39	1655	78	79	Dorothea	Brosch	–	7
40	1629	80	81	Mathäus	Blaschke	–	55
41	1665	82	83	Susanne	Kahlig	–	HV1
42	1660	84	85	Georg	Gromann	–	56

(continued)

Table 3.1 (continued)

Number	Year of birth	Father's number	Mother's number	Given name	Family name	Inferred ethnicity	Address
43	1678	36	37	Katharina	Weiss	–	24
44	1658	40	89	David	Blaschke	–	48
45	1656	0	0	Anna	Anon.	X	?
46	1666	92	93	Mathäus	Münster	–	HC
47	1658	94	95	Anna	Kunz	–	MK17
48	1661	96	97	Paul	Schwirtlich	–	3
49	1665	98	99	Juditha	Münster	–	MK49
50	1637	100	101	Paul	Kuntschig	–	23
51	1668	76	103	Justine	Ertel	–	52
52	1630	104	105	Paul	Anders	–	39+HV2
53	1638	106	107	Katharina	Kasper	–	39
54	1630	108	109	Johann	Kunert	–	38
55	1648	110	111	Marina	Stach	–	?
56	1650	112	113	Michael	Kasper	–	13–19
57	1655	114	115	Anna	Fussel	–	MK22
60	1665	92	93	Johann	Münster	–	HC/OD
61	1667	68	69	Susanne	Blaschke	–	53
62	1660	124	125	Lorenz	Schwirtlich	–	58
63	1661	50	127	Susanne	Kuntschig	–	23
64	1615	128	129	Georg	Mendele	–	VI25
65	1615	130	131	Katharina	Schinke	–	VI58
66	c1620	0	0	Michael	Wellert	G	DK
67	1634	134	135	Katharina	Brosch	–	?
68	1634	136	137	Thomas	Blaschke	–	53
69	1635	138	139	Marina	Schwirtlich	–	HV22
70	1614	140	141	Georg	Ertel	–	2
71	1636	142	143	Marina	Graf	–	47
72	1605	144	145	Georg	Weiss	–	26
73	c1605	0	0	Barbara	Anon.	X	?
74	c1610	148	0	Martin	Popp	–	45
75	1616	150	151	Dorothea	Kasper	–	3
76	1627	140	141	Jakob	Ertel	–	52
77	c1630	154	155	Sabine	Schreiber	–	52
78	c1615	156	157	Paul	Brosch	–	7
79	c1615	158	159	Anna	Hanuschke	–	?
80	c1605	160	161	Gregor	Blaschke	–	55
81	c1605	162	163	Dorothea	Schreiber	–	55
82	1635	164	165	Johann	Kahlig	–	HV1
83	c1635	166	167	Anna	Weiss	–	BA
84	1628	168	169	Martin	Gromann	–	56
85	1632	170	0	Katharina	Rohleder	–	MK46
89	1633	178	179	Susanne	Nitschmann	–	MK34

(continued)

Table 3.1 (continued)

Number	Year of birth	Father's number	Mother's number	Given name	Family name	Inferred ethnicity	Address
92	1632	184	185	Johann	Münster	–	47
93	1633	186	187	Justina	Brosch	–	50
94	1634	188	0	Matthäus	Kunz	–	OD/MK17
95	1633	190	191	Susanne	Steffan	–	MK17
96	c1630	192	193	Georg	Schwirtlich	–	43
97	c1630	0	0	Katharina	Anon.	X	?
98	1635	196	197	Johann	Münster	–	MK49
99	1640	198	199	Barbara	Schindler	–	MK15
100	c1605	200	0	Andreas	Kuntschig	–	1
101	c1605	0	0	Judith	Anon.	X	?
103	1637	136	137	Ursula	Blaschke	–	53
104	c1605	208	0	Gregor	Anders	–	HV2
105	c1605	210	0	Anna	Kahlig	–	HV4
106	c1605	150	213	Martin	Kasper	–	39
107	c1600	214	215	Marina	Anders	–	39
108	c1600	216	0	Mathes	Kunert	–	38
109	c1600	0	0	Anna	Anon.	X	?
110	1620	220	221	Mathes	Stach	–	LO
111	c1620	0	0	Marina	Anon.	X	?
112	c1615	150	213	Georg	Kasper	–	13–19
113	1624	226	227	Katharina	Anders	–	?
114	c1600	228	0	Martin	Füssel	–	MK45
115	1624	230	231	Marina	Schlosser	–	MK16
124	1635	248	249	Adam	Schwirtlich	–	58
125	1639	250	251	Ursula	Pohanke	–	37
127	1631	254	255	Marina	Schwirtlich	–	?
128	c1585	256	257	Martin	Mendele	–	VI26
129	c1585	258	259	Anna	Greger	–	VI35
130	c1585	260	261	Andreas	Schinke	–	VI58
131	c1585	0	0	Anna	Anon.	X	?
134	c1605	268	0	Markus	Brosch	–	KA30
135	c1605	0	0	Anna	Anon.	X	?
136	c1605	160	161	Valentin	Blaschke	–	53
137	c1605	274	0	Marina	Gromann	–	OD
138	c1610	276	277	Georg	Schwirtlich	–	HV22
139	c1610	278	279	Barbara	Futschig	–	JO21
140	c1585	0	0	Valentin	Ertel	G	2
141*	c1585	0	0	Unknown	Anon.	–	?
142	c1590	284	0	Bartl	Graf	–	TO
143	c1590	286	287	Esther	Pietsch	–	47
144	c1585	0	0	Georg	Weiss	G	26
145	c1585	0	0	Dorothea	Anon.	X	?

(continued)

Table 3.1 (continued)

Number	Year of birth	Father's number	Mother's number	Given name	Family name	Inferred ethnicity	Address
148	c1590	296	0	Urban	Popp	–	8
150	c1580	0	0	Gregor	Kasper	G	3
151	c1580	302	303	Unknown	Anders	–	HV2
154	c1600	324	325	Mathes	Schreiber	–	52
155	c1600	0	0	Esther	Anon.	X	?
156	c1590	312	0	Michael	Brosch	–	6
157	c1590	302	303	Katharina	Anders	–	HV2
158	c1590	316	0	Paul	Hanuschke	–	7
159	c1590	318	319	Dorothea	Schenk	–	KA
160	c1575	0	0	Johann	Blaschke	C	3
161	c1580	286	287	Margarete	Pietsch	–	47
162	c1580	324	325	Martin	Schreiber	–	55
163	c1580	0	0	Anna	Anon.	X	?
164	c1600	328	0	Johann	Kahlig	–	HV1
165	c1600	0	0	Marina	Anon.	X	?
166	c1600	0	0	Johann	Weiss	G	BA
167	c1600	0	0	Susanne	Anon.	X	?
168	c1600	336	337	Andreas	Gromann	–	56
169	c1600	0	0	Ursula	Anon.	X	?
170	c1600	340	0	Andreas	Rohleder	–	MK46
178	c1600	356	357	Georg	Nitschmann	–	MK34
179	c1600	358	359	Unknown	Jünger	–	?
184	c1600	0	0	Johann	Münster	G	HC
185	c1600	370	0	Dorothea	Rohleder	–	MK8
186	c1610	156	157	Georg	Brosch	–	50
187	c1600	0	0	Katharina	Anon.	X	?
188	c1600	0	0	Simon	Kunz	G	OD
190	c1590	380	0	Blasius	Steffan	–	LY+MK17
191	c1590	382	0	Katharina	Anders	–	MK17
192	c1590	0	0	Bartl	Schwirtlich	C	43
193	c1595	386	0	Dorothea	Heigel	–	MK61
196	c1600	392	393	Christoph	Münster	–	MK49
197	c1600	394	395	Anna	Schlosser	–	MK16
198	c1605	396	0	Gregor	Schindler	–	MK15
199	c1605	0	0	Katharina	Anon.	X	?
200	c1580	400	0	Andreas	Kuntschig	–	1
208	c1580	0	0	Simon	Anders	C	HV22
210	c1580	420	421	Gregor	Kahlig	–	HV4
213	c1590	160	161	Unknown	Blaschke	–	3
214	c1570	0	0	Matthäus	Anders	C	39
215	c1570	0	0	Marina	Anon.	X	?
216	c1570	0	0	Wenzel	Kunert	–	MV10–13

(continued)

Table 3.1 (continued)

Number	Year of birth	Father's number	Mother's number	Given name	Family name	Inferred ethnicity	Address
220	c1590	0	0	Andreas	Stach	C	LO
221	c1590	0	0	Katharina	Anon.	X	?
226	c1600	452	453	Peter	Anders	–	13–19
227	c1600	0	0	Barbara	Anon.	X	?
228	c1560	0	0	Martin	Füssel	X	VE
230	c1600	394	395	Jakob	Schlosser	–	MK16
231	c1600	462	0	Ursula	Schindler	–	MK55
248	c1610	192	193	Bartl	Schwirtlich	–	43
249	c1600	0	0	Susanne	Anon.	X	?
250	c1600	500	0	Johann	Pohanke	–	37
251	c1600	286	287	Marina	Pietsch	–	47
254	c1600	0	0	Bartl	Schwirtlich	C	HC
255	c1600	0	0	Margarete	Anon.	X	?
256	c1555	0	0	Stantke	Mendele	X	VI26
257	c1555	0	0	Dorothea	Vögler	G	VI59
258	c1555	0	0	Jakob	Greger	C	VI35
259	c1555	0	0	Anna	Anon.	X	?
260	c1555	0	0	Valentin	Schinke	C	VI58
261	c1555	0	0	Johanna	Anon.	X	?
268	c1575	0	0	Christoph	Brosch	C	KA30
274	c1575	0	0	Hans	Gromann	G	OD
276	c1575	0	0	Johann	Schwirtlich	C	HV22
277	c1575	0	0	Eva	Bayer	G	MK61
278	c1590	0	0	Mathes	Futschig	C	JO21
279	c1590	0	0	Marina	Walzel	G	JO
284	c1570	0	0	Martin	Graf	G	LO
286	c1555	0	0	Kaspar	Pietsch	G	47
287	c1555	0	0	Unknown	Kuntschig	G	59
296	c1570	0	0	Albrecht	Popp	G	8
302	c1560	0	0	Martin	Anders	C	HV2
303*	c1560	0	0	Unknown	Anon.	X	?
312	c1560	0	0	Martin	Brosch	C	6
316	c1560	0	0	Gregor	Hanuschke	X	7
318	c1560	0	0	Georg	Schenk	G	KA
319	c1560	0	0	Dorothea	Anon.	X	?
324	c1560	0	0	Jakob	Schreiber	G	55
325*	c1565	0	0	Unknown	Anon.	X	?
328	c1570	420	421	Martin	Kahlig	–	HV1
336	c1570	0	0	Kaspar	Gromann	G	56
337	c1570	0	0	Ursula	Anon.	X	?
340	c1570	0	0	Georg	Rohleder	G	MK46
356	c1570	0	0	Jakob	Nitschmann	G	MK34

(continued)

Table 3.1 (continued)

Number	Year of birth	Father's number	Mother's number	Given name	Family name	Inferred ethnicity	Address
357	c1570	0	0	Unknown	Erler	G	MK22
358	c1570	0	0	Matthäus	Jünger	G	MK5
359	c1570	0	0	Ursula	Schindler	G	MK12
370	c1570	0	0	Wenzel	Rohleder	G	MK8
380	c1565	0	0	Markus	Steffan	C	VE
382	c1565	0	0	Martin	Anders	C	MK17
386	c1570	0	0	Georg	Heigel	X	MK61
392	c1570	0	0	Michael	Münster	G	MK49
393	c1570	0	0	Unknown	Schenk	G	MKD57
394	c1570	0	0	Thomas	Schlosser	G	MK16
395	c1570	0	0	Unknown	Rohleder	G	MK51
396	c1575	0	0	Michael	Schindler	G	MK12
400	c1560	0	0	Gregor	Kuntschig	G	1
420	c1550	0	0	Valentin	Kahlig	C	HV4
421*	c1550	0	0	Unknown	Anon.	X	?
452	c1570	0	0	Georg	Anders	C	13–19
453	c1570	0	0	Unknown	Schmitzer	G	13–19
462	c1570	0	0	Georg	Schindler	G	MK55
500	c1570	0	0	Simon	Pohanke	C	DO

Numbering of individuals follows that used by Schindler, while an asterisk following the number indicates individuals not listed by Schindler. The year of birth is either as given by Schindler or inferred (preceded by the letter "c"). "Unknown" indicates that the given name is not known by Schindler, while "Anon." indicates the absence of information on the family name. The ethnicity of founder ancestral individuals is determined from the family names as either German (G), Czech (C), or unknown (X), while descendant individuals are indicated by a hyphen (-). The address refers either to birth (preferred) or residence address. Addresses from Hynčice consist of numbers only. Addresses from other villages are indicated with initial letters indicative of the village's name followed, where available, by a number indicating the address: BA, Blahutovice; DK, Dolejší Kunčice; DO, Dobišov; HC, Hynčice; HV, Horní Vražné; JO, Jakubčovice nad Odrou; KA, Kamenka; LO, Loučky; MK, Mankovice; OD, Odry; TO, Tošovice; VE, Véska; VI, Veselí. The location of the villages is shown in Fig. 3.1

years between the two registers. The records were made in the Odry registers because Veselí did not get its own church until the end of the eighteenth century and rectory only in 1809. In the seventeenth century, it belonged to the Odry parish. Similarly, land holdings in Veselí of the seventeenth century were recorded in Odry land registers from the years 1636, 1650, 1683, and 1688. A land register was called *urbarium* or *urbar* because it was kept in the city of the domain (from Latin *urbanus*, pertaining to a city or *urbs*). It contained lists of field, meadows, woods, etc. and of their owners.

The different pastors and priests spelled the names of the oldest known Mendels differently. Slavic pastors preferred to end the family name with an "a" (i.e., Mendela, Mendula), as would be more natural in their language, whereas Germanic pastors used an "e" at the end (i.e., Mendele, Mandele, Mendtle). Schindler argued that originally the name did have an "e" at the end, the "–le" ending being a characteristic of many family names in Swabia,[21b] which he believed was the country from which the Mendels originated. According to him, they left Swabia before, during, or shortly after the so-called Peasant Wars in the sixteenth century. A prelude to the Peasant Wars was an uprising that erupted on Easter of 1514 in the valley of the Rems River but spread rapidly through the area north of the present-day city of Stuttgart in southern Germany.[29] The rebellion was ignited by the devaluation of currency introduced by the local overlords as a measure aimed at improving their financial situation. The duke talked most of the rebels into abandoning the insurrection and then executed some of them, flogged others, and expelled the rest from the land. Schindler believed without providing any evidence in support of his speculation that the Mendels were in the group of the expelled peasants. If so, the Mendels could have arrived in Kravařsko as early as in 1514. Another possibility is that the Mendels left Swabia during the Peasant Wars some ten years after the uprising in the Rems River valley. The incident that sparked these wars occurred in 1524 in the town of Stühlingen, which is right at the border of present-day Germany and Switzerland. The revolt then spread throughout the Germanic provinces, with the exception of Bavaria, and leaped over into Austria.[29] The insurgents demanded, among other things, abolition of serfdom, guarantee of basic human rights, and reform of religion. Although Luther's teachings inspired the uprising, the reformer distanced himself from it at the beginning and later turned sharply against it. In the many battles of the wars, more than 100,000 peasants fell or were executed brutally by the victorious lords. The Mendels might have fled the devastation of the wars or the wrath of the lords.

Consistent with Schindler's speculation about the Swabian origin of the Mendel lineage is the apparent Protestantism, more specifically Calvinism,[30] of the earliest known representatives of this lineage in Veselí. Both southern Swabia and the Odry region in Silesia of the sixteenth century were of a Calvinist leaning, and it would therefore have been natural for Calvinist émigrés from the former region to seek refuge in the latter. Since a number of other Germanic names in the Odry region ended with –el, Schindler speculated further that the Mendels arrived in Kravařsko with at least 20 other refugees from southern Swabia.[21b] The apparent prosperity of the oldest known Mendels of Veselí suggests that they must have been settled in the region for some time before their names first appeared in the historical records. If they did arrive after the uprising in the Rems River valley or in Stühlingen, they could had been in Kravařsko for more than 100 years or three to four generations before Konstantin (Standtke) Mendel's—enough time for Konstantin's ancestors to work themselves up to the farmsteads that he and his sons held. We must emphasize, however, that in the absence of a direct genealogical link between Kravařsko and Swabia, the Swabian origin of Mendel's ancestors remains a mere speculation. Others have used a similar indirect argumentation based on the distribution of family names to posit that the Mendels came from Bavaria.[31] The distribution of the name "Mendel" in present-day Germany (Fig. 3.9) is consistent with Schindler's speculation.

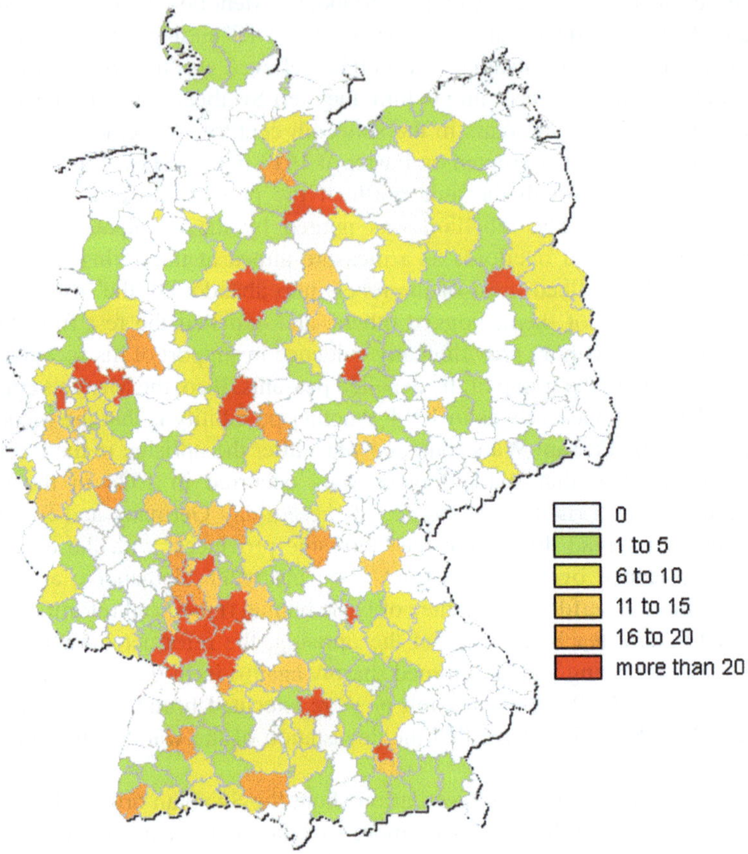

Fig. 3.9 Distribution and frequency of the name "Mendel" in present-day Germany calculated from telephone-book entries. The name is most frequent in southern and southwestern Germany, in the region from which Mendel's ancestors are believed to have originated. The place with the highest frequency is the city of Germersheim near the French border, where there are registered 74 persons under this name (source: *Bundesamt für Kartographie und Geodäsie*, http://www.bkg.bund.de)

To summarize the Mendel line of descent, Konstantin Mendel had a son Martin, who had a son Georg, who had a son Wenzel (a German version of the Czech name Václav).[25] In 1683, Wenzel Mendel moved from Veselí to Hynčice, where he acquired the farmstead No. 6. The following year he married Marina Wellert, the daughter of the *Erbrichter* in Kamenka, a village north of Odry. Subsequently, the Mendel line of descent died out in Veselí, but continued in Hynčice. There, Wenzel Mendel had a son Andreas, who was a mere cottager, but his son, Anton I, rose again to the farmer status when he bought the farmstead No. 58. His son Valentine had a son Anton II, who became Johann Mendel's father (Fig. 3.10).

Of course, by concentrating on the Mendel lineage in the pedigree of Johann Mendel's ancestors, we are making the same mistake as the genealogists of the past

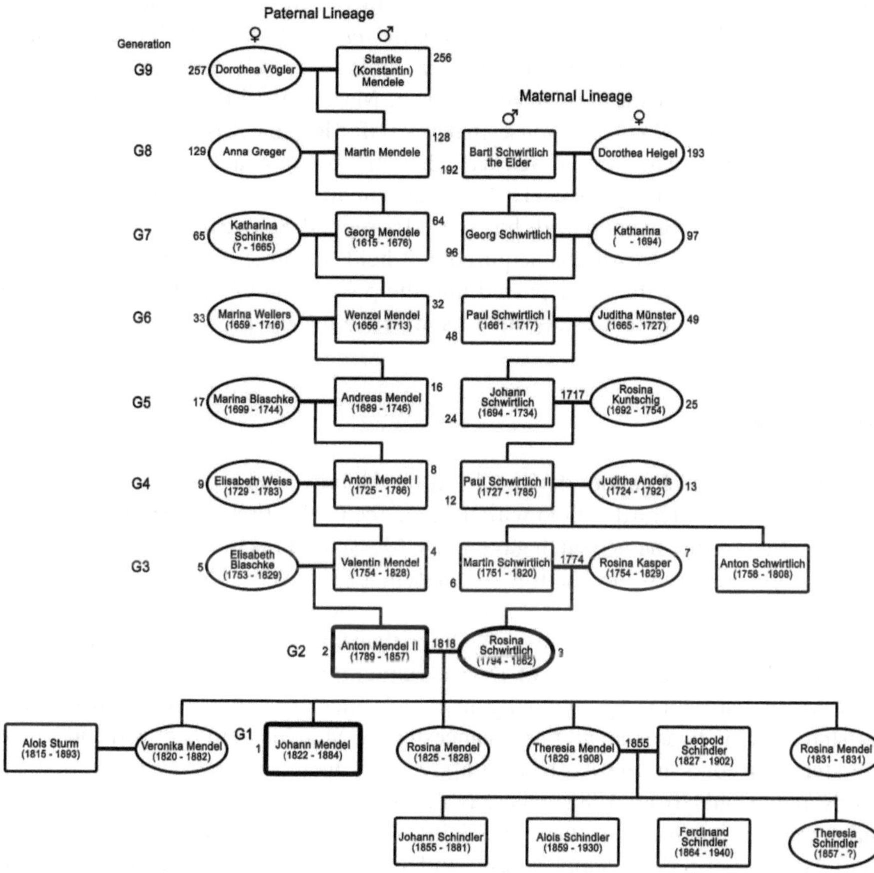

Fig. 3.10 Simplified version of Mendel's family tree

had made. The only thing special about the contribution of this lineage to the founder of genetics is the family name and the mother tongue, neither of which has anything to do with biological inheritance. From a genetic perspective, the Mendel lineage is no more important than any other line of descent in the pedigree. It is the pedigree as a whole that tells us something about Johann Mendel's genetic endowment. We cannot analyze here the pedigree in any detail[26] but must at least outline the maternal (Schwirtlich) line of descent of our protagonist. Alois Schindler was able to follow the Schwirtlich line back to the eighth generation, one generation less than the paternal line. The dates for the oldest known Schwirtlich, Bartl Schwirtlich the elder, are not available, but we estimate that he might have been born around 1590. He was a *Bauer* (farmer) who lived in Hynčice No. 43 and married Dorothea Heigel from Mankovice No. 61. They had two sons, Georg and Bartl the younger. Georg Schwirtlich (estimated birth at around 1630) married Katharina and with her started a line of descent that through Paul I, Johann,

Paul II, and Martin Schwirtlich led to Rosina Schwirtlich, Johann Mendel's mother. The first three Schwirtlichs of the line were *Bauern*, who lived in Hynčice No. 3. The fourth, Paul Schwirtlich II, ended up in debt, apparently through mismanagement, and had to sell the farmstead. His son, Martin, Rosina's father, became a *Gärtler* in Hynčice No. 13, a house that his wife, Rosina Kasper, inherited. The brother of Martin was Anton Schwirtlich, who became a teacher in the first, improvised school in Hynčice. The second son of Bartl Schwirtlich, Bartl the junior, married Susanne and with her started a second Schwirtlich line of descent in Hynčice. These Schwirtlichs acquired and held for three generations the farmstead No. 58, the one that later passed into the possession of the Mendels. A third Schwirtlich line of descent established itself in Horní Vražné.

The Surname

"Mendel" is a Jewish name, derived from the Hebrew *immānūēl* (i.e., *im*, with + *ānū*, us + *el*, God; literarily "God with us") via the Greek Emmanouēl, which in the Old Testament is the name Isaiah gave to the Messiah of his prophecy.[32] In Yiddish the name derives from the Hebrew *Menashsheh* (*Menashe*, *Menashem*), meaning "comforter" or "causing to forget." Because of the name, a Jewish origin of the Mendel lineage has been proposed.[33] Other than the name, however, there is no evidence for G. J. Mendel being of Jewish extraction; on the contrary, Schindler[21b] mustered at least four arguments against such ancestry. First, the records in the parish register indicate that all known ancestors of Mendel were baptized and hence Christians. Second, in the Austrian Empire, Jews were required by law to have a family name only from 1782, whereas the name "Mendel" appears in the Odry parish register much earlier than this. Third, very few Jews were peasants, most of them being innkeepers, shopkeepers, and merchants, whereas nearly all G. J. Mendel's ancestors were land cultivators of one kind or another. And fourth, there are several possible ways of deriving the name from a German, rather than Hebrew, word. One is that the original name was *Mandel*, which in German means "almond," but in Austrian German something else. In Austria at harvest, the cut stalks of grain used to be bound into bundles or sheaves, and 15 of such bundles were laid crosswise upon one another into layers, which were then left on the field to dry. The set of 15 sheaves, and by extension of any kind of 15 items, was called a "Mandel." The second possibility is that the original name was either *Mantel* (an overcoat) or *Mangel*, which signified either a shortage of something or a wringer—a device for squeezing out water from wet clothes. The third, and in Schindler's view the most likely, possibility is the transformation series Mann→Mannle→Mandele→Mendele→Mendel, where "Mann" means "man" or "husband." Schindler pointed out that in Kuhländchen the ethnic Germanic people used Mannle, Mandele, and Mendele as designations for a man of short stature, which some of the Mendels, including G. J. Mendel, seem to have been. Incidentally, Germans in general and Swabians in particular are rather fond of diminutives, which they form through the –el or –ele endings. Thus, in Swabia Bauer becomes

Bäuerle, Meyer Meyerle, Schnabel Schnabele, Birk Birkle, and so on. Notwithstanding all this, at the St. Thomas Abbey in Staré Brno, at least some of Mendel's colleagues ("brothers") held him for being of Jewish origin (see Chap. 5).

Surveying Mendel's family tree, one notices several Slavic names among Mendel's ancestors (supplementary Fig. S1, see http://extras.springer.com/2013/978-3-642-35253-9), a telltale sign that intermarriages had occurred between the Germanic immigrants and the indigenous inhabitants of Veselí. Schindler[21b] regarded the following five to be Slavic names (Germanic transcriptions are in parentheses): Fučík (Futschik), Švertlík (Schwirtlich), Říha (Greger), Šimek (Schimke = Simon), and Ondra (Anders). He does not give the sources on which he based these assignments. The sources we have consulted[34] attribute Czech origin to the names Blaschke (Blažek, Blažka), Brosch (Brož, Brožek), Kahlik (Kahlík, from kahlovati = hudlovati, meaning "to bungle"), Schwirtlich (Švertlík, from old Czech šverkati/švertati = švitořiti, meaning "to twitter" or "chatter"), Stach (Stach, Stašek, Stoš), Pohanke (Pohanka, meaning "buckwheat," *Fagopyrum esculentum*), and Steffan (Štěpán, Štefan), with Czech names given in parentheses. There are altogether 38 different surnames in Schindler's complete list of ancestors.[26] Of these, at least seven are of Czech origin (or 11, if Schindler's putative Czech names are taken into account). Among the eleven are some names which occur at high frequency among the ancestors. Thus, the two most common surnames in the Mendel family tree are Blaschke and Schwirtlich (17 and 16 individuals, respectively), both of which are of Czech origin. All these observations indicate that frequent intermarriages took place between the Germanic and Slavic peoples in the *Kravařsko/Kuhländchen* region and that G. J. Mendel bore about 33 % Slavic genes in his genome.[26] According to Schindler,[21c] Schwirtlich, the maiden name of Mendel's mother, could have been derived from "Světlík" and the Czech root "světlo," meaning "light." A related name, Švrtlík, had been once common in parts of Silesia.[35] Hence, Mendel's maternal lineage must have been of Czech origin. The fact that Rosine Schwirtlich was German speaking could be explained by assuming that in her lineage a Czech male ancestor married a woman of Germanic descent and, to make life easier for her, spoke German with her and with their children at home. With time, the original Czech name of the male lineage became Germanized and converted from Švertlík to Schwirtlich. If this interpretation is correct, Mendel would be of Germanic descent on the paternal side (Fig. 3.10) and of Slavic origin by way of his mother. The contribution of the Slavic gene pool to the genomes of Mendel's ancestors is therefore undeniable.

Parents

Anton Mendel was born on April 19, 1789. We surmise that at the age of six or seven years he went to school but that the education he received was quite elementary. The two surviving letters he wrote to his son[28] reveal him struggling for words as he tries to express himself in *hochdeutsch* (high or standard German), which, to a *Kuhländer* that he was, was almost like a foreign language. The grammatical errors disclose that the writer was not a person who read newspapers at breakfast and a chapter of a novel before falling asleep. His handwriting, nearly undecipherable at places, contrasts with the neat, calligraphic script of his son. It reveals a hand used to holding the reins of a horse team or a hoe rather than a quill. At the same time, however, he must have considered himself lucky for being able to communicate in writing with his son at all and to sign his name on official documents. Many Hynčicians could not do even this, for until the end of the eighteenth century most of the peasants in the Austrian Empire were illiterate. The overlords and the state officials were of the opinion that the less the peasants knew, the better. Only toward the end of the century did Maria Theresia and especially her son, Joseph II, realize that freed and properly educated peasants were better for the country's economy than serfs held in ignorance. Only then did they issue the *Allgemeine Schulordnung* (general school regulation), a decree that took education out of the hands of the clergy (the church) and placed it under the control of the state. The specific directives of the decree were the following: All children will receive compulsory education between the ages of six and 12 years. The pupils will be taught German, reading, writing, arithmetic, and religion. Every region in which there lived 90–100 school-age children will have an elementary school (*Volksschule* or *Grundschule*) to which no child will have to walk for more than half an hour. The schools will employ only teachers who have passed a qualifying examination. The fixed teachers' remuneration will be in cash and in kind and will be the responsibility of the community. Parents who use children for labor instead of sending them to school will be prosecuted. And the enforcement of these directives will be the responsibility of special regional committees and commissioners.

Not all these directives were implemented immediately and everywhere, but slowly the situation began to improve. In Hynčice, the peasants embraced the reform eagerly but had a request to make. The nearest school for their children was in Dolní Vražné, but since a new school was planned there, which would be even farther from Hynčice than the old one, they applied for a permit to have a school in their own village. In the application they argued that the trip to the new school in Dolní Vražné would not only be too long, but on rainy days floods could also make it perilous. After some haggling, the authority granted the permission and the school was completed in 1796.

The idea of having their own school was planted into the minds of the Hynčician peasants by one of Mendel's ancestors, Anton Schwirtlich, an uncle of G. J. Mendel's mother. By all accounts Anton Schwirtlich was a remarkable man.[36] Born in 1758 as one of ten children in a family that went bankrupt, he earned his daily bread by livestock herding for other farmers and later as a farmhand. He must have been highly intelligent, for although he never received any schooling, he taught himself to read and write and other elements of primary education. At the age of 20, he was drafted into the army for 14 years, fought in the "Potato War"[37] over the Bavarian succession, and then, when the war ended, was sent home on "vacation" to wait for Austria's involvement in yet another war. During this interim period, he decided to make himself useful by teaching the children of Hynčice, whose parents did not want them to make the long daily trips to Dolní Vražné. Effectively, in the years

Fig. 3.11 House No. 12/13: the first school in Hynčice

from 1780 to 1788, he opened a private school in his native village. In one room of his brother's house No. 12 or 13 (Fig. 3.11), he taught 15–20 children how to read and write and perform basic arithmetical operations. The "tuition" for attending the school was whatever the pupils' parents could afford to contribute to the teacher's livelihood, mostly in kind. The school closed when Anton Schwirtlich found himself a wealthy wife and purchased a smallholding in the newly founded colony of Emauzy near Hynčice. When the new war came, he did his duty and returned home to die of tuberculosis in 1808. His legacy, however, lived in Hynčice. His pedagogical success with the children made the peasants realize the value of education and inspired their desire to have a permanent elementary school in their village.

Anton Mendel reached school age in 1795. Hence, if he went to school in that year, as decreed, it would have been to the one in Dolní Vražné. From 1796 until 1801, he then apparently attended the new elementary school in Hynčice. Once out of school, he presumably worked on his father's farm, which he, as the eldest son, was expected to take over one day. At the age of 19 years, he was drafted into the army. Originally, the Austrian Empire had a professional army, but with the many wars the empire waged, the recruiting of mercenaries had to be supplemented with conscriptions.[38] In theory, the only difference between these two ways of procuring *Kanonenfutter* (cannon fodder) was that in the former, but not in the latter, the young men volunteered to join the army. Toward the end of the eighteenth century, all men between the ages of 17 and 40 years were made eligible for military service. There were many exemptions, however—nobles, priests, clerks, physicians, free peasants and their eldest sons, and burghers of free cities. Moreover, every eligible man could buy himself out for the stiff price of 300 guldens,[39] which few could scrape together. Until 1802, the military service was for life, but later it was reduced to ten to 14 years, depending on the arms (infantry, artillery, cavalry) to which the recruit had been assigned. Anton Mendel must have started his active duty in 1808. During his service, the Austrian Empire had been engaged in the last two of the five so-called Coalition Wars against France, in which Austria was joined by other countries, mainly Prussia, Russia, and Britain (some of these, however, switched sides during the conflict).[40] The Fourth Coalition War started on April 9, 1809, and was fought on three fronts, in Bavaria, Italy, and Tyrol, under the commands of Erzherzog Karl von Österreich, Erzherzog Johann von Österreich (the brothers of the reigning Emperor Franz I), and Feldmarschal Chasteler de Courcelles, respectively. After some initial victories, the Austrian armies were forced to retreat, and on May 10, Napoleon's army besieged and then took Vienna, the capital city of the empire. After an additional defeat in the Battle of Wagram on July 6 and the loss of some 100,000 men, Austria signed a peace treaty with France on October 14, 1809. A few years later, however, on August 11, 1813, it declared yet another war on France, the Fifth Coalition War. By this time, Napoleon had suffered a humiliating defeat in Russia in 1812, his army was severely depleted, and France exhausted. Although his troops put up fierce resistance to the coalition forces in two decisive battles, at Leipzig in "The Battle of the Nations" on October 13, 1813, and then at Waterloo on June 18, 1915, the French were defeated. In the end, though, Austria was in no better shape at the end of the hostilities than it was when it first started them in 1792.

Anton Mendel must have participated in at least some of these hostilities. Since the stage of the war theater was all of central Europe, his regiment undoubtedly did a lot of marching through the different lands, and the soldiers must have gotten good lessons in geography. They also had an opportunity to acquaint themselves with different ethnic groups and their lifestyles, cultures, and customs. Anton Mendel must have had a good supply of stories to tell on long winter evenings, from both the battlefield and encounters with various civilian populations. After eight years of service, he was discharged from the army, two years earlier than the shortest term called for. Assuming that he enlisted at the age of 19 years, he must have returned home in 1816 as a 27-year-old man. Why his term was shortened is not known, but in1816, his father, Valentine Mendel, was already 62 years old and perhaps he petitioned the authorities for an early discharge so that his son could take over the farm. On the other hand, it is also possible that when the hostilities ended and with no other war in the making, the army discharged some of the veterans earlier than scheduled. Upon his return, Anton Mendel took over the farmstead No. 58 upon which stood a wooden house with a thatched roof, perhaps resembling one of those depicted in Fig. 3.12. Anton tore the house down and built in its place a new, stone and brick house with a slated roof (Fig. 3.13). Although he did as much of the destruction and construction as he could himself, the undertaking not only swallowed up all his savings but also put him in debt until his retirement.

With the new house nearly finished, his thoughts turned to marriage and his choice fell on Rosina Schwirtlich[41] from house No. 13 in Hynčice. It was certainly not a choice motivated by the expectation that she would improve his financial situation through her dowry, for Rosina was a daughter of a *Gärtner*, a smallholder, who had a hard time keeping his head above water himself. The claim that she was a daughter of a gardener and that it was she who instilled a love of flowers in her son is a myth that arose through mistranslation of the German word *Gärtner*. There is no evidence that Rosina's parents, Martin Schwirtlich (1751–1820) and Rosina

Fig. 3.12 Scene from a Kravařsko village at the beginning of the nineteenth century. The house Anton Mendel tore down to build a new one might have resembled one of the depicted wooden, thatched-roof houses

Fig. 3.13 Johann Mendel's birthplace: house No.58 as it looked about a century after Anton Mendel built it. The drawing shows only one part of the entire estate

Schwirtlich née Kasper (1754–1829), were growing plants for sale. Actually, Rosina's ancestors were farmers, but her grandfather, Paul Schwirtlich II, had to sell half of his land because of debts, and this reduced the descendants to a smallholder status. Rosina's dowry might not have amounted to much, but her genetic endowment invigorated the Mendel lineage. Anton Mendel and Rosina Schwirtlich got married on October 6, 1818, on the same day as Anton's brother Johann was marrying Rosina's sister Judita and moving into the house No. 13.

Most of what we know about the physical appearance and characters of Anton and Rosina Mendel is based on oral tradition recorded primarily by Alois

Schindler.[27] No picture of them has been preserved if any ever existed. Schindler himself was only four years old when Anton Mendel died and nine years old when Rosina Mendel passed away, so his own memories of his grandparents could have been only dim at best. The memories of the persons who were in a more mature age when they knew the old Mendels were nearly half a century old by the time Schindler interviewed them and so no longer entirely reliable. We get an additional glimpse of Anton Mendel from his only preserved brief letter to his son, written in 1852, but no letters written by Rosina Mendel have reached us. It has been claimed that the elder daughter, Veronika, resembled her father and the younger, Theresia, her mother and that Johann displayed a mix of traits inherited from both parents (see below). Surely, these were subjective judgments, however. For all these reasons, the characterization that follows must be treated with circumspection.

The Mendels were not a race of giants. Both the father and the son were of short, heavyset stature, as were apparently many of their ancestors. Schindler[21b] saw in this physique an adaptation to a highlander lifestyle in Veselí, but he forgot to enlighten us on the advantages a short person might have at an altitude of 500 meters, as compared to one living at 300 meters above sea level. More likely, the short stature might have been a heritage from much older ancestors—those in Swabia if that was where the Mendels originated. There might have been some giants in the Schwirtlich lineage, but apparently they were destined to die a hero's death.[21b]

Anton Mendel was a serious man. Somehow, it is hard to imagine him in the company of the "merry haulers of Heinzendorf," described by Felix Jaschke.[22] He might have, now and then, joined his neighbors in the tavern to drink a jug of beer and brood over local affairs and world politics. And once in a while, at weddings and other special occasions, he might have even had one too many and then laughed and sang with the others. But most of the time, the worries would have been written on his face, the precariousness of peasant life weighing heavily on his shoulders. He was a man of principles, respectful of the laws, customs, and traditions. He dutifully fulfilled his obligations to the overlords in both payments and in robota, paid taxes to the state and tithes to the church, even when it meant that he and his family had to work like slaves to make ends meet. He also was respectful to members of his own family if convention required it. When Johann was ordained a priest, he acknowledged his son's new status by addressing him in a letter[42] *Hochwürdigster Herr Sohn* (Reverend Sir) and in third person, signing the letter formally *Anton Mendel Bauerausgediengster in Nr. 58 in Heinzendorf* (retired farmer settled in No. 58 in Hynčice). In the same letter, he requested formally the son's consent to the planned marriage of Johann's sister Theresia, as required when the brother was a cleric. He was an able farmer, managing his land judiciously and with great skill. But his passion was the orchard attached to the house. In it he planted different varieties of apple, pear, cherry, and plum trees, the shoots of which he collected from as far away places as Opava and Olomouc. The trees responded to his tender care, and the orchard developed into a showpiece of his estate.

About Rosina Schwirtlich, we know even less than about her husband. Apparently, she attended the elementary school in Hynčice, where she learned to read and write. Although no document written by her has been preserved, Alois Schindler assures us that her handwriting was "beautiful... well readable, simple, and without any embellishments"[43] and compares it to that of her son and her uncle, Anton Schwirtlich. Schindler also characterized her as being good-natured, quiet, calm, and modest. If Anton Mendel was the gloomy cloud hanging over the family, Rosina, true to her maiden name, was the bright light ray piercing the cloud.

Anton and Rosina Mendel were peasants. The stereotype of a peasant, current then, but surviving at some places to this day, has been that of a somewhat dim-minded person. Indeed, the word itself has a slight or strong pejorative connotation to it, depending on who uses it and in what context. The truth is, however, that to be a good farmer requires not only hard work but also considerable intelligence. It requires not only practical knowledge, experience, and good judgment but also the ability to think rationally all the time. A farmer must be mindful of decisions made every day, most of which influence the success or failure of his undertaking. When and where to plant what? When to begin with the harvest? How to prepare the soil? When to take what to the market? And so on. Each of these decisions must take into account many variables, which must be weighed against one another. The variables come from observations, in which the farmer must be able to differentiate all the time the essential from the trivial. He must then draw the right conclusions based on the weighing of the variables and have the confidence to base his actions on his deductions. If all this sounds like a veiled description of scientific work, it is because the essence of the mental process underlying the activities of a farmer and of a scientist is similar. Both involve observation, planning, decision-making, weighing of possibilities, and rational judgment. In short, they both require *special* intelligence. There is every reason to believe that Anton and Rosina Mendel had that special "peasant" intelligence, which enabled their son to venture into a realm others before him and, for a while even after him, did not dare to enter.

Teachers and Philanthropists

Historians call the eighteenth century the Age of Enlightenment, a century in which a group of philosophers made "knowledge," "reason," and "progress" the catch words of Western civilization. It was the rare period in human history, when philosophers had the ears of not one, but three enlightened rulers, Yekaterina II of Russia, Friedrich II of Prussia, and Joseph II of Austria. One principal tenet of the movement was that humanity could be changed for the better through the acquisition of knowledge—through education. To effect such a change, three things had to be accomplished: Education had to be reformed, it had to be made accessible to all people, and a new breed of teachers had to be raised. Reforming education was the task of pedagogues, the philosophers concerned with methods of knowledge dissemination. Accessibility of education had to be mandated by persons in power. And the transmission of the theory into practice required enlightened teachers. Moravia-Silesia, specifically the region of Mendel's birth, was one of the few places in Europe, where all three ingredients exigent for the educational reform

Fig. 3.14 Jan Amos Komenský (Comenius, 1592–1670), educator and philosopher, who was active in Fulnek from 1614 to 1621

were present. Here, Komenský laid the theoretical foundations for the reform, Countess Maria Walburga took the first steps toward bringing education to the peasants, and Schreiber and Makitta had been notable examples of enlightened teachers.

Jan Amos Komenský (Comenius, 1592–1670; Fig. 3.14) actually lived in the century preceding the Age of Reason but belonged to a group of philosophers who through their writings and actions prepared the soil from which Enlightenment would spring.[34] He knew from his own experience how dreary learning could be. The first thing his Latin teacher might have told him could have been "*Repetio est mater studiorum*. Repetition is the mother of learning. Repeat! And repeat again and again!" Words, the six cases of each noun, all the possible tenses of a verb, and all the other delicacies of the Latin grammar, as well as everything else the pupils were taught in the other subjects, had to be learned by heart—endless hours of mind-numbing memorizing. And when all was firmly deposited in the mind, the pupils were still unable to actually speak Latin, because their brains simply could not compute all the cases and tenses fast enough for the fluency of a speech. The inducements to learning were a painful tug on the pupil's ear and a blow of a switch on an outstretched hand. Komenský thought that there had to be a better way of learning Latin and all the other required subjects. To prove his point, he himself prepared a textbook of Latin for beginners. He called it *Orbis Sensualium Pictus*

(The Visible World in Pictures) because it consisted of engravings depicting scenes from daily life, nature, and the universe, each picture accompanied by one or two descriptive sentences in Latin and in the pupil's native tongue, followed by the identification of the depicted objects in the two languages. The success of the book was phenomenal. It was translated into 16 languages and remained in use for over two centuries. Apparently, the children enjoyed the pictures and through them found it easier to associate words with objects and concepts in both languages. Komenský made learning through enjoyment and play one of the principles of his educational reform, which he proposed in his *Didactica Magna*, The Great Didactic. The other principles included the coeducation of boys and girls, abolishment of corporal punishment, inclusion of science into the syllabus, emphasis on useful and practical knowledge, and the dictum that failure in learning is not the pupil's but the teacher's fault. Komenský had the opportunity to apply these principles in practice in his homeland, but for a short time only, because at the age of 30 years he became a man hunted for his religious beliefs. He was a pastor whose allegiance was on the losing side of the religious strife that tore apart his country; it was derived from the teachings of Jan Hus. After Hus' death, his followers split into two large factions and several smaller groups. Of the two factions, the more moderate one, the Utraquists (from the Latin *sub utraque specie*, under each kind, referring to the sect's tenet that laity should receive both the cup and the bread in the Eucharist), prevailed and became the Czech version of Protestantism. Of the smaller groups, the most successful one became *Jednota bratrská* (the Unity of Brethren, also called the Czech or Moravian Brethren or the Moravian Church), which advocated a simple life, opposition to wars, and harmonious relationships among people. The small city of Fulnek and the adjacent village of Kunín in Moravia, near the Silesian border and within a one hour walking distance from Hynčice, became the center of the Moravian Brethren's activity. Komenský became the last bishop of the brethrens and the teacher of their children at the school in Fulnek. After the Battle of the White Mountain, the Moravian Brethren, like all other non-Catholics in the Czech lands, were given a choice, convert or leave the country. Thousands chose the latter. Because of his activities, Komenský had to leave clandestinely to spend the rest of his life in exile in Poland, Sweden, Hungary, England, and the Netherlands, where he wrote and published most of his philosophical works.

Stirred by the novel *Émile, ou de l'éducation* (Emile, or on Education) by Jean Jacques Rousseau, several educators in the second half of the eighteenth century and the beginning of the nineteenth century revived and extended Komenský's ideas and founded institutions in which they attempted to bring them into practice. Most prominent among them were Johann Heinrich Pestalozzi (1746–1827) in Neuhof of present-day Switzerland, Johann Heinrich Basedow (1724–1790) in Dessau, and Christian Gotthilf Salzmann (1744–1811) in Schnepfenthal near Leipzig, both in present-day Germany. In Moravia, a bold attempt to improve the education of peasant children was undertaken by Countess Maria Walburga Josepha Cajetana Truchsess-Zeil (1762–1828; Fig. 3.15). As her title and long name suggest, she was born into two old noble families and married into a third. Her father was Count Franz Xaver von Harrach und Rohrau (1732–1781), who

Fig. 3.15 Countess Maria Walburga Truchsess-Zeil (1753–1817), enlightened philanthropist active in Kunín. She was probably responsible for Johann Schreiber's transfer to the parish church in Dolní Vražné

could trace his ancestry back to the thirteenth century and to the ownership of the village Harachy in southern Bohemia.[45] Countess Maria's mother, Countess Maria Rebekka, came from the von Hohenems family and her husband was Count Clemens Alois Truchsess-Zeil (1753–1817). Walburg was a castle near Ravensburg in Swabia, which the original Walburg family owned in the early twelfth century. Hohenems and Rohrau are villages in present-day Austria. The former is in the state of Vorarlberg, and the latter, the birthplace of Joseph Haydn, is near the Czech border. And Zeil am Main is a village in Bavaria. The attachment of the names of these different places to the names of the nobilities served to distinguish the various lineages within a given family. The "Truchsess" is a title, one of the four highest in the medieval peerage system, signifying that the person's ancestors were in charge of the king's or emperor's table, obviously a position of trust, considering how many rulers had been disposed of by being served a poisoned meal. The marriage of Countess Maria to count Clemens was not a happy one, not in the least because all their children died at a very young age. Deeply depressed, the countess separated from her husband and retired to her Kunín estate, which she inherited from her father and which included villages Butovice and Suchdol, in addition to Kunín itself. To come out of the depression, she plunged into a project, the aim of which was to provide proper education to the children living on her estate. In 1792, following the examples of Salzmann and Basedow, she founded an institute of education for talented boys and girls, whom she provided with clothing, board, and lodging. She also equipped the institute with all the necessary teaching aids and hired a person to run it and teachers to educate the children; she herself taught some of the classes. Among the teachers were Christian Carl André, of whom we will have more to say later in the book, and the chaplain Johann Edmund Schreiber, to whom we will return shortly. Schreiber was also the first director of the institute. The discipline at the institute was strict, almost militaristic, but the learning was innovative, with emphasis on practical subjects. The boys were taught principles of rational farming and horticulture; the girls had, in addition to the basic courses, also courses in sewing and knitting. The countess also lectured the village women on women's missions and on the new methods of education. Obviously, she was an exceptional woman in more than one respect.

On the estate, Schreiber established a fruit-tree nursery, for which the countess imported varieties from France and other parts of Europe. In an attempt to upgrade the farmers' orchards, she distributed some of the new varieties to the villagers, but the peasants, distrustful of everything that came free of charge from the nobility, let them wither away.[46] Schreiber, more familiar with the peasant mentality than the countess, advised her to resort to a trick. Following his suggestion, she let it be known in the villages that anybody caught taking seedlings from the nursery will be punished. At the same time, however, she instructed the watchmen to make enough noise when making their rounds of the garden at night to alert any thieves. The ruse worked: After a few nights, the plot with the new varieties was plundered, and after a few years, French apples were the adornment of many a peasant's orchard.

At first, authorities had only praise for the institute, but later, when the mood in the country began to change from enlightened to absolutistic, they began to view it with different eyes. Acting on repeated denunciations from an informer within the institute (a disgruntled priest of a despicable character), they began to investigate charges that the countess had failed to attend Catholic services regularly; that she allowed children to be taught natural history, including lessons on animal reproduction; and that she had bad influence on the peasants' family life by lecturing the wives on immoral issues.[46] Although most of the charges had a factual basis, from today's perspective they only showed how progressive the countess was. From the viewpoint of the authorities in the Austrian Empire, however, they were indications that the institute had become a place from which notions alien to the monarchy were being spread. Ultimately, Schreiber was forced to leave and the institute itself was closed in 1814, when in addition to the problems with the authorities, it ran into financial difficulties.

At about the time when Schreiber became the lightning rod for criticism of the institute, a clerical position involving also teaching duties became vacant at the church in Dolní Vražné, and he was appointed to it, possibly with the help of the countess. The office Johann Schreiber (1769–1850) assumed in 1805 at the church of Saints Peter and Paul in Dolní Vražné is called *fara* in Czech and *Pfarre* in German (from Greek *paroclus*, "the one who holds out a hand," in the present context, "who administers sacraments"). It can be translated into English as *parish*, *parsonage*, *vicarage*, or *rectory*, none of which is precisely equivalent to the Czech and German terms because the organizations of the churches are somewhat different in England and in the Czech lands. The same applies to the term for the priest who holds this office: *farář* in Czech, *Pfarrer* in German, and a choice of English terms *parish priest*, *parson*, *vicar*, or *rector*. Here we shall use the terms *rectory* and *parish priest* for the office and its holder, respectively, and reserve the term *parish* for the group villages (towns) administered by a given rectory. The rectory is the building in which the office is located, and at the same time it also houses the living quarters of the parish priest, as well as his assistant, the chaplain (*Kaplan* in both German and Czech), the cook, and other servants. With the building comes also a garden and perhaps even a small field. The parish priest is the spiritual leader of the community. He and the chaplain perform all the rites and services connected with the church, keep the parish register, and also teach religion at the local schools.

In the Middle Ages, there used to be a rectory at the church in Dolní Vražné, but it became defunct during the Thirty Years War, and the village came under the jurisdiction of the rectory in Mankovice.[13] In 1781, however, the Tolerance Decree issued by Joseph II called for a reorganization of the system of rectories and the closure of some of them. The rectory in Mankovice was included among the latter, while the rectory at Dolní Vražné was resurrected and the Mankovice rector moved to it in 1785. The church in Dolní Vražné was, however, in a sorry state. Not only was it all wooden and too small for the enlarged congregation, but it also was in danger of collapsing. A new church was therefore built in 1798 and consecrated in the following year. A chronicler[13] found it worth recording that for the construction, the builders used 240,000 bricks burned on one of the local farms and that the kilns used for the burning were heated with coal—so novel was the use of burned bricks and coal at that time! Johann Schneider persuaded Countess Maria Walburga to accept the patronage over the church and so attain influence over the appointments of its rectors. It was a tactical move, which paid off a few years later. The patronage was the formal tie of the countess to Mendel's birthplace, since Hynčice belonged to the Dolní Vražné parish. Some biographers of Mendel[47] claim that the countess actually owned Hynčice, but this is incorrect. Hynčice belonged to the Odry domain, which had never been a possession of the countess.

In 1802, the parish priest of Dolní Vražné died, and Schreiber was appointed as his successor. By then, not only the church but also the rectory was rebuilt, the latter consisting of seven rooms and some adjacent cubicles. One of the first things Schreiber undertook after moving to Dolní Vražné was to transform part of the rectory garden into a fruit-tree nursery. In it he would grow the varieties he brought with him from Kunín and also begin a quest for new varieties by growing plants from seeds, which local children collected for him. Schreiber, the experienced horticulturist that he was, knew of course that the fruit of a tree grown from a seed would not necessarily be the same as the fruit from which the seed originated. Many fruit-tree varieties were hybrids which, when propagated sexually, might produce different types of offspring and the original type may all but disappear. To propagate a promising variety, Schreiber, like generations of horticulturists before him, resorted to asexual propagation by *grafting*. It consisted of apposing the cut surface of a *scion* (a short piece) from the variety to be propagated onto the cut surface of a *stock* (root or rooted stem) from another tree to produce a chimera in which tissues of disparate origin were joined together.

Fruits and vegetables were an important part of villagers' menus. Fruits alone could not sustain hardworking men and women, but they could still the hunger of a child. In those days, fruits could also be something that they have since ceased to be for most of us—a delicacy. The supermarket fruits of our times may contain all the prescribed vitamins and "fibers," for which we now consume them, but their early harvesting, artificial ripening, cooling, and all the other treatments they undergo in the hands of the suppliers all but deprive them of their taste, smell, and frequently also their visual appeal. Only the old, village-born of us may still remember what a ripe apple just fallen off a tree tasted like and what it was like to enter grandmother's pantry stored with fruits. In Mendel's village, undoubtedly, every family with even the smallest piece of land grew fruit trees on it, while the children of those parents, who did not, found ways of getting their share of the harvest. What a thrill it was to steal the forbidden fruit! As a priest, Herr Pfarrer Schreiber had to tell them that it was a sin, but as an enthusiastic pomologist, he probably could not reprimand too harshly those who confessed to it.

Herr Pfarrer taught the children religion, but more than anything else he liked to spend the time with them in the garden, showing them how to prepare the scions and the stocks for grafting; how to locate the cambium, the thin layer of tissue between the bark and the wood, which had to be apposed between the two grafted surfaces; how much wax had to be applied to the surface of the joint to keep it from drying out or getting infected; how to prepare the bast; and how to wrap it around the graft. *Herr Pfarrer* knew so much about fruit trees! No wonder he was one of the founding members of the Pomological Association in Brno in 1816[48] and attended regularly its annual meetings. No wonder, too, he was elected a corresponding member of the Agricultural Society in Brno and was undoubtedly forgiven by the Lord for being proud of this honor.

We know much less about the second teacher at the Hynčice elementary school, Thomas Makitta. He was a farmer's son from Klokočov who became a certified teacher. He was born in 1774, married Rosina Schenk in 1797, started at the Hynčice elementary school either in 1795 or 1796, and taught there until 1836 (not 1839, as claimed by Iltis[8]). Schreiber must have joined him in 1806, at the earliest, and taught religion as well as principles of horticulture and beekeeping, presumably until 1850. Makitta must have therefore taught the entire Mendel family—father, mother, son, and both daughters, although Theresia probably for one year only (the remaining five years she must have been taught by Georg Schramm, Makitta's successor). Johann and his sister Veronika overlapped for four years in their school attendance. Schreiber, too, taught the whole family, except the father, who finished school one year before the priest's arrival at Vražné.

The teaching must have been a great challenge even for these experienced educators. The school had only one classroom, and there were up to 80 children in attendance in some years. Presumably, the pupils were divided into two groups, one consisting of the first through third graders and the other of the fourth through sixth graders, one group attending the school in the morning and the other in the afternoon. Even this division obviously left some 40 pupils and three grades together in one class. The teachers must have applied all their pedagogical skills

Fig. 3.16 The part of Hynčice in which the farmstead No. 58 is located. In the background is the Veselský kopec on the top of which lies the village Veselí. To the right of it is the Pohoř Hill and in the valley between the two hills is the town of Odry. The drawing is based on a photograph presumably taken in 1931

just to maintain discipline, not to mention instructing the children. Presumably, while one grade was instructed orally, the other two were engaged in writing compositions, calculating, or drawing on a slate. Rapidly advancing pupils like Johann Mendel might have been engaged as teacher's assistants in tutoring the lower grades. In addition to the standard subjects, Makitta and Schreiber strived to acquaint the children with basic natural history. When the news about the added subjects reached the authorities, they ordered the teachers to stop this *Unfug* (garbage) and restrict themselves to the standard syllabus.

Farmstead No. 58

The house Anton Mendel built stood on an estate at the lower end of Hynčice, not far from Horní Vražné (Figs. 3.16 and 3.17). The estate consisted of three parts: the farmyard surrounding the house and its auxiliary buildings, an orchard in front of the house, and the fields behind it. The picture of the house No. 58 (Fig. 3.13) is based on a photograph taken in the 1920s, about a century after Johann Mendel's birth, but with the exception of minor changes it may have retained its original appearance. It nevertheless gives a false impression about the size of the estate since it represents only the living quarters of the Mendel family. In reality the building was only one section of a complex arranged into a quadrangle like a fortress (Fig. 3.18). The buildings of the complex enclosed an inner courtyard, accessible from the outside through a gate. Beside the house in which the family lived, the other structures of the complex served functions associated with farming. There were stables for horses, cow houses for the cattle, and various lesser

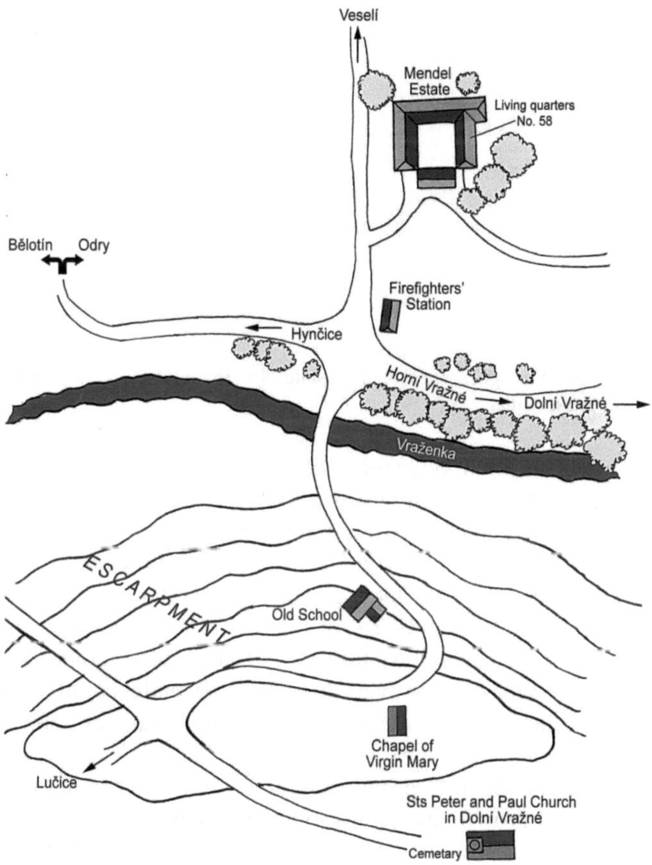

Fig. 3.17 Landmarks of Mendel's Hynčice—Vražné

structures for the rest of the livestock: pigsties, goat and sheep dens, coops for the poultry (hens, ducks, and geese), and rabbit warrens. The rest of the buildings served as storage facilities of different kinds. There were barns for storing harvested crops and for the straw after threshing, haylofts, rooms for short-term storage and processing of green fodder, granaries, and underground cellars, in which potatoes, beet, and fruits were stored. Other parts of the complex served for storage and repair of farming machinery such as plows, harrows, rollers for flattening larger clods of earth, wagons, and an assortment of small equipment and tools. Wood for heating was stored in wood bins and a wooden outhouse served for the natural needs of the whole family. In the central farmyard were a well, a dunghill, and a cesspool for liquid manure.

The orchard encompassed about 40 acres together with the land extending from the house down to the road and the stream passing through the village. On the opposite bank of the stream, the land rose precipitously in the form of a low, wooded escarpment. The distance from the stream and the elevated location

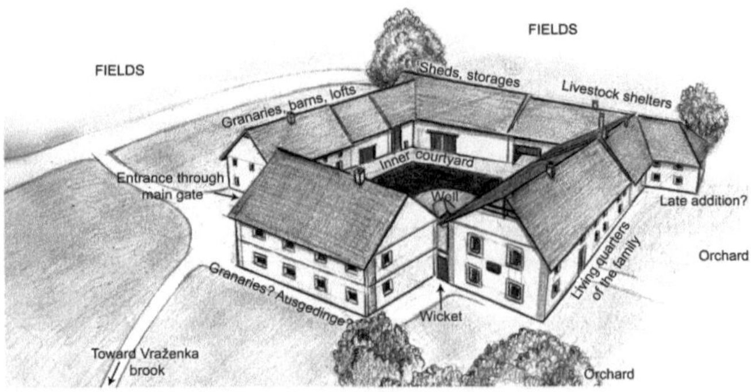

Fig. 3.18 The present-day appearance of the estate No. 58 (now 69). The farmstead has been converted into a cultural center

protected the house from floodwaters, when they came after heavy rains. The orchard contained varieties of apple, pear, cherry, and plum trees, all of them thriving under Anton Mendel's loving care. Johann liked to assist his father by his work in the orchard, and so the two of them spent many happy hours together absorbed in pruning the trees, protecting them against hares in the winter, hoeing and fertilizing the ground around each of them, and grafting them in an effort to grow new varieties. This last activity fascinated Hans (as the father probably called his son), and he was eager to learn everything Anton Mendel could teach him about it. Undoubtedly, he began experimenting with the technique as soon as his father allowed him to do so and must have been excited when some of the grafts actually took and new varieties began to grow on some of the branches of "his" trees. The fascination with plant growth and plants in general stayed with him for the rest of his life. Several biographers attribute to Gregor Mendel the statement that plants were his children. If he really said this, he might have been expressing something deeper than a mere fondness of trees and flowers. Growing plants might have been Gregor Mendel's subconscious substitute for watching his own children to grow, which was denied to him by the fact of being a Catholic priest. At any rate, the basis for the fondness was laid in his father's orchard. And something else probably originated there, too. In one corner of the orchard stood a group of beehives (Fig. 3.19), which supplied the Mendel family with the main sweetener for their drinks and meals—the honey. Here, too, Johann must have been an eager assistant to the elder Mendel in transferring the hives close to trees or fields that were just in bloom, cleaning the hives, collecting the honey, pursuing the swarm that left the hive with the new queen, catching the swarm and settling it down in a new hive, and in the fall setting up the bees for their winter sleep. And here, too, Johann became infected with his father's enthusiasm and grew up into a master apiculturist. A small part of the orchard was converted into a vegetable garden in which Rosina Mendel grew lettuce, cabbage, cauliflower, carrots, radishes, turnips, kohlrabi, onion, garlic, cucumbers, pumpkins, rhubarb, parsley, fennel, and caraway. There may have been strawberry beds, raspberry patches, and flowerbeds intermingled

Fig. 3.19 A farmer catching a swarm of bees

among the vegetable beds. The rest of the orchard ground was covered with grass, which was harvested for hay.

The house was simple in design. Although it had two stories, only the ground floor was used for dwelling, whereas the upper floor may have served as a spare granary.[49] The actual living space consisted of three small-to-modest-size rooms, approximately 17, 10, and 4 square meters in floor area. In the largest of them stood an imposing tiled stove, which may have served to heat all three rooms. The base of the L-shaped structure was made of firebricks and faced with ornamented tiles.[50] The rest of the structure was constructed from tiles joined together on the inside with clay and wires. The foot part of the "L" formed a cooking range, consisting of a furnace covered with iron plates and equipped with a grate at the base and an underlying ash-collection chamber. Fireclay-lined ducts led hot air and gases from the furnace into the upright part of the stove, where it heated two ovens, and then into metal ducts connected to the chimney. The bricks and the tiles absorbed the heat from the fire and released it slowly into the rooms. Thus heat continued to emanate from the stove long after the fire in the furnace extinguished. On cold winter days, family life centered on the stove, the only warm place in the house. To warm up their beds on cold nights, the Mendels might have placed a heated, cloth-wrapped brick under the comforter. The rooms were outfitted with simple, functional furniture, made by the local cabinetmaker and decorated with folk-art motifs. A table with stools, an assortment of chests, a cupboard, a modest wardrobe, and beds were probably all the Mendels could afford. The walls were decorated with tiles and icons and the ceiling plastered. Attached to the house was a small unit into which Anton and Rosina planned to retire once their son would marry and take over the farm. Viewed from the outside, the house had an austere appearance, the façade lacking any ornamentation, the gable being of a simple design, and non-protruding windowsills (Fig. 3.13). The gray, slated roof and the absence of a decorative entrance to the farmyard underscored the sobriety of the building. The message the house proclaimed to the world a message: Here lives a hardworking family, which cannot afford any frills.

Attached to the farmstead were until recently 16 hectares (39.5 acres) of land, mostly fields and some meadows. This may seem too large an area for a single family to manage without any outside help or any sophisticated machinery and burdened with a duty to work three days per week on allodial land. Indeed, older documents[51] assessed the size of the land as mere 20 *Joch* or 11.6 hectares (28 acres), which would have been more malleable, considering that one third of the land was actually not cultivated (see below). Thus, land may have been added to the estate at later time. The strip of arable land extended from the house all the way to the forest covering the slopes of the Veselský kopec.

Following century-old traditions, the farmers of Hynčice grew on their fields five basic crops—four cereals and one pulse (legume). The cereals were rye, wheat, barley, and oats; the pulse was either the garden pea or lentil. (It was therefore already in his childhood, on the fields of Hynčice, that Mendel first encountered the plant that later became his icon, his coat of arms—the garden pea.) Each of the five crops was grown for a different purpose. Rye was used to make bread from the flour obtained by grinding the grains; wheat flour for making dumplings, noodles, and pastry; barley to feed the livestock and to make grouts, semolina, and porridge, as well as for malt used by local breweries to produce beer; oats to feed horses and for making oatmeal; and the pulses as meal supplements, for soups, salads, and casseroles. The straw of the grains was used as bedding in the stables and the cowsheds. Rye and wheat were generally sowed in the fall and harvested in the summer of the following year, whereas barley and oats were sowed in the spring and harvested in late summer of the same year. The farmers divided their fields into three sections and sowed rye and wheat in the fall in one section, the rest of the crop in the spring in the second, and let the third section lay fallow. In the following years, they rotated the crops in such a way that every three years any given section would lay fallow and so be allowed to regenerate nutrients and moisture for another two seasons of crop cultivation. At the end of the fallow period, the farmers burned or plowed in the weeds, thus enriching the soil with the ashes or organic material. We can assume that when Anton Mendel started farming on his own, he too followed these traditions, both in regard to crop selection and the three-year crop rotation system. During his time, however, farming practices in central Europe underwent dramatic changes. New crops were introduced, the crop rotation system was altered, and new farming tools and equipment were invented. The combination of these three factors influenced the productivity of the farms—it ushered in an agricultural revolution. It is unthinkable that Anton Mendel was not affected by at least some of the changes. The three main new crops were potato, lucerne (alfalfa), and sugar beet. Potatoes first appeared in the Kravařsko region in 1760 and, like elsewhere in Europe, were at first regarded with distrust. However, this attitude changed rapidly during the famine of 1772, when potatoes became the staple food of the starving population.[38] Lucerne and clover were known to farmers for an even longer time but came into wide use as animal fodder in connection with the introduction of the four-year crop rotation system. In this system, the fallowing of a section had been eliminated, and the rotation was extended from three to four years in which the crops followed one another in the sequence wheat—rye/barley/oats—lucerne/clover—potato/beet. As the different crops took out from the soil and added to it different nutrients, there was no need to leave any land fallow, and crops could be grown on all sections every year. Thus, the system in effect added one third of land to the farmer's holdings. Sugar beet for sugar production may not have been introduced in the Kravařsko region until the time close to Anton Mendel's retirement, but cultivation of the common beet for feeding the livestock may have started during his farming years.

Johann Mendel Is Born

If there is anything certain about a person's life, it is that he or she was born and that he or she is going to die. One might expect, therefore, that the dates of these two framing events would also be indubitable, but this is often not the case, even when the man is a celebrity. Take Beethoven, for example. Biographers of the great composer have been arguing for over a century whether he was born December 17, December 16, or even at some earlier date.[52] The only certainty seems to be that he was baptized on December 17, 1770. (Although Beethoven himself insisted that he was born in 1772, biographers discount this claim as mystification, perhaps a carry-over from his childhood, when his father, in an attempt to present him as another *Wunderkind* à la Mozart, made him two years younger.) The reason for the uncertainty about the birth date of many persons is simple: Generally, a child's birth was not recorded in the parish register on the actual day of birth but on the occasion of baptism, which did not need to coincide with the former. The delay provided an opportunity for communication errors between the child's parents and the keeper of the register, commonly the parish priest. Johann Mendel, too, might have been a victim of such a misunderstanding. A priest ordinarily performs the baptismal ceremony in the church within ten days after the birth of the child, either before or after the mass. During the ceremony, the priest pours holy water (i.e., water blessed by the sign of the cross by a bishop or priest) on the infant's head pronouncing the words: "I baptize thee in the name of the Father and the Son and the Holy Spirit."[53] The person holding the baby is neither the father nor the mother but one of the godparents. They are persons—usually neighbors or friends of the parents—who sponsor the child, their godson or goddaughter, for its admission into the Christian fraternity and at the same time assume responsibility for its faith when it comes to age. In the case of an emergency, when there is a danger that the infant might die, the church allows the ritual to be performed by any person anywhere, even with ordinary water if holy water is not available.

It is at baptism that the child is given a name, the choice being influenced by two preferences: Favored are names of saints and of relatives. Saints are persons who, through their exemplary life (holiness), sacrifice (martyrdom), achievements (theological, scholarly, missionary, political), altruism, and miracle working, are veneered by the Roman Catholic Church. There are more than 20,000 saints, but of these only a few thousands have been canonized (i.e., included in an official list or canon). Some saints have a *feast day* in the liturgical calendar on which they are remembered by prayers, flowers, and votive gifts. Catholics believe that saints, being in heaven and so having immediate access to God, can be persuaded by prayers to intercede on someone's behalf and so act as sort of supernatural lobbyists. Catholics believe further that by choosing a saint's name for their child, they provide it with a heavenly guardian or patron. Often parents choose a name of a saint, whose feast day is close to the child's birthday or day of baptism. In addition to celebrating annually a birthday, a person may also observe a "name day" on the feast day of the saint whose name he or she bears.[54] Especially popular saints have many persons named after them, who celebrate their name day on the saint's feast day. Thus, in the Czech lands, on Saint Joseph's day, an appreciable proportion of the male population spends a good part of the day in the beer halls in the company of their namesakes. Mendel, however, was baptized Johann not because he was born a few days before Saint John's day but because that was the name of his uncle, Anton Mendel's younger brother. The people in Kravařsko often named their children after their relatives. Because of this practice and because of frequent marriages between related persons in that region, the repertoire of first names among these people had been rather limited. The repeated occurrence of names such as Johann, Andreas, Marina, Martin, and others has confounded the work of Mendel's genealogist Alois Schindler. It explains the ambiguities in the oldest generations of Mendel's pedigree (supplementary Fig. S1, see http://extras.springer.com/2013/978-3-642-35253-9).

Having completed this theological detour, we return to the question of Mendel's birth date. Two dates must be considered: July 20 and July 22, 1822. The former is the date recorded in the parish (baptismal) register and indicated on Mendel's birth certificate; the latter is the date transmitted to us through family tradition. Since the register records Johann's baptism on the same day as his birth, we have a unique case here of a child being baptized (according to the register) before it had been born (according to the family tradition). Obviously, one of the sources of information, either the register or the tradition, must be wrong—but which? Let us first have a closer look at the entry in the parish register. The page on which the entry appears is inscribed at the top left corner "1822," and so it presumably records all the births that occurred in that year in the villages Hynčice and Dolní Vražné (recall that Horní Vražné, lying between these two villages, belonged, at the time of Mendel's birth, to the Mankovice parish). From the page we learn that in 1822, seven children were born in the two villages (the page fails to specify which of the seven were born in Hynčice and which in Dolní Vražné), five boys and two girls, five legitimate and two illegitimate, and all to Roman Catholic parents.[55] (It may appear somewhat unusual that a parish register of a Roman Catholic Church would contain a column for registering the birth of Protestant children. The simple explanation is, however, that there was no Protestant church in the area, and since the law required the registration of all children born, the Catholic Church had to register also children born to the few Protestants living in the area.) In addition to our Johann Mendel, another Mendel was born on May 12, also a boy, and he was also named Johann. That, too, may appear strange, until we realize that he was the son of Anton Mendel's younger brother Johann and was therefore named after his father, as was Anton's son after his uncle. According to the register, Johann Schreiber, the *Pfarrer* of the Saint Peter and Paul church in Dolní Vražné, baptized the two Mendel boys, whereas a priest identified as Gabriel Enz… (the rest of the entry is illegible), presumably Schreiber's assistant, baptized the remaining five children. Some biographers[55] regard the baptism by the Pfarrer as a sign of Schreiber's respect for the Mendel family.

Another peculiarity of the register is that all seven children born in 1822 were baptized on the day of their birth, which is rather unusual. It seems unlikely that every father rushed an infant that has just been through the stress of delivery into the church (which for the residents of Hynčice was not exactly around the corner) to have water poured on its head. The custom in most of Silesia of that time had been to have the child baptized on the first Sunday after the day of birth. (Of the seven babies, all but one were born on a weekday.) Furthermore, babies have the bad habit of choosing the most inconvenient hour of the day for their entrance to this world, and one might not expect the parish priest or his assistant to be ready at any time of the day or night to perform the ceremony immediately after birth. It has been argued[55] that the birth of Anton Mendel's son was an emergency, which required accelerated baptism. The argument is based on a note scribbled across the columns in the individual entries in the parish register. The note is nearly illegible, but it seems to read *Kath. Tauf…* etc (Catholic baptism… etc) in all but one of the entries, which has been interpreted to mean that the ceremony was performed in the church as prescribed. The one exception is that in the entry recording the baptism of Anton Mendel's son, the "etc" is replaced by "Nr. 35." This deviation has prompted the speculation[55] that Anton's son was baptized neither in the church nor at home at No. 58, but in a house No. 35. Interestingly, one of the two illegitimate children, Anna Brenner, is listed as coming from the same house. The explanation offered for this odd coincidence is this:[55] Number 35 was the house of the midwife. Anton Mendel's son was born at home but was then taken to the midwife's house, because he appeared weak and his parents worried that he might not survive. They thought that the midwife would be more qualified to take care of him for a few days. Because of this emergency situation, the infant was baptized in the midwife's house, either by the priest or the midwife herself. Anna Brenner was born in the midwife's house because the "parents of a girl in trouble were usually not eager to have the child born in their own house."[55] These conjectures have, however, several weaknesses. First, transferring a weak newborn into another house would not appear to have been a good idea. Also, the mother would have had to be transferred with the newborn if for no other reason than to feed it. But was she in a condition for such a transfer? Second, what could the midwife do for the newborn at home that she could not do at Mendels' house? Third, in a small village like Hynčice, the birth of an illegitimate child could not be kept secret, so what would be gained by having it delivered in the midwife's house? Fourth, the register gives the *address* of the illegitimate child as house No. 35, so its mother had to come from that house, which makes it unlikely that it was a midwife's house.

The 1822 page of the register gives the impression that its entries were not made one by one on the day of baptism, but rather all at one time, by one person, perhaps at the end of the year. If so, the information entered might not have been fresh, but recalled from memory or from notes. Furthermore, for an official document, the entries in the register are made rather carelessly. Not only was there any effort spent to make the entries legible, but also there seems to have been little regard for their accuracy. The latter conclusion is suggested by the two Mendel entries. In both is the family name of the sons spelled correctly as "Mendel," whereas that of the father incorrectly as "Menndle." Under these circumstances, it cannot be excluded that some of the dates recorded may have been off by a few days and that the reference to house No. 35 in Johann's entry is in error.

But is the family tradition more reliable than the parish register? Certainly to the parents, the birth of a child is a much more profound event than to a parish priest, especially if the father anxiously awaits the birth of his successor on a farm. It is hard to believe that such an event would not be writ firmly into the parents' memory if not recorded in the family bible. On the other hand, it is also true that 200 years ago, and especially on a farm, it was more difficult to keep track of the flow of time than it is today. There were no wristwatches and the massive pendulum clock, if a family owned one, did not show dates. Wall calendars, produced *en masse* today, were a rarity then. And of course, there was no radio nor television to remind people what date it was. The villagers had, however, a way of breaking up the monotony of the time flow: They used major feast days of the saints to date events. Saint Nicolas' Day, Saint Martin's Day, Saint Katherine's Day, and many others served as reference points into which the year was divided. According to the Mendel family tradition, Johann Mendel was born on Saint Magdalene's Day, which comes on July 22.[55] (Mary Magdalene was a repentant prostitute who washed Christ's feet with her tears.[56]) The question is, however, can we be sure that the tradition has been passed on to us undistorted. We know of the tradition from two sources. First, contemporaries of Mendel reported that the abbot always celebrated his birthdays on July 22. And second, Alois Schindler recorded[43] that his mother Theresia, Johann's younger sister, as well as other relatives, insisted that in the family Johann's birthday was always celebrated on Saint Magdalene's Day. If so, both the day of birth and the day of baptism recorded in the parish register would have to be incorrect. Mendel must have become aware of the discrepancy, at the latest when he entered the monastery in Brno in 1843 and had to present his birth certificate. The latter was issued on September 12, 1834, as one of the documents required for matriculation in the Gymnasium in Opava. Since the document was based on the information in the parish register, it gave the date of his birth as July 22, 1822.

Since, however, it was mailed to the school,[55] Mendel may not have seen it until later. The reason why he made no attempt to correct the error may have been quite simple: By that time, he must have realized that it would be nearly impossible to convince the responsible bureaucrats that a mistake had been made. Anyway, he probably thought it unimportant whether he was regarded a couple of days older and went on celebrating his birthdays as he always had. He was right, of course: Plus or minus two days does not make much difference in the story of his life. If we vented here the controversy in some detail, it was not because we consider its resolution important but because we think that it provides an interesting insight into the circumstances of his birth.

The Sisters and Other Relatives

It is commonly claimed that Gregor Johann Mendel had two sisters, the older Veronika and the younger Theresia. In reality, however, he had four sisters, two of whom died shortly after their birth. The firstborn was Veronika who came to the world on May 5, 1820. After her came Johann on July 20 or 22, 1822. Next was Rosina, who was born on March 17, 1825, but died November 4, 1828, at the age of three years and one month. After Rosina came Theresia, born on May 4, 1827. And the last was another girl, who was born on September 7, 1831, and since by then the first Rosina was already dead, the fifth addition to the Mendel family was baptized Rosina again, in memory of the first Rosina and in honor of her mother. But the name seemed to bring bad luck for the second Rosina died hardly one month after her birth, on October 10, 1831. The cause of the two infants' demise is not known, but an infection, a common cause of infant mortality in the pre-Pasteurean era, would seem a likely possibility. Johann was six and nine years old when the two Rosinas died, and the passing undoubtedly left a deep imprint on his memory. According to most biographers,[43,51] Johann was closer to Theresia than to Veronika, an assertion that rests on two circumstances, one being a fact and the other an inference. The fact is that it was Theresia and not Veronika who later renounced her right on a dowry to enable Johann the continuation of his studies. While this could be interpreted as a sign of special fondness of Theresia for Johann, one must not forget that Veronika was nine years older than Theresia, and hence of the two, only she was of marriageable age at the time the decision was made. The decision might have, therefore, been based primarily on practical rather than any other considerations. It has also been suggested that it was her disposition toward cheerfulness that pulled Johann closer to Theresia than to Veronika. Here, however, we must realize that this inference is in essence based on the interpretation of a single photograph on which the younger of the two sisters has a more cheerful expression than the elder one (Fig. 3.20). Well, maybe so, but we all know how photographs can give totally false impressions. Also, one must keep in mind that because of the dowry arrangement, Gregor Mendel later enabled Theresia's sons to study and that it was through them that most of the information about the Mendel family has reached us. It is only natural that the sons focused primarily on their own

Fig. 3.20 The Mendel sisters. The drawing is based on a photograph presumably taken some time after 1852

side of the family history. We know little about Veronika's side because her lineage did not produce any dedicated reporters like the Schindler brothers. On the other hand, it must have been Johann who took care of and played with Theresia in her first few years, because all the other members of the family had other things to do, and this circumstance might have influenced their close relationship. A recently published letter[27] written by Alois Schindler on December 22, 1822, and addressed to Hugo Iltis provides a most direct account of the relations among the Mendel family kins, and for this reason we print here the translation (our own) of its pertinent second half:

Veronika had a gloomy expression on her face and in that she resembled her father, Anton Mendel, who was a very serious man. Gregor Johann Mendel inherited from his father his small stature. But his beautiful mild smile and his high forehead of a thinker originated from his grandmother, Rosina Kasper (Josef's daughter), who married Martin Schwirtlich, Hynčice 13. My mother resembled her prelate brother in her face, but was taller than Gregor Johann Mendel. Her tallness was the heirloom of the Schwirtlichs. One Schwirtlich (one of Mendel's uncles) fell in the battle at Leipzig as an infantryman and for the infantry only tall men were chosen. [...] This year I had the opportunity to lookup my mother's and my prelate uncle's final certificates from the elementary school. My mother was a very good student, she received four "very good" marks (reading, writing, counting, language). The prelate uncle earned two" very good" and two "good" marks. This, however, does not mean anything. Hans Mendel certainly performed as well as his sister Theresia but the teacher, Makitta, was a very cautious man and he took into account that Johann's certificate will go to Lipník. We children of Teresia have generally performed very well and how poorly was I graded when I decided to continue my studies! Of the Sturm children, only the eldest one performed very well *in the elementary school. Uncle Mendel wanted him to study in Brno and to cover his expenses, but the parents refused to give their consent because they wanted to save Mendel's offer for their younger children. Strangely enough, their performance at school was so mediocre that further study did not come in consideration. Perseverance, diligence, and untiring striving are the heirlooms of the Mendels. Had Wenzel Mendel not married Marina Wellert and does not Wellert or Willert mean a strong will? Anyway, Gregor Mendel's and Theresia Schindler's cheerful disposition and lightness in the way of looking at things go back to Rosina Schwirtlich. After all, the self-educated Anton Schwitlich, the first teacher in Hynčice, was a very close relative of Gregor Mendel. Veronika Sturm had remained lean all her life and had not gained weight even after her climacterium, By contrast, Gregor Mendel and Theresia Schindler always tended toward well roundedness of their forms and this inclination toward corpulence came from the Schwirtlich hereditary material. As prelate, Gregor Mendel made multiple attempts to shed some fat, getting up at 4 o'clock in the morning and giving up soup and fluids. It helped, however, always only for a short time. My mother did not take a single drop of beer the whole year long, was always very agile and hard working, and was the heaviest woman in her native village. Although plagued by gout, she lived to the age of 78 years. The Sturm aunt had not lived that long. Alois Sturm senior, Gregor Mendel's brother-in-law, was the son of an Erbrichter in Mankovice and a great penny-pincher (*Sparmeister*). Recently I received from home records of several purchases (*Käufe*) from Alois Sturm. I deduce that from the estate Johann (Gregor) Mendel was paid yearly 10 fl. Coventionsmünze as a contribution to his studies and 100 fl. Con. as his share in inheritance. Ten fl. contribution to study is really not much.*

The State of the Empire

In 1822, the year of Johann Mendel's birth, Franz I celebrated the thirtieth anniversary of his rule over the Austrian Monarchy, first as the emperor of the Holy Roman Empire of Germanic Nations (which he dissolved in 1806) and then, from 1804, as the emperor of the Austrian Empire.[57] He succeeded his father, Leopold I (see Fig. 3.3), who in turn was the successor of his older brother, Joseph II. Leopold's rule was short, a mere two years, but during this time he managed to stop the implementation of most of the innovations that Joseph pushed through. Franz went even farther and actually reversed most of Joseph's decrees. As a result, not much had changed for Anton Mendel under any of these three emperors in that the *corvée* remained in place. The first 23 years of Franz's rule had been turbulent, as during that time five war episodes were separated by short periods of shaky peace. The wars exhausted the empire economically and brought it on the verge of collapse in 1811. Ultimately, however, the monarchy emerged victorious from the war and recovered from the crisis. From 1815 on, it embarked on an extended period of peace, recovery, and modest prosperity. The main reason why the prosperity was not better than modest was primarily Franz himself. The country needed reshaping, modernization, receptivity to new ideas, and innovations. But Franz was neither willing nor able to lead the empire in this direction. On the contrary, his orientation was exactly opposite. His credo was "Hold onto everything that's old, for our predecessors knew that old things were good. I will not approve or sanction new ideas... I do not need scholars, I need good citizens...Those who serve me, must teach what I order. Who does not know how to do this or comes with new ideas, can go, or I will get rid of him."[38] With a sovereign holding such views, the monarchy was bound to stagnate. The parole of the entire period had been *status quo*—keep things the way they are and be content, don't stir up the calm waters. To retain the *status quo*, Franz turned the monarchy into a monstrous bureaucracy. Nearly all Habsburgs displayed a tendency to rule bureaucratically, but under Franz Austrian bureaucracy achieved new heights. He himself was the personification of a fossilized civil servant. Since he thought that he knew everything better than everybody else and that he could not trust anybody, he insisted on handling all petitions, even the most trivial ones, himself. Consequently, he spent long hours in his office, shuffling papers from one pile to another, unable or unwilling to heed major issues, which should have been the focus of his attention. He governed primarily through two persons, Metternich and Kolowrat, who in their turn ruled over an army of civil servants. Prince Klemens Wenzel Nepomuk Lothar von Metternich (1773–1859) was in charge of foreign affairs and internal security, and Graf Franz Anton Kolowrat-Liebsteinsky (1778–1861) was responsible for just about all other business of the state, but for finances in particular. Franz, an Austrian; Metternich, a German; and Kolowrat, a Bohemian represented the acme of bureaucracy in their respective nations. Although they were as different in character and disposition as three persons could be, they had three things in common: dread of revolution, conservatism, and resentment of nationalism. The French Revolution of 1789 and the spread of its ideas across Europe scared the wits out of them and led them to a firm resolve not to let anything like it happen in the Austrian Empire. The enlightened rule of Joseph II revealed to them the adverse consequences liberalism could have for the welfare of their own class, the aristocracy. And the mixed ethnic composition of the Austrian Empire (27 % Germans, 18 % Hungarians, 16 % Italians, 13 % Czech, and the remaining 26 % divided among nine other nations) made them realize that their government was perched on a powder keg, and this made them nervous about anyone bearing a torch, or even a match, of nationalism. All three statesmen were therefore continuously on alert for signs of emerging revolution, liberalism, and nationalism. Together, they turned the Austrian Empire into a police state over which Metternich spread a cobweb of spies, undercover agents, and paid informers. They committed a substantial part of the bureaucratic apparatus to gathering information about people, operating under the motto: Everybody is a suspect until proved innocent. The agents busily compiled thick dossiers on the activities of citizens by any means, including reading people's private correspondences, and not even bothering to make a secret out of it.[38] At seven o'clock every morning, a messenger picked up all the incoming and outgoing mail at the post office and brought it back exactly two hours later. During those two hours at the police station, agents sifted the mail, opening letters written by or addressed to known suspects, as well as some others chosen at random, and recorded passages of interest or simply confiscated the letters. Another popular means of controlling what people read and thought was censorship. Any critique or unfavorable comments addressed at the ruling dynasty, state politics, or the Catholic Church and any comments smacking of liberalism or nationalism were blotted out from newspapers and books. To prevent that a reader might get information from materials published abroad, importation of foreign newspapers and other printed matter was strictly forbidden. Although similar measures were in effect in other European countries, in the Austrian Empire, the fear of new ideas progressed to paranoia.

A Day at Home with the Mendels

As far as is known, the Mendels did not employ any servants or farmhands, so the two adults and the three children—Hans (Johann), Broni (Veronika), and Resi (Theresia), as the parents called them affectionately—had to manage the household and the fieldwork all by themselves. Considering the size of the property, this was no small task accomplishable only if the entire family, including the children, worked very hard. From a tender age, the children were led to contribute their share of work, starting with simple duties and gradually progressing to more complex chores. On a farm, there is always some work for everybody, and so the children were kept busy. The more they had to work, however, the more intensely they savored the rare moments left for play—Sunday afternoon, holidays, and days on which the weather restricted activities outside the house. They did not know boredom, the mother of mischief, because there were thousands of things they wanted to do in the precious free time they could get. From the earliest time, they learned that every slice of bread had to be earned and that the earning entailed hard work. The work in the fields and in the garden taught them that one can harvest only what one had sown and cultivated carefully and that any attempts at shortchanging the crops in care may have disastrous consequences. All these and many other experiences were important practical lessons that shaped their characters.

In the absence of contrary evidence, we can assume that the dependence on one another and the sharing of happy moments, as well as misfortunes, bonded the family together tightly. In a farmer family, there is neither time for nor the inclination toward any overt expressions of affections. Fondness of one another is shown by deeds rather than words. This is, indeed, the impression one gets from reading between the lines of the few known letters Gregor Mendel wrote to his parents.[28] Undoubtedly, like in any family, there must have been strains also in the Mendel family, but on the whole, it seems, the relationships between the parents and the children and among the children themselves were harmonious. No documents have survived that might have allowed us to reconstruct the first ten years of Mendel's life in any specific detail. All that we can do, therefore, is to use the available general information about life in that period of time in that corner of Silesia to sketch out in broad strokes the daily and annual routines Johann presumably had experienced. Hopefully, the vignettes we are about to paint will give the reader the flavor of the environment in which Johann Mendel spent his preteen years. We begin with a description of what *might* have been one day in the life of the Mendel family. In the next two sections, we will follow the annual turnover of seasons reflected in the family's work and feast days.

For the parents, the day began at dawn in the summer. Before they sat down to breakfast, they first took care of the livestock—Anton of the horses and the cattle and Rosina of the rest. The stables and cowsheds had to be cleaned; the manure removed and replaced with fresh bedding; the animals themselves had to be cleaned, watered, and fed; and the cows milked. The pigs, goats, and the poultry also had to be attended correspondingly. By then, it was time to start a fire in the stove, prepare breakfast, and wake up the school-age children. The breakfast consisted usually of cold or warm milk and bread smeared with homemade butter. Although the habit of drinking coffee began to spread through the empire as early as 1685,[38] it may not have reached Hynčice before the 1820s, but even if it had, it would have been too expensive for the Mendel family to indulge in on a daily basis. They might have, however, reached occasionally for the cheaper substitute, the chicory. They may have also drunk tea every now and then, probably with milk, but for regular consumption, it too may have been beyond their means. However, they might have used a substitute from different herbs or dried linden flowers. So, mainly milk and water from their own well were their staple beverages. After breakfast, they went about their individual duties. The children walked to school; the father would hook up the wagon and ride off to work in the fields, either his own or those of the lord; and the mother would either go with him, in which case she would take with her also the preschool children, or she would attend to the many chores in and around the house and the children would have to help her. The family would come together again around noontime for lunch, which would be the main meal of the day, generally soup and a main course. The former could be a decoction from the meat of the main course, potato, noodle, vegetable, or a semolina soup. All the meat of the main course came from home production, mostly from a slaughtered pig or poultry. One or two pigs supplied enough meat to feed the family the whole year. To last this long, the meat had to be canned, smoked, or salted, while other parts of the pig were processed into different kinds of sausages, salami, *Presswurst*, bacon, and lard. All these varieties could become part of the main course. The poultry meat came chiefly from chicken, but on special occasions, it might have from a duck or a goose. The meat was served with potatoes, dumplings, cabbage, or other vegetables. Friday was a meatless day for religious reasons, and there were other such days throughout the year, and on these days the main course usually consisted of semolina prepared in different ways or of various vegetables. Pickles or canned fruit was usually eaten with the main course. During the harvest, when the entire family worked the whole day on the field, they had a cold lunch, usually bread with butter, topped with homemade cheese, yogurt, marmalade, or plum jam, and milk. It could, however, also be a piece of bacon, a slice of *Presswurst*, or aspic, with bread and beer (adults only). Similar selections served also as snacks during the morning and afternoon breaks.

After lunch, the family went after their tasks in the fields or at home and worked until dusk. In the evening the animals had to be taken care of, more or less in the same way as in the morning, before the family could sit down to a supper and the evening prayer. The supper was often no more than a bowl of boiled potatoes, peeled hot and eaten with butter and milk or buttermilk.[42] The prayer was ordinarily short, *Pater Noster* (Our Father) and *Ave Maria* (Hail Mary), but at certain times of the year, the family assembled for the rosary, five or 15 sets of prayers of much longer duration. Each set consisted of one Pater Noster, ten Ave Marias, and one Gloria Patri (Glory to the Father), the final set being followed by the so-called Apostles' Creed, a brief statement of the Christian faith, which began with the words "I believe in God the Father Almighty, Creator of Heaven and earth…" To count the prayers, the worshipers used the rosary, a string of small and large beads, the former for the Ave Marias and the latter for the Pater Nosters and Gloria Patris. The circular string had a pendant with the crucifix for the Apostles' Creed. The worshipers knelt in front of a crucifix or an icon mounted on the wall.

After the prayer, the children went to bed, while the parents attended to a few remaining chores, and then they, too, retired. After dark, the only sources of light were from candles and oil lamps. The former were expensive and so were used sparingly. The oil lamp, in its latest improved version, was a vessel filled with vegetable (rapeseed) oil, from which projected a tubular wick enclosed between two metal cylinders. A glass cylinder protected the flame at the end of the wick. The more efficient kerosene lamps became available only in the second half of the nineteenth century and electric lamps only in the twentieth century.

The routine just described varied somewhat depending on the type of work that the seasons of the year brought with them. It also changed on Sundays and holidays. On Saturday afternoon, the children cleaned and swept the farmyard and washed the floors in the house. On Sunday morning, they all put on festive dresses and walked to the Saints Peter and Paul church to attend the service. Although the church was modest compared to other churches Johann would see later in his life, as a child he must have been awed by what he encountered there—the size of the building, the height of the bell tower, the spaciousness of the interior, and the embellishments of the altars and of the pulpit. No less awe-inspiring was the solemnity of the rite; the atmosphere, permeated by the fragrance of flowers; and the scent of the burning candles and smoldering myrrh in the censer. But by far most impressive was the ceremony of the mass itself, with the priest, the acolytes, and the altar clad in ceremonial vestments, the priest chanting in an incomprehensible language (Latin), the crowd singing hymns (often out of tune), the pipe organ booming like an approaching storm, and on special occasions the sound of a soprano soloist descending from the choir onto the congregation as if from heaven. For a long time he marveled at the mysteries of the ceremony, for which he could get only fragmentary explanations from his parents. Only later, when he began to learn Latin did he realize, for example, that the phrase *Dominus vobiscum* repeated by the priest several times during the ceremony did not mean *Dominus wo bist du?* (Lord, where are you?), as he thought, but the greeting "The lord be with you."

And only later still, did he begin to grasp the origin and meaning of the ceremony. He learned then that, supposedly, on the eve of his immolation, when he sat down to the Last Supper with his 12 disciples, the Apostles, Christ took a loaf of bread, broke it into pieces, and with the words "Take and eat, this is My Body," gave it to them. He then took wine in a chalice and with the words "All you drink of this, for this is My Blood" he passed it onto them.[58] Afterward he directed them "Do this in remembrance of Me."[59] With time, the Catholic Church developed around this simple act an elaborate ceremony, the mass, in which the priest and the assembled laity said prayers, sang hymns, chanted prescribed phrases, and performed ritual movements. In the center of the ceremony remains, however, a reenactment of the events at the Last Supper and on the following day.[53] According to the Christian creed, Christ, representing one version ("the Son") of a single tripartite God, came to this world to give humanity a second chance to save itself for God's Kingdom, after it forfeited the first one through Adam and Eve's original sin. For reasons incomprehensible to any non-Christian, God the Son had to become human, suffer a cruel death, and by this sacrifice appease God the Father to lift the curse imposed on humanity for the original sin. At the Last Supper, Christ transubstantiated his body into bread and blood into wine, while on the day of his crucifixion he offered his real body and his real blood to God the Father on behalf of all Christians. In the part of the mass called the *Communion*, the priest repeats the act of transubstantiation, turning bread in the form of a wafer called the *host* (from Latin *hostia*, victim) and wine into the actual body and blood of Christ, even though the physical properties of the bread and wine do not change. He then takes both the bread and the wine but offers to the congregation only the former. This type of Communion is called *under one kind*. Originally, however, both the priest and the laity received bread and wine in a *Communion under both kinds*, and this practice has been resumed in some non-Catholic churches. In the Catholic ritual, only those people of the congregation, who prior to the Communion confessed their sins to the priest and were absolved of them, can receive the host.

One can assume that two masses were offered for the Vražné and Hynčice congregation on Sundays, a shorter one early in the morning and a more elaborate one at 10 AM. Rosina Mendel probably attended the morning service, so that she could prepare lunch while the rest of the family was attending the High Mass. It was at this service that Johann Mendel first encountered Johann Schreiber, who would later become his teacher and play a decisive part in his life. It is doubtful, however, that while watching *Pfarrer* Schreiber celebrate the mass and deliver sermons from the pulpit, the thought might have crossed Johann's mind that one day he, too, will embark on a similar career.

A Year of the Natural Cycle on Anton Mendel's Farm

For a farmer, a new year did not begin in January, as it did for other people, but in October, when everything planted in the preceding cycle had been harvested and the time came to prepare the soil for the growth in the new cycle. The fields were empty, and only the stubbles remained after the grain reaping, and the dry stems left after the potato harvest, were burned. The hectic time of the summer's agricultural activities gave way to a calm period of preparation for the long winter sleep. To help the soil to regenerate, fertilizers had to be provided, primarily manure. The farmer had to cart the manure from the farmyard, distribute it over the fields, and then plow it in. In Anton Mendel's time, the original way of plowing, in which the ground was merely broken up with an arrowhead-shaped implement, had been substituted by a new, more efficient method. The method was based on the use of a new model of the plow with a blade or plowshare that cut under the soil and a moldboard that lifted the sliced soil, turned it over, and pulverized it. A pair of horses or oxen hitched to the plow pulled it, and the plowman kept it in the soil at the desired depth and in the right direction through shafts for both hands. A skilled plowman working with well-trained draught animals could handle them and the plow alone. If the animals were not well trained, he needed a boy to lead one of the horses along the furrows. An older, heavier version of the plow had two wheels in front of the cutting section to bear a part of its weight; the later versions were without wheels. The plowman would start either in the middle of the field and proceed toward the sides or start from the sides and move toward the center, furrow after furrow. Flocks of birds, crows in particular, would follow him to feast on the unsheltered worms and grub. Hawks might hover above the furrow to sweep down on any escaping rodents.

Plowing completed, the farmer hitched his horses to a wooden roller and standing on the roller's frame for greater weight drove it over the furrows to break down the clods. Then he hitched the horses to a wooden or metal harrow with a spiked underside to complete the breakdown of the clods, smooth the ribbed field, and collect uprooted weeds for burning. The seedbed thus prepared, the plowman turned into a sower. We can assume that at least initially, Anton Mendel sowed his grain crops by hand and that only later, if at all, did he get a horse-drawn sowing machine, the drill. As a sower, he would walk the field up and down, holding with one hand a folded apron filled with grain and with the other hand casting the grain evenly over a broad area in a gesture immortalized by the French artist Jean Francois Millet in his famous painting. (Millet, better than anybody else, captured the life and activities of western European farmers in the nineteenth century in a series of drawings and paintings. Because of their realism, the pictures evoke farmers' toils accurately and expressively.) The task of the sower is more difficult than it might seem. Of course, anybody can scatter seeds over an area, but to scatter them so that a uniform covering of crop will sprout from them is rather tricky. A high density of crop in one place requires thinning later and is wasteful, whereas seeds scattered too thinly provide space for the growth of weeds. Only a skilled sower can scatter them correctly. To prevent birds from eating the scattered seeds, the field has to be harrowed immediately. The introduction of the sowing machine alleviated these problems. The machine consisted of a wheel-mounted seedbox with an agitator dispersing the seeds into furrows made by the drill, which also covered the grain.

On rainy days, when the farmer could not work in the fields, his attention turned to the crops harvested in the preceding season. In the case of the cereals, the grain had to be released from the ears by threshing. There were no threshing machines in Anton Mendel's time; the grain had to be separated from the rest of the plant by beating the latter with a flail, a wooden tool consisting of a club (swizzle) fastened by a leather stripe to a long handle (staff). The farmer untied several sheaves and spread them on the bare floor of the barn and then, swinging the club of the flail, struck them repeatedly in rhythmical motions to knock out all the grain from them. For greater efficiency, neighbors helped each other and worked in teams, striking the crop in sequence, one after the other. From the rhythm of the thumping sounds coming out of a barn, a passerby could guess how many threshers were at work. The threshers worked systematically from one end of the barn to the other, and when they reached it, they raked the straw away and swept the grain with the chaff into bins or sacs. To separate the grain from the chaff, the farmer had to wait for a dry day with mildly blowing wind, and, working outdoors, he would cast handfuls of the mix into the air and collect the heavier grain in a basket while letting the lighter chaff blow to the ground a short distance away to be used as horse fodder.

The field plowed, the winter crops sowed, and most of the grain harvest threshed, the farmer could devote his time to a variety of minor outdoor activities: clearing ditches and drains; cutting birch tree branches for brooms; removing molehills on the meadows and exterminating the moles and voles; repairing farm buildings; pruning and stalking trees in the orchard; collecting, sawing, and chopping firewood; and repairing or rebuilding fences. Weather permitting, he might continue some of these activities through the winter months and into the spring, until the fields would require his full attention again.

In winter, when the weather confined him indoors, there was no shortage of work either: repairing and sharpening farm tools; sawing timber; plaiting baskets; carving wooden spoons, platters, bowls, and mugs; repairing carts and wagons; forging new plowshares in the village smithy; and of course every day in the morning and evening feeding, watering, and cleaning the livestock. The highlight of the winter months was, however, the slaughtering of the pig. The one selected for the sacrifice had been pampered the whole year, getting the best fodder and extra rations, making it the fattest animal on the farm. Ultimately, however, the day came, when it had to pay with its life for all the special favors it received. Pigs are intelligent animals, and judging from its heartbreaking squeals when hauled out from the pigsty on the fateful day, it apparently apprehends what is coming. Although the executioner was a professional, the act itself could sometimes be botched. Roped by a hind leg, the pig had to be knocked unconscious by a blow on its head with a maul and then quickly turned on its side and bled from the heart to collect as much of blood as possible to make blood sausages (such as *Presswurst*) and also to assure high quality of the meat. Occasionally, however, the pig woke up while being bled and got away, and the whole family then participated in chasing it around the farmyard with much ruckus. The carcass was then singed with boiling water, pulled up by its hind legs to hang from a crossbeam in a barn, scraped, and slit open on the belly side. For some reason, this moment of the pig slaughter was a favorite subject of seventeenth-century Flemish, French, and Dutch painters, including Rembrandt. The butcher then kept the whole family on its toes for much of the day, for every step had to be timed and coordinated and everybody had to help. The bulky butcher, fortified by generous amounts of plum brandy (*slivovice*), worked, short sleeved, outside, where the temperature was deep below the freezing point, and inside in the steamy, overheated room with the cauldron of boiling water, without any apparent ill effects on his health. They all paused for a short while when the first assortment of boiled meat came out from the cauldron. Joined by neighbors, relatives, and friends, they then delighted in what some would swear represented the culinary highlight of the year. After this short break, the work resumed and continued until late into the night and, at a slower pace, for several days afterward, for not just the meat but all the organs and tissues were utilized in one way or another. If the pig was male, the one exception to this thrift was the testicles, which were hung on a tree in the orchard for the birds to feast on. As the meat had to last for up to one year, it had to be preserved by salting or smoking. To salt the meat, the Mendels steeped it in brine for several days and then hung it in the chimney; before eating it, they had to simmer it in water to remove most of the salt. For curing the meat, they might have used a communal smoke house.

As soon as the weather permitted it, but no later than in March, Anton Mendel resumed his work in the fields—plowing, followed by the harrowing, sowing of the spring crop, and planting potatoes and vetches. Planting was backbreaking work, because the planter worked the whole day bent to the ground dispensing sliced tubers of potatoes, saved from the previous season, into the furrows at a regular distance of one foot, which then had to be covered by plowing. Once the potatoes sprouted, they had to be hoed to aerate the seedling and to remove weeds. Peas and beans, too, had to be planted individually, into holes made with a pointed stick.

By June, it was time for the hay harvest, the first harvest of the cycle. Meadows near the brook, fertilized by silt deposited during occasional floodings, yielded lush grass. The yield of other meadows depended on the year's precipitation. The farmer cut the grass with the scythe, which had the blade at nearly the right angle to the long handle. The handling of the scythe required certain skill, for the grass had to be mowed close to the ground, but at the same time care had to be taken not to hit the ground with the tip of the blade and break it. A dulled blade had to be sharpened with an abrasive whetstone, moistened by spitting on it, and then moved along the blade's cutting edge, altering the sides with each rapid move. Unevenness of the cutting edge had to be straightened with a hammer, knocking the blade positioned on a small metal anvil. The persistent monotonous knockings coming from the farmsteads and carrying through the village were familiar sounds of early summer evenings. Women followed the reapers, scattering the cut grass to dry. For several days afterward, they turned it over periodically until it was ready to be gathered into haystacks and carted into the barn or stored in the attic for the winter. The scent of the coumarin emitted by the fresh hay was one of the delights of rustic life, and some considered it a special treat to take a nap on Sunday afternoon in the hay just harvested and stored in the attic.

No sooner was the hay in the barn than it was time to reap the cereals and the vetches. Like the grass, the barley, oats, peas, and beans were cut with scythes. For reaping the wheat, farmers originally used sickles, cutting high up on the stalks near to the ears or low near to the ground. In the former case, the straw had to be cut subsequently with the scythe, the purpose of this inefficient method of harvesting being to lose as few of the ears as possible. It is likely, however, that Anton Mendel, at least in later years, reaped wheat and rye with a scythe that had a special attachment on the handle minimizing the loss of grain. Women following the reaper collected the stalks and bound them into sheaves, which they then assembled into rows of stacks or *mandels*. Later, when the stalks dried up and the ears completed their ripening, horses led by one of the children would draw the wagon along the row, from one stack to the next. At each stop Anton would pick one sheave after the other with a pitchfork and lift it onto the wagon, on which Rosina would organize them neatly into a boxlike arrangement extending high above the wagon's base. The arrangement would be such that it would withstand, without a loss of a single sheave, the perilous swings it would be subjected to during transport. In the barn, the sheaves had to be unloaded and again one by one stacked and stored until there was time for threshing. The last sheave, as well as the horses drawing the last harvest wagon, were decorated with flowers to show to the village that for their family the grain harvest was over. On Sunday, after all farmers had completed their grain harvest, the village celebrated with music, dancing, drinking, and feasting. Once the last stack disappeared from the field, it was time for the gleaners to move in and gather any remaining ears. Women and children from families which did not own any land went through the field and systematically, bent over the stubble, picked the ears with one hand, and with the other passed handfuls into a bag on their backs. Millet captured this act in the second of his three best-known paintings (*The Gleaners*, the third being *The Angelus*). At home the gleaners loosened the grain from the ears by hand and ground it into flour. Though meager, the yield could save a family from starvation.

The last to be harvested were potatoes, turnips, and beets. Potato tubers were uncovered by plowing, collected into baskets, and carted for storage in the cellar or on an earth-covered pile in the garden behind the house. The wilted potato stems were piled up and burned. The smell of burning potato stems drifting over the landscape was a sure sign, if any was needed, that autumn had arrived. Children delighted in this time of the year, for not only did it signal the end of the hard work on the fields but also because burning the piles offered various diversions. The most treasured one was baking potatoes in the hot ashes of a burned out fire. Peeling the blackened crust and consuming the inner part while still hot was one of many unforgettable experiences of a childhood on a farm.[60]

We can assume that only for the first few years of his life was Johann Mendel a mere passive observer of all these activities on the farm. As soon as he learned to walk and speak, he became a participant in these activities—such was the law of the farm in those days. Since his mother had to work in the fields during the planting and harvesting seasons, the younger children had to take care of many of the house chores, while the older ones had to work in the fields with their parents. When everybody was out in the fields, the younger children were probably left in the care of Anton Mendel's parents (Rosina's parents were already more than 30 years dead when Johann was born), who might have lived in a small room in the house No. 58. Valentin Mendel died in 1828 at the age of 74; his wife Elisabeth followed him one year later at the age of 75. Johann was therefore six and seven years old, when he lost his grandparents. But until this happened, he probably spent much time with them.

The repertoire of Johann's duties must have been quite variable. It may have included sweeping the yard; driving the ducks to the brook and keeping an eye on them; grazing the goats, sheep, and cattle; feeding the rabbits goutweed and grass he had cut in the orchard or at the brook; feeding the young ducks and geese with a special fodder prepared from cut and chopped nettles; fetching water from the well; churning butter; and hoeing vegetables in the garden. As he grew bigger and stronger, he had to join his parents working in the fields. He picked stones uncovered by plowing (they could ruin a scythe during reaping), uprooted thistles and other weeds in sprouting wheat fields, planted and hoed potatoes, turned over drying grass during hay harvest, and helped collecting tubers during potato harvest, as around the house, in the fields, too, the list of things to do seemed inexhaustible. Hardly was one task completed, when a dozen others were already waiting for him. Seen from a present-day perspective, it may appear as child slavery, but Johann's parents were no slave drivers. No doubt, they would have preferred to see their children play, rather than toil, but if they did not want them to starve, they had no other choice than to seek their help. Does this mean that Johann's childhood was an unhappy one? By no means, it was hard, but not unhappy, for the more limited was his free time, the more intensely he lived it.

A Year in a Village

In his works, Millet depicted just about all of the peasants' labors mentioned in the preceding section and many more. As he was a contemporary of Mendel, even though in a different part of Europe, his art offers a much more telling peek into the toilsome life of the nineteenth-century European peasant than any narrative might hope to provide. In one way, however, the view purveyed by Millet is misleading, in that it creates the impression that the peasant's days were filled only with drudgery, from daybreak to dark, the whole year around. In reality, the year was punctuated by festivities that let the farmer forget for a while all his worries and hardships. Some of these festivities, such as weddings, were occasional affairs, but others were regular occurrences. The majority of the latter were religious holidays, celebrations, or commemorations of either events mentioned in the sacred text of Christianity, the New Testament, or feast days of prominent saints. The church frowned upon the few surviving profane customs rooted in pre-Christian paganism and made a great effort to replace them by its own traditions.

Starting the year—as in the preceding section—from the postharvest time, the first widely commemorated occasion was All Saints Day on November 1, followed immediately by All Souls Day. The former was dedicated particularly to those saints, known or anonymous, who did not have a special feast day marked in the calendar. On All Souls Day, people remembered their deceased relatives, friends, and acquaintances. They decorated their graves with spruce and fir branches, white berry twigs, chrysanthemums, and candles. They said prayers privately at the graves and at special services and liturgical ceremonies in the church. According to the Christian doctrine, prayers addressed to the saints and masses celebrated on those two days were particularly efficacious in shortening the time souls had to suffer in purgatory. The two holidays were an occasion on which dispersed relatives met at the graves of their loved departed ones. The melancholy of the late, fog-enshrouded autumn days, the somber mood of the people crowding the cemetery, the sea of lights flickering in the dusk, and the special mix of scents emanating from the burning candles and the coniferous lopping all created a special atmosphere, which subdued even the children, making even them to think of the terminableness of human life and of the mysteries looming beyond the grave.

As the days were getting shorter and colder and the nights longer, the two holidays that were on the children's minds, and also on their tongues, were Saint Nicholas Day and Christmas. The former was not really a holiday, but that mattered little to the children, because the event that concerned them did not happen on that day anyway but on its eve, on December 5. On that evening, the children expected, not without some trepidation, to be visited by Saint Nicholas personally. They were told that if they wouldn't eat anything the whole day and be pious, they would actually see in the evening Saint Nicholas descending on a golden rope from the sky. Alas, if they were lucky, they saw only a spectacular sunset, but no rope and no heavenly messenger, obviously because they were not pious enough! The reason for the trepidation was that Saint Nicholas would not come alone, but in the company of—the devil! The origin of this tradition pairing two opposite poles of Christian mythology is uncertain, as is also the historicity of Saint Nicholas himself. Not disturbed in the least by such theological nuances, on the eve of Saint Nicholas Day, two adult villagers disguised themselves as the legendary pair: Saint Nicholas wearing a long white robe, long white beard, a bishop's miter on his head, long staff in one hand, and basket with presents (contributed by the children's parents in each house) in the other and the devil sporting a black or blackened goat skin, a mask with horns, an oxen tail, a hoofed foot, birch broom in one hand, and rattling chains and a bag with black coal in the other. Thus attired, they went from house to house, accompanied by a crowd of curious onlookers, acting as the chorus in a Greek tragedy. The youngest children received the couple with a mixture of pious awe, fright, and agitation, not doubting for a moment about the genuineness of the representatives of heaven and hell. The older children had some doubts, noticing that the white robe was made of linen, the miter of paper, and the voices of the two were somehow familiar, but they, too, were frightened enough to think of all this only after the couple's departure. The oldest children, of course, knew better. Saint Nicholas inquired about the children's good or bad behavior, asked some of them to say a short prayer or sing a song, and then rewarded the well-behaved ones with a bag of gingerbread, apples, and walnuts. The devil took care of the misbehaving children, whipping them with the broom, of course to an accompaniment of heartbreaking howls of the sinner, trying to hide behind mother's shirts and promising under duress that he will behave. He received only black coal from the devil to remind him of his promise. We have grounds to believe that in Mendels' house the devil never had any reason to be called on to support the parents' educational process.

No sooner had the children taken in this exciting event, than they began counting days to the next: Christmas. The church counted with them, starting on the Sunday nearest to November 30, the feast day of Saint Andrew, one of the 12 Apostles. From this first Advent day (the word meaning "coming"), Christians had four weeks to prepare themselves for the celebration of the Nativity—the birthday of Christ. Never mind that nobody knows the year, not to mention month and day, on which Christ was born. The founder of Christianity had to have a birthday, and so one was chosen, but not quite arbitrarily. The chosen date, December 25, happened to be close to the date of winter solstice, an occasion widely celebrated by many pagan peoples. On that date (around December 22), the sun, which since the summer solstice around June 21 had been peaking its daily path across the sky at progressively lower points every day, reached the lowest point. It appeared to remain at this point (which is what the word "solstice" means) for a few days, before it began to rise again, peaking every subsequent day a little bit higher in the sky. To the ancient people, the rise appeared as if it was the birth of a new sun, and this was what they celebrated with festivities and ceremonies. Knowing how hard it was to uproot old customs, the church cleverly placed the birth of Christ into the period in which people were used to celebrating the birth of a new god (sun) and allowed them to integrate into the Nativity festivities many pre-Christian customs and traditions.

Most of the traditions were ancient, indeed. They went back to the time when animism—the belief that spirits inhabited all things in nature, animate or inanimate—was widespread. A part of this belief was that on special occasions, the spirits could be persuaded, first, to help people, especially the farmers, whose prosperity was very much influenced by nature's caprices and, second, to use their extraordinary powers to reveal one's future. Winter solstice, the birth of a new god and by extension the birth of Christ, was obviously a magical moment, in which it was possible to communicate with the spirits in nature and ask for favors. Hence, in Anton Mendel's time, when one spoke of "Christmas magic," one did not refer to the sound of ringing cashier registers, as the merchants do in our times, but to the true meaning of the word "magic": the performance of rituals with the intention of controlling supernatural forces to achieve a desired effect. It is only when viewed from this perspective that the Christmas Eve customs make sense. Sharing a Christmas cake with the cattle and horses and garlic with the rooster, gander, and drake; burying fish bones from the Christmas Eve dinner under the fruit tree in the orchard; and dropping apples and walnuts into the spring well in the fields and honey into the farmyard well—all these acts were meant to ensure that the animals, trees, and water sources would stay disease free, and remain productive. Similarly, pouring melted wax or lead into a water bowl and reading from the emerging shapes one's future, peering into the depth of a farmyard well to glimpse the likeness of one's future husband, or shaking an elder bush and listening from which direction a dog would bark to indicate the place to which a girl will move as someone's wife—all these signs could be believed or half believed only under the assumption that some magical force operating on Christmas Eve was capable and willing to reveal things to come.

Onto this animistic heathen background, the Catholic Church imposed its own mythology. The story of Christ's birth is presented in two parts of the New Testament, the Gospels of Saint Luke and Saint Matthew. The former tells how an angel appeared to the shepherds outside of the town of Bethlehem informing them that the Messiah prophesied in the Old Testament was born. The latter relates how a group of magi (wise men) following a bright star they noticed on the night Christ was born came to a stable in Bethlehem, where they found the Christ child in a crib. Later the church expanded this story by turning the group of magi into three kings, identifying them as Caspar, Melcher, and Balthazar and having them present gifts of gold, frankincense, and myrrh to the infant. Starting from the Middle Ages, it became a tradition to portray the events described by these two stories in a display of figures and articles—the Nativity scene, also called *crèche* because the center of it was the crib (which is what the French word means) with the infant Christ laying on straw or hay in a stable. The other subjects and objects in the display were Christ's mother Mary, her husband Joseph, the shepherds, the Three Kings, oxen, sheep, and other animals, with the star of Bethlehem (depicted as a comet) and a suite of angels in the sky above the scene. It is quite possible that Johann's father cut the figures for a crèche out of plywood and let the children color them and display them on the top of a chest in a natural setting achieved by using moss, twigs, grass, stones, and other objects. The children must have delighted in assembling the Nativity scene every Christmas. A much more elaborate crèche was probably on display in the Saints Peter and Paul church, with actual statues of the Holy Family and all the witnesses of the event.

The crèche was one familiar Christmas symbol adorning the Mendels' home at this time of the year; the other, the Christmas tree, was probably also on display, beginning on Christmas Eve. The custom of trimming a spruce or a fir tree for Christmas originated in the Germanic countries in the seventeenth century and by Mendel's time most likely reached also Silesia. The impetus for the custom was probably a tree decorated with apples to symbolize the Garden of Eden in the German medieval mystery plays. When the plays were abolished, the tree found its way into the homes as a symbol of the Christmas season, and its trimming grew gradually more elaborate. In Mendel's home, the decorations might have included apples, gilded walnuts, homemade sweets and paper-wrapped candies, angel figurines cut out of plywood, and small candles. The children no doubt admired their tree and probably fought every day with the temptation to steal a cookie or a candy from it.

It was, however, not just the tree that the children looked forward to during the entire Advent period. It was also the near certainty that under it would be a present or two for each of them. The guessing what it might be that year grew in anxiety as Christmas was approaching—ever so slowly, it seemed to them. The parents of the three Mendel children could not afford buying expensive presents for them, but they always managed to scrape some extra cash to get them what they knew they wanted most. The time at which the children were allowed to open their presents varied from region to region, and we do not know what the custom was in the Mendels' home. We may assume, however, that, like in many other parts of Silesia, after all the work was done on Christmas Eve, the livestock were served special treats and the prayers were said and the Mendel family sat down to a dinner of pea soup and fish, usually a carp, as the main course. During the meal, the tension among the children was visibly rising, but there was still no sign of a Christmas tree. The children were told that the Christ child, who might also bring presents for them, would trim the tree, and so they listened intently for the sign that the tree had been delivered. In reality, one of the parents or a relative secretly decorated the tree in the afternoon already in one of the bedrooms that remained under lock and key. Finally, the signal came, the ringing of a small bell, and they were allowed to enter the bedroom. And there it was, already lit and glittering in its full beauty, the most wonderful Christmas tree. And under it wrapped presents for each of them, and when they then discovered that they got just what they wanted, their joy became boundless.

The children played with their new toys until they fell asleep and had to be taken to their beds. Later at least one parent with the eldest child, but in some years the whole family went to the church to attend the midnight Christmas mass. If the weather was fine, the trip to and from the Saints Peter and Paul's church was a memorable experience. Usually, by this time of the year, large amount of snow had fallen already, and the frost was biting at the pilgrim's noses, as they forged their way up the hill toward Dolní Vražné together with other villagers. In the cold night, nobody felt like talking and so they must have appeared like dark, moving shadows, of which only the crackling of the snow under their feet betrayed their realness. What made the trips most memorable, however, was the sky above the travelers' heads. On a moonless, clear, frosty night, it offered a sight that nowadays, in the light-polluted world, fewer and fewer people encounter. If not the frost, the view of the starry heaven would have taken their breath away and lifted their spirits to euphoric heights. In the church, their spirits remained in the heavenly spheres, kept there by the music to which they were treated. Specially composed for these occasions, it contained folk tunes reflecting the adoration of the shepherds and other country folk, to which the villagers were best attuned. At the end, the congregation joined in by singing Christmas carols.

On Christmas Day, the family feasted on a special midday meal, usually a liver dumpling soup, goose or duck, and Christmas cake full of raisins as a desert. Home-baked cookies were also available on that day and during most of the Christmas season, which ended on January 6, the Three King's Day. On that day, three village boys, dressed as the Three Kings (one of them, the king from Africa blackening his face with soot) and holding a large star, went from door to door, caroling a special Three Kings' song at each. In each house, they scented the living room and the stables with a censer, sprinkled the rooms with holy water, and inscribed on the door with chalk the letters C + M + B, which everybody thought stood for the initials of the Three Kings, but the church insisted that it meant *Christus Mansionem Benedicat*, May Christ bless this house.

The long winter evenings in the post-Christmas period were occasions for the farmers to get together for neighborly chats. Even on these occasions, however, the people were rarely just sitting around idly and talking. For the mistress of the house, they were an opportunity for a feather stripping session. On a farm, little was wasted and feathers were an important commodity used to stuff comforters and pillows. A farmer's wife would therefore pluck all her geese in the summer months and also collect all the feathers of the birds killed for the meat. To be used, however, the feathers had to be stripped to remove their prickly parts. To do this, one had to pick the feathers individually; hold each at the tip in one hand; rip off the soft parts, the vanes, from the shaft with the other hand; save the vanes; and burn the rest. Combining a palaver of friendly neighbors with communal feather stripping was a way of doing useful work and having a good time together. Having stripped all the feathers in one house, the party moved to the next house. Although the men folk rarely participated in the stripping, they went along with their wives because they enjoyed the lively, entertaining conversation, as well as the cake the hostess baked for that occasion. The children of the house had to help with the work, which they, for once, did not mind, because they were eager to listen to the conversation of the adults. They were actually quite reluctant to obey, when sent to their beds as the conversation turned R-rated. If there was a gifted storyteller in the group, usually an older person who "had seen the world," as they said, he could spellbind the audience. Those who heard a ghost story told on a winter night in a cozy room, illuminated only by the flickering light of an oil lamp, with the wind howling outside and strange cracking noises coming from the ceiling as if somebody were walking in the attic, would not easily forget the experience.

With the Christmas season over, it was soon time for the Christians to prepare themselves for the next festive period, the Holy Week, culminating in Easter Sunday. As with Christmas, the church timed the Easter festivities to coincide with those celebrated by pre-Christian people. After the new sun festival, the next deeply rooted pre-Christian tradition was to hail the awakening of nature from the winter sleep. It coincided roughly with the spring equinox, the date (March 21 or 22), when the night was of equal length as the day (which is what the word "equinox" means). Rather than trying to eradicate the heathen custom, the church reinterpreted it and gave it a new meaning. Instead of celebrating the rising of new life in nature, it used the occasion to institute a feast of the resurrection, the rising of Jesus Christ from the dead. To do so and so fit events described in the New Testament and to some extent also in the Old Testament, into the astronomical cycle, it had to make Easter a movable feast. The Holy Week became a commemoration of the events that preceded immediately the death of Christ on the cross, which coincided with the Jewish Passover (Pesach). The latter celebrated God's rescue of Israelites from Egyptian bondage, as described in the Exodus, the second book of the Old Testament. To strengthen the links between the diverse traditions, the church drew parallels between the rescue of the Israelites and the rescue of Christians from eternal suffering in hell through the death and resurrection of Jesus

Christ. After a period of incertitude about the method to be used in determining the date of Easter every year, the church declared Easter Sunday to be the first Sunday after the full moon following the first day of spring, which in the Northern Hemisphere is the day of the spring equinox. Consequently, Easter Sunday could fall on any Sunday between March 22 and April 25, depending on the year. Because of the mobility of Easter Sunday, several other holidays preceding and following it became also movable feasts.

The Holy Week was a time of sorrow. To prepare themselves for it spiritually, Christians made the seven-week period preceding Easter Sunday, the Lent, the time of prayers, penitence, and fasting, starting on Ash Wednesday. But before prevailing on its faithful to feel sorry for their sins, the church gave them an opportunity to commit more transgressions against the Lord's Commandments, by sanctioning the carnival, which culminates on Shrove Sunday, Monday, and Tuesday before Ash Wednesday. Carnival (a word derived from *carne vale* or "meat adieu"), *Fasching*, *Fastnacht*, or *Fasnet* in German (all being terms derived from *fasten*, to fast) and *Masopust* (abstinence of meat) in Czech, the time of weddings, dancing, singing, drinking, clowning, and general merrymaking, echoed the ancient Roman Saturnalia and Lupercalia, the former orgiastic celebrations of Saturn, the god of agriculture, and the latter the fertility rites honoring Lupercus, the god of the shepherds. The highlight of the last three days of carnival was a procession of fools led by a bear. As generally no live animal was available, a man dressed in bearskin had to do. What Johann would, however, remember best of those days were probably the special Shrovetide donuts his mother made on that occasion. Yellow from all the eggs and butter that went into them, filled with marmalade, and hot oil dripping from them as they were served directly from the pan, they were one of the unhealthiest and most wonderful delicacies he had ever tasted. But on Ash Wednesday the party was over. The villagers, waking up with big heads, went straight to the church to ask God for forgiveness. The sign of the cross the priest made with ash on their foreheads, accompanied by the words "for dust thou art and unto dust shalt thou return," launched them into the dreary period of Lent.

Although Easter was a commemoration of events reported in the New Testament, remnants of pagan traditions remained alive throughout its entire period and also in the time preceding it. On the fifth Sunday of Lent, girls dressed a thatch, taken from a dilapidated roof, with rags of woman's clothing, pinned it on a pole, and carried it in a procession, as an effigy symbolizing death associated with the winter months, out of the village and to the river. There they threw it in the water to be carried away and then cut a young tree, which they brought into the village while singing that they took death from the community and brought in its stead new life. On the sixth Sunday of Lent, Christians celebrated Christ's entry into Jerusalem, described in the Gospel of Saint Matthew as a triumphant event during which people spread palm fronds and garments before him. In the church, the priests in Mendel's time commemorated the event by a special ceremony, in which, carrying imitations of palm leaves, they walked in a procession around the church and blessed sprigs of willow trees and catkins brought by the people. Back home, the people then placed some of the blessed twigs behind the icons on the wall and stuck others into the sprouting grainfield to protect their homes and fields from disasters.

During the entire Holy Week, the priests held services every day, each day in connection with a different ceremony reminding the believers of the last days of Christ. The most elaborate ceremonies were reserved for the second half of the week. On Maundy Thursday,[61] the occasion of the Last Supper and of Christ's arrest and imprisonment, the bishop washed the feet of 12 poor men, as Christ washed the feet of his 12 disciples. After the Gloria part of the mass, all the church bells rang and then fell silent (they flew to Rome, the believers were told), not to be heard again until nearly two days later. During those two days, boys in each village gathered at six o'clock in the morning, at noon, and then again at six o'clock in the evening to walk in a procession through the village, stopping to say prayers at the wayside crosses and the chapel or church. During the walk, they made rhythmical clip-clip-clap-clap-clap sounds with wooden, homemade instruments, the clappers, usually one in each hand. Each instrument consisted of a rectangular platform pierced through by a handle to which was attached, on the other side of the platform, a swiveling hammer. By jerking their hands rhythmically, all simultaneously, they produced a wooden sound that carried through the village and alerted people to the time of the day and to the occasion.

On Good Friday, according to the New Testament, Christ died on the cross at three o'clock in the afternoon. The account that several unusual phenomena accompanied his death probably gave rise to the belief that when the Passion, the description of the agony and suffering of Christ prior and during the crucifixion, is read or sang in the church, the earth opens at places to reveal treasures buried in it. Mendel's contemporary, the Czech poet Karel Jaromír Erben, captured beautifully the spirit of Good Friday in a legend based on this belief. The muted atmosphere of the day, the mournful music performed at the church, and the invocation of the last hours of Christ in word, music, and display in the church must have affected Johann deeply.

The mood changed gradually during Holy Saturday. Although it commemorated Christ's burial, an anticipation of his resurrection pervaded already the day. At noon the clip-clap boys made their last round through the village, and in the evening, during the service, the bells began to toll again, a sign that the climax of the Holy Week was near. On Easter Sunday, the day on which, according to the Scriptures, Christ rose from the dead, the primeval urge to celebrate the return of the spring came to the fore in a mix of Christian and pagan customs. In ancient Judea, the Jews observed Passover by sacrificing a lamb and feasting on it. Christians took over the tradition and made the lamb a symbol of Christ sacrificed to save humanity. In Mendel's time, however, most Christians no longer feasted on a real lamb on Easter Sunday, but ate instead a cake baked in the form of a lamb. A Christian tradition was to decorate the church and sometimes also the homes with Easter lilies, which were viewed, presumably because of their large, pure white blossoms, as a symbol of pure new life beginning through the resurrection of Christ. The heathen symbols included colored eggs, rabbits, switches, and pure water. An egg was a symbol of spring, of new life, and of a hidden life force readying itself to burst into the open.

Coloring the eggs with bright colors symbolized the sunlight of the spring. The Mendel children probably colored hard-boiled eggs by simply dipping them into colored solutions such as those made by boiling onionskins. But in some Moravian homes, gifted women and men developed egg coloring into a folk art. In some families, the parents told the children that a rabbit would bring the eggs and hide them at different places in the orchard. The children then had great fun searching for the eggs. Rabbits, one of the fastest multiplying mammals, were brought into the picture as a symbol of fertility and so also of spring.

An Easter switch could be a single, freshly cut willow sprig or a bunch of sprigs artfully entwined into a flexible, braided rod with bows attached to its tampered end. The art of entwining the sprigs into a rod passed from one generation of boys to the next, but not every boy could master it. On Easter Sunday, boys armed with the braided rods went from house to house, whipping gently (or not so gently depending on the boy's disposition) all the female members of their household, while singing the prescribed song, asking for a donation of eggs and a piece of a cake. The whipping supposedly transferred some of the vigor of the new life contained in the sprigs onto the lashed person and so made the girls more energetic, healthier, and merrier and rejuvenated the mature women. The ancient Slavs attributed similar power to springwater and from this superstition probably arose a tradition of boys trying to surprise girls in their sleep on the morning of Easter Monday by splashing cold water on them. On Tuesday, the girls were supposed to return the favor, although they rarely did. But on Easter Monday, there was a lot of chasing, squealing, shrieking, and giggling in the early morning in the village.

After Easter the spring celebrations continued on May Day. Once quite extensive, in Mendel's time the May Day festivities were already reduced to the erection of the maypole in the village square. The pole consisted of a long shaft, at the end of which was attached a small tree decorated with multicolored ribbons. In some villages, the youth used to dance in circles round the pole. In Silesia a boy in love with a girl would place a May Day tree secretly on the roof of the house in which she lived on the night before May Day. This night, known as Walpurgis Night, was also an occasion for the boys to carry out many pranks, some of which could be quite ingenious and these would then become part of the village lore for many years. Thus, a farmer might wake up on May Day morning to realize that a wagon was missing from his farmyard. Thinking that the rascals had pulled it to one of the neighbors' yards, he would go looking for it. But at the gate he would encounter a group of villagers staring at the roof of his house and laughing. Looking up he would get the shock of his life: There his wagon was, straddling the crest of the roof! At night, the boys took it apart and piece by piece hauled it on the roof and assembled it there, while all the inhabitants of the house were sleeping soundly. At another house, the boys would capture all the sleeping hens and force them through the single small opening of a cistern. A great racket coming from the cistern would then bring the farmer out of the house in the morning. Realizing what happened, he would face a big problem: How to get the chicken out? The affected farmers would get mad, but that was just about all they could do, because the perpetrators were unknown and anyway, on Walpurgis Night, such pranks would have to go unpunished. In pre-Christian times, May Day Eve was also an occasion to light great bonfires and to celebrate the advent of a new season around them. But the church Christianized this custom also and moved it to June 24, the feast day of Saint John the Baptist.

For the months of May and June, the church also instituted a series of movable religious holidays, the dates of which depended on that year's date of Easter. Forty days after Easter Sunday came the Ascension Day marking the end of Christ's earthly existence and celebrating his ascension to heaven. Ten days later, or 50 days after Easter Sunday, followed the Pentecost or Whitsuntide, the former name (preferred by the Roman Catholics) deriving from the Greek word for 50 and the latter name (espoused by certain other churches) referring to the white garments worn by the children baptized on that day. The two days, Whitsunday and Whitmonday, commemorated the descent of the Holy Spirit upon Christ's disciples gathered to observe Jewish thanksgiving at the end of the wheat harvest. The presence of the Holy Spirit empowered the apostles to preach the Gospel of Christ. The actual roots of both the Jewish and Christian Pentecost festivities reach deep into the prehistory of humankind, when people celebrated the victory of the sun god. Right after Whitsunday came Trinity Sunday, a celebration of the mystery of single God in three versions. And the Thursday after Trinity Sunday was Corpus Christi honoring the presence of Christ in the host, the bread taken at the Communion. On that day the priests, under a baldachin, carried the host, enshrined in a gold receptacle, the monstrance, in an opulent procession, in which everybody who was somebody in the village participated, while the "nobodies" stood on the sides. The route was decorated with young birch trees and flowers. Beautifully dressed young girls marched before the priests and cast flowers from their baskets on the road. The procession stopped for prayers at makeshift altars placed facing each one of the four corners of the earth and then returned to the church for a solemn *Te Deum* (Thee, O God, we praise) service.

Corpus Christi was the last major feast of the liturgical calendar. By then it was harvest time, the busiest time of the farmer's year, a period during which all festivities had to be suspended to get the crop under the roof safely and as quickly as possible. Soon after the harvest home, the days began to shorten, the sun peaked lower and lower in the sky with each succeeding day, flocks of swallows and starlings could be seen preparing for their long flight south, and before long it was All Saints Day again. The liturgical cycle punctuating life in the village with feasts went into a new round. For the Mendel children, each of the feasts meant a reprieve from the tedium of hard work on the farm. The festivities were a rich source of memorable impressions and valuable experiences, and the holidays and Sundays were highly treasured stretches of time for play, among themselves at home or with other children in the village, at the stream, in the woods, or other places they longed to explore.

The Birth of a Naturalist

The description of the various religious rites in the preceding section might have left the impression that Johann Mendel was a particularly pious boy. In reality, however, there are no grounds for believing that this was the case. His family was religious but apparently not overzealously so. If Johann had felt predetermined to become a priest or if he had been an altar boy, surely the Schindler brothers would have heard about it from their parents and mentioned it, but they didn't. We can assume, therefore, that Johann was like most other village boys in his free-time activities. Like the other boys, he made exploratory expeditions into the thickets, copses, and hedgerow in the region; climbed trees to fetch twigs of mistletoe or young, red spruce cones; spied on birds and sometimes raided their nests and hunted sparrows and other birds with a sling; scoured the streams for fish and crayfish; and in the winter tracked hares, pheasants, and foxes. Like other boys, he could judge when willow branches got enough sap in the spring to make whistles out of them; he knew how to forecast weather from the flight of swallows, the croaking of the green tree frog, the biting intensity of gadflies, the distance over which the air carried the mooing sound of the cattle, or the time the dew in the meadows required to evaporate. He knew where to search for edible mushrooms and where to find the sweetest wild strawberries, the largest blueberries, and the juiciest raspberries. He could tell a house martin from a swallow and recognize a hawk or peregrine high in the sky from its silhouette.

All of these activities, observations, and experiences brought Johann Mendel into contact with nature, which was the first requirement for the development of a naturalist. But all the other boys had the same opportunities for contact, yet not one of them turned into a naturalist. Therefore, there must have been an additional factor involved in the conversion of a mere contact into a life-long infatuation. This other factor was what interested Thomas Makita and Johann Schreiber in Johann Mendel. It is not too difficult to imagine what might have happened in the Hynčice school. Both the teacher and the priest discerned a gifted child in the mass of souls struggling to master the multiplication table. They began to nurture the talent and the child responded positively to their attention. In the eyes of a peasant child of that period, the priest and the teacher were the most respected and revered personages, second only to God and the child's parents. The intensity of the veneration took a quantum leap when the child became aware of having attained the status of their favored pupil. Noting that both the teacher and the priest were fond of trees, plants, and animals, cultivated and wild, the child's interests, too, were steered in that direction. He began asking questions and brought plants and other natural objects for identification. The child's interest pleased his instructors, and so they paid even more attention to him. In the end, this mutual psychological amplification of the bond between the instructors and the pupil led to the fixation of the child's interests: The foundation for the development of a future naturalist was laid.

A Momentous Decision

Recognizing Johann's intelligence and eagerness to learn, his teachers thought he should be given the opportunity to receive a higher education. But to be sure that they were not misjudging him, they wanted him to transfer for the final, sixth year of his compulsory education to a school in Lipník, a town some 20 kilometers south from Hynčice. The school they had in mind was specifically set up for gifted children of underprivileged parents and consequently was more competitive than the ordinary village school. They reasoned that if Johann would do well there, they could recommend with good conscience his transfer to a *Gymnasium*. It is somewhat strange, though, that their presumably best pupil went to Lipník with a report card on which he had "very good" notes in two subjects, but only "good" in two others. One might think that a teacher would only recommend a pupil who had "very good" in all subjects. Alois Schindler,[27] as noted earlier, suggested that the teachers were covering their backs just in case that Johann would not do so well in Lipník. If so, the incident would not speak well of the teachers: If the boy was outstanding in all subjects, they should have given him the corresponding grades; if they were not certain, they should not have raised his hopes for further study.

Before Johann, Makitta and Schreiber sent at least two other pupils to the Lipník school, Winkler from house number 51 in Hynčice and Krist from Horní Vražné. Johann knew both of them well and listened wide-eyed, when they came home for vacations, to their enthusiastic accounts of the school and of life in Lipník. The teachers therefore did not have to persuade him: He was ready for the test year in the town school. The teachers realized, however, that the boy alone would not be able to convince his parents to allow him to study there; they would have to help him. Who actually talked to the parents is uncertain. Most biographies claim it was Makitta, which seems to make sense since Makitta was Johann's main teacher, whereas Schreiber taught only religion and extracurricular practical horticulture. Also, since the information apparently stemmed originally from Alois Schindler who must have had it from his mother, it seemed reliable. Nevertheless, doubts have been raised about this assertion;[46] indeed, several arguments can be raised in favor of Schreiber's part in making the momentous decision. Schreiber, as the spiritual leader of the community, was better acquainted with the individual families in his parish than Makitta. With his Kunín experience, he was also the person with wider contacts, including contacts to the Lipník school. He could, therefore, lay the case better before the parents than Makitta. Also, Schreiber, as the main person behind the drive to teach the Hynčice children elementary natural history, may have been particularly fond of Johann because of his interest in this subject, and the feeling seems to have been reciprocated by the pupil. The fact that later, in his letters home, Gregor Mendel added frequently greetings to *Herr Pfarrer*,[28] but never to Makitta, has been interpreted as a sign of Gregor's gratitude to the former for his help in convincing the parents.[46] But there is another explanation for his omission: All these letters date from a period when Makitta was already dead.

Schreiber may have proposed to Johann's parents: Send him for one year to Lipník; if he does not do well, take him home and make a farmer out of him, and nothing will be lost, because the school will not cost you anything. If, on the other hand, he excels there, too, it would be a sin to waste your son's God-given gift. Undoubtedly, the parents wanted to know what Johann would become if he did study, and Schreiber would have told them that, realistically, there were only two possibilities: Johann could become a priest like Schreiber himself or a teacher like Makitta. Biographers generally assume that Johann's mother was easily swayed by Schreiber's arguments, for all she wanted was a better life for her son no matter what. Anton Mendel needed more time to mull over the proposition. Like nearly all farmers, he was deeply attached to the land on which he toiled. He felt that since he inherited it from a Mendel, he should pass it on to a Mendel again. He had to take into account the possibility that since Johann was his only son and since there was a good chance that both his daughters might marry husbands who would inherit farms from their fathers, the Mendel farmstead might end up in the hands of a stranger. If that happened, he and his wife would have to move out, and the question would be: where would they go? (As it turned out, Viktoria Mendel married a man, Alois Sturm, who was prepared to take over the farmstead No. 58, and so, although the Mendel possession of the farm ended, at least the new owner was not a total stranger. It seems, however, that Anton and Rosina moved out of the farm anyway.) On the other hand, since there had been no sign that the *robota* would be abolished any time soon, the prospect that at least his son would escape the injustice of the peasant's lot was just as appealing to Johann's father as it was to his mother. For Anton Mendel it was a difficult decision, but in the end he too consented. And so, in the fall of 1833, the 11-year-old Johann said goodbye to his family, his home, and his childhood, since a new phase of his life was about to begin, a phase in which he would be forced by circumstances to gradually take more and more responsibility for himself. He packed his meager belongings and with tears in his eyes mounted the horse-drawn carriage in which his father was going to take him to Lipník.

Inevitability and Serendipity: Part 1

A favorite question historians and biographers ask is "What if…?" What if Grouchy, instead of pursuing a detachment of Prussian corps, had turned west, as his general urged him, and hasted to join Napoleon against Wellington at Waterloo? What if father Einstein had not given his four-year-old son the compass that supposedly arose Albert's interest in science? What if the student Charles Darwin had not met John Stevens Henslow and the latter had recommended somebody else for the position of a naturalist on the HMS Beagle? As for Mendel, the little that is known for a fact about his childhood provides many occasions to ask the what-if question. Instead of wasting a page on idle speculations, however, we will simply highlight six major occasions in which the course of his life had been determined by chance events and would have enfolded differently, had the events taken a different form.

First, emigrants from a place somewhere in Germany settled at the Silesian-Moravian border and by intermarrying with the local Czech population created a gene pool with the potential of assembling a combination necessary for the birth of a genius. Second, the relatively frequent intermarriages within the local group increased the chances of the right genes coming together in this unique combination. Third, Johann's birth into a farmer family brought him in close contact with nature, a condition necessary for stimulating his interest in natural sciences; taught him to work hard, a characteristic he would need during his studies; and instilled in him perseverance, without which he could not keep on going later, during his work on an eight-year-long experiment, but above all it infused in him the ability of parsimonious, commonsensical reasoning that would allow him to view things differently than all his learned predecessors and contemporaries had. Fourth, Johann's childhood among simple village people, most of them poor in terms of material means but rich in customs and traditions, contributed to his becoming a humble person, respectful of people's faith, sensitive to their suffering, and compassionate with their fates. All these attributes may have facilitated his decision to become a priest. And, as we shall elaborate later, without becoming a priest, almost certainly Mendel would not have had the opportunity to carry out the experiments that made him the founder of modern genetics. The experience of village life also provided him with a rich store of memories onto which he could fall back subsequently and bear life's adversities easier than he might have otherwise. Fifth, it was Johann's lucky turn of events to grow up in a pocket of enlightenment existing in a country darkened by the cloud of absolutism. Growing out of the tradition founded by Jan Amos Komenský, in Mendel's time the pocket had been maintained by the Countess Maria Walburga Truchsess-Zeil and her emissaries. Of the latter, Schreiber and Makitta had the greatest influence on the direction Johann's life would take. But the chain of serendipities reaches deeper into the past. Had Anton Schwirtlich, and this is the sixth serendipity, not whetted the appetite of the inhabitants of Hynčice for indigenous education and spurned their desire to have their own school, Johann Mendel would have attended the school in Dolní Vražné, where the teacher probably would not have noticed his talent or, even if he would have, done nothing to promote it. Similarly, if Hynčice had not the fortune of having Makitta assigned to its school and Schreiber to its parish, almost certainly Johann Mendel would have ultimately taken over his father's farm and posterity would have never heard of him.

Postscript: Mendel's Nationality

Had Johann Mendel taken over the farm, as his father expected him to do, no nation would have cared about claiming him as one of its own. But since he became a figure known worldwide, three or four nations now compete for the honor of claiming him as their famous representative: the Austrians, the Germans, and the Czechs. Before we consider the grounds for their claims, let us specify what we

Fig. 3.21 Commissar: *Tell me, Herr Mendel, what are you actually, Austrian, German, or even Czech?* Mendel: *Silesian, Herr Commissar*

mean by "nationality." The dictionary[18] gives several definitions of this term, of which the one pertinent to our considerations is this: "the status of belonging to a particular nation by birth or naturalization." And it defines a "nation" as "a stable, historically developed community of people with a territory, economic life, distinctive culture, and language in common." By this definition, the Austrian Empire was a conglomerate of multiple nations, only one of which remained Austrian after the disintegration of the empire in 1919. Since Johann Mendel was neither born in Austria itself nor was he a naturalized member of that nation, he does not qualify for Austrian nationality. For the same reasons, his nationality is not German. What is now considered to be the German nation came into existence only in 1871, and it did not include the part of Europe in which Mendel was born. In fact, since the middle of the eighteenth century, the Austrian Empire was an enemy of Prussia, under which leadership the German nation ultimately came into being. This leaves us with the Czechs. Mendel was born in one of the hereditary lands of the Czech crown, which were then included in the Austrian Empire but which even the Habsburgs continued to recognize as separate nations, at least formally. The land in which Mendel was born was Silesia. Politically, therefore, Johann Mendel was of Silesian nationality (Fig. 3.21), and he was identified as such in official documents.

The situation is complicated, however, by the fact that most of Mendel's ancestors spoke German rather than Czech, as did the rest of the Silesian population (and some of his ancestors, as well). This circumstance led Alois Schindler to the conclusion that the ancestors were immigrants from Germanic states. Even if this supposition were true (and there is no hard evidence that it is), it would still not make Mendel a German national, either politically or genetically, for all the reasons stated earlier in this chapter. And anyway, this dispute about Mendel's nationality has little relevance to his standing. He is no longer a figure whose legacy pertains to one nation only. His contribution has long since become a treasured possession of all humankind.

References and Notes

Abbreviation: FM *Folia Mendeliana Supplementum ad Acta musei Moraviae*. Moravské zemské muzeum Brno; published since 1966

[1]"Happy, happy days of youth which can never be recalled! How is it possible not to love it, to cherish memories of it? Those memories refresh and elevate my soul, and serve me as the fountain of my best enjoyment." Leo Tolstoy: Childhood, boyhood and youth. Translated by Isabel Hapgood, Lear Publishers, New York 1949

[2]John Locke (1632–1704) in his *Essay Concerning Human Understanding* argued against the concept of innate ideas and compared instead the mind of a newborn child to *tabula rasa*, the clean slate. He coined the phrase *Nihil est in intellectu nisi quod prius fuerit in* sensu. (There is nothing in the mind what was not first in the senses.) His inspiration was the French philosopher Pierre Gassendi (1592–1655)

[3]Dvorský F (1907) O starožitném panském rodě Benešoviců. Musejní spolek, Brno

[4]There used to be another village called *Kravařov* near Opava, on the left bank of the Opava River and on the main road connecting Opava with Hlučín. Its existence is first mentioned in documents from 1569; See[14]. It is now part of Suché Lazce

[5]Váša P, Trávníček F (1946) Slovník jazyka českého, 3rd edn. Fr Borový, Praha

[6](a) Kuhn W (1954) Siedlungsgeschichte Oberschlesiens. Oberschlesischer Heimatverlag, Würzburg 1954 (German interpretation of Silesian history.) (b) Macůrek J. Slezsko a jeho význam v českém státě 14.–18. století. (c) Čáda F. Právní začlenění Slezska do zemí českých. These two articles (b and c), which give the Czech interpretation of the events, were published in *Slezsko, český stat a česká kultura*, pp. 42–64 and 65–74, respectively, Matice opavská, Opava 1946

[7]German interpretation: Matzura J. Das Kuhländchen, seine Grenzen und Gröse. Unser Kuhländchen 19: 365–373, 1913 (reprinted in[28], pp. 160–168). For Czech interpretations, see Kravařsko, vlastivědný sborník východní Moravy, vol. (1932–1938), published in Nový Jičín

[8]Iltis H (1924) Gregor Johann Mendel. Leben, Werk und Wirkung. Julius Springer, Berlin, An English translation by E. and C. Paul was published under the title *Life of Mendel* by George Allen & Unwin, London 1932

[9](a) Die österrechisch-ungarische Monarchie in Wort und Bild: Mähren und Schlesien, Vol. 17. Eduard Beyer, Wien, 1897. (b) Scholz FE (ed.) (1998) Kuhländchen—unvergessene Heimat. Verlag Gerhard Rautenberd, Leer. Germany

[10]Květ R (1996/1997) Mendel's birthplace Hynčice from the old trails and historical borders of Moravia and Silesia. FM 31/32: 53–57

[11]Wright WE (1966) Serf, Seigneur, and Sovereign. Agrarian Reform in Eighteenth century Bohemia. University of Minnesota Press, Minneapolis, MN

[12]In this count are included small but independent units, which may not qualify for being called domains. They may be classified as estates or farmsteads (*statky* in Czech)

[13]Rolleder R (1903) Geschichte der Stadt und des Gerichtsbezirkes Odrau. Private print of the author. Steyr

[14](a) Hosák L (1967) Historický místopis Moravy a Slezska v letech 1848–1960. Profil, Ostrava. (b) Černý F, Váša P (1907) Moravská jména místní. (Výklady filologické). Matice moravská, Brno. (c) Hosák L, Šrámek R (1970, 1980) Místní jména na Moravě a ve Slezsku. Vol. I and II. Academia, Praha. (d) Schwarz E (1961) Die Ortsnamen der Sudetenländer als Geschichtsquelle. 2nd edn. Verlag R. Lerche, München

[15]Ottův slovník naučný (1888–1909) Illustrovaná encyklopaedie obecných vědomostí. Vol. 1–28. Jan Otto, Praha

[16]The Czechs (and Germans) distinguish between *hrad* (*Burg*) and *zámek* (*Schloss*), the former being fortified with thick walls and the latter usually not. Both terms are often translated into English as "castle," but this is confusing. Here we reserve the word "castle" for *hrad/Burg* and translate *zámek/Schloss* as "manor house." In Odry, there was both a castle and a manor house. See also http://moravskekravarsko.cz/encyklopedie/objekty1.phtml?id=135332&menu=4

[17]Brunner O (1965) Land und Herrschaft: Grundfragen der territorialen Verfassungsgeschichte Österreichs im Mittelalter. R.M. Rohrer, Wien

[18]Merriam–Webster's Collegiate Dictionary (1996) 10th edn. Merriam—Webster, Springfield, MA

[19]Brandl V (1876) Glossarium illustrans bohemico-moravicae historiae fontes. C. Winker, Wien

[20]Lutterer I, Majtán M, Šrámek R (1982) Zeměpisná jména Československa. Slovník vybraných jmen s výkladem jejich původu a historického vývoje. Mladá Fronta, Praha

[21](a) Schindler A (1911) Gregor Johann Mendel und seine Ahnen. Unser Kuhländchen 1; 347–352. (b) Schindler A (1914) Die Ahnen Gregor Johann Mendels. Unser Kuhländchein 3; 37–49. (c) Schindler A (1922) Die Ahnen des Naturforschers Gregor Johann Mendel. Das Kuhländchen 4; 89–91

[22]Jaschke F (1913) Beschreibung ganz sicherer und unzweifelhafter oder im engeren Verstande gehörenden Ortschaften zu dem sogennanten Kuhländel nach der Meinung des mährischen Wanderes. Manuscript from the year 1817 cited by Josef Matzura in Das Kuhländchen, seine Grenzen und Grösse. Unser Kuhländchen 19:365–373. It must be pointed out, however, that Felix Georg Jaschke (1756–1831) is now regarded as not an entirely reliable source of information (see Jan Hanák, in *Moravské Kravařsko*. Region Moravské Kravařsko, Nový Jičín, 1998)

[23]Alois Schindler was born on July 22, 1859, in Hynčice to Theresia, Johann Mendel's younger sister, and her husband Leopold Schindler. Alois was the couple's second of three sons, the first being Johann and the third Ferdinand. All three sons studied with G. J. Mendel's support in Brno and in Vienna. Alois and Ferdinand became practicing physicians in Zlaté Hory and Butovice, respectively; Johann worked as an assistant at the *Technische Hochschule* in Brno, but he died early of tuberculosis. Alois Schindler died suddenly on June 1, 1930, on a train to Olomouc. Of the three, he was the closest to Mendel's heart

[24]A. Schindler's letter of November 28, 1928 to P.A. Matoušek; reprinted in[28]

[25]Schindler A. Ahnentafeln bekannter Sudetendeutscher. 4. Ahnentafel: Prälat Gregor Johann Mendel, Naturforscher, Entdecker der Vererbungsgesetze. *Sudetendeutsche Familienforschung* 1; 186–187, 1928, and 2: 76–80, 1929–1930

[26]O'hUigin C, Mrhačová E, Klein N, Klein J (2012) Mendel's ancestors: The Schindler's list. FM 48(1):5–23

[27]Simunek MH, Thümler U, Sekerák J (eds.) (2011) The Letters on G. J. Mendel. Correspondence of William Bateson, Hugo Iltis, and Erich von Tschermak-Seysenegg with Alois and Ferdinand Schindler. Studies in the History of Sciences and Humanities vol. 28. Prague 2011

[28]Kříženecký J (1965) Gregor Johann Mendel 1822–1884. Texte und Quellen zu seinem Wirken und Leben. J.Ambrosius Barth, Leipzig

[29]Blickle P (1985) Der deutsche Bauernkrieg von 1525. Wissenschaftliche Buchgeselschaft, Darmstadt

[30]Collison P (2004) The reformation: a history. Modern Library, New York, NY

[31]Lenz F (1922) Gregor Mendel. Münchener medizinische Wochenschrift 69:1349–1350

[32]The Holy Bible Containing the Old and the New Testament. Authorized King James Version. Camex International, New York, NY 1989

[33]Bateson W (1909) Biographical notice of Mendel. In: Bateson W (ed) Mendel's principles of heredity. A defence, 2nd edn. Cambridge University Press, Cambridge, pp 319–316

[34](a) Svoboda J (1964) Staročeská osobní jména a naše příjmení. Nakladatelství Československé akademie věd, Praha. (b) Kotík A (1897) Naše příjmení. Praha. (c) Moldanová D (2004) Naše příjmení. Mladá fronta, Praha. (d) Beneš J (1998) Německá příjmení u Čechů. Vols. 1 and 2. University of JE. Purkyně, Ústí nad Labem

[35]Vašek AE (1922) Kolem pana Řehoře Mendela. Moravské noviny 43(163)

[36]Schindler A (1922) Anton Schwirtlich, der erste Lehrer von Heinzendorf. Das Kuhländchen 4:137–139

[37]When the Wittelsbach lineage died out in 1777, Joseph II, the coruler with Maria Theresia, tried to annex Bavaria. This move, however, led to a new war, a fourth one in a series, between Prussia and the Habsburgs. Since the armies did not fight much, but consumed all the potatoes in Silesia and Moravia, the encounter came to be known as the *Potato War*

[38]Hora-Hořejš P (1998) Toulky českou minulostí. Vol.7 Od konce napoleonských válek do vzniku Rakouska-Uherska (1815–1867). Via Facti, Praha

[39]The silver-based unit was called a *Thaler* in German, an abbreviation from *Joachimsthaler* deriving from *Joachimsthal*, the German name for the Czech *Jáchymov*, and *Thal*, a valley, hence Joachim's Valley in which silver was mined and coins minted since medieval times. The Czech name of the Thaler was *tolar,* and one of the many derivatives of this designation became the US dollar. In Mendel's time, the Thaler was commonly referred to as *Conventionsthaler* (alternatively spelled with "K" instead of "C"). Although gold was generally valued more than silver, the gold-based monetary unit had a lower denomination than the silver-based one, one Thaler equating to two guldens. Correspondingly, one conventional Thaler had the worth of 120 or 200 *Kreuzer*. Banknotes ("paper money") or *Bancozettel*, as the Austrians called them, were introduced in 1762 and were denominated in guldens only. For the coins, the term Florin was preferred over the gulden. Prior to 1762 in the Habsburg Empire, one understood under "money" coins containing certain amounts of silver or gold. Since there were many states in Europe and all of their rulers insisted on having their own currency, a great variety of coins were in circulation, so that converting one into another was a challenge for both merchants and customers. In 1750, Maria Theresia decided to overhaul the monetary system in her empire, taking into account the situations in the neighboring states as well. Since the starting point of her reform was an agreement reached at a convention between Austria and Bavaria, the new monetary system came to be known as *conventional.* At this convention, the relation of gold-to-silver values was established at the then current international prices of the two metals at the ratio 1:14. On this foundation, two principal currency units were introduced, one based on gold and the other on silver. The gold-based unit was called *Gulden* in German or *florin* in Latin (both these terms being shortenings from "golden florin," coin originally used in medieval Florence and imprinted with a figure of a lily), *forint* in Hungarian, and *zlatý* (meaning "golden") in Czech. As an illustration of Gulden's worth in Mendel's time, here are the salaries in Guldens per year in professions at three different levels: higher office holder 500–700, teacher 130, and worker 100–200. The *Gulden* was divisible into copper subunits. A single copper coin was called *Kreuzer* in German (so named because the coin bore a figure of a cross, or *Kreuz*), *krajczár* in Hungarian, and *krejcar* in Czech. Originally one *Gulden* was divided into 60 Kreuzer, but after the decimalization of the currency in 1857, one Gulden became 100 Kreuzer. One Gulden amounted 60 Kreuzer, one Kreuzer to four Pfennig, and one Pfennig to two Heller. Four and half Gulden amounted to one gold *Ducat*. The Gulden was retained as the currency of the Austro-Hungarian Empire until 1892, when the *Krone* in German, *korone* in Hungarian, and *koruna* in Czech replaced it. The *Krone* bore a figure of a crown, hence the names in all three languages. The exchange rate was two crowns for one gulden

[40]Grab AI (2003) Napoleon and the transformation of Europe. Palgrave Macmillan, New York, NY

[41]The first name of G. J. Mendel's mother is commonly spelled as "Rosine." This spelling probably derives from the printed version of Schindler's List[21]. Iltis[8], however, spells the name as Rosina, with an "a" at the end. We think this is the correct spelling, for two reasons. First, in the parish register, in the entries we had an opportunity to check, the spelling is consistently with an "a" at the end. The second reason is etymological. In German, "Rosine" means a "raisin"; now, which mother would want to call her newborn daughter "Raisin"?! To call the daughter "Rose" is of course a different matter, and indeed "Rosina" may be a corruption of the Czech *Růžena* or *Růženka*, which clearly derives from *růže* or rose. This interpretation is in keeping with the Slavic origin of G.J. Mendel's mother

[42]Letter of August 23, 1852; see[28]

[43]Schindler A (1902) Gedenkrede auf Prälat Gregor Joh. Mendel anlässlich der Gedenkenthülung in Heinzendorf am 20. Juli 1902. Private print of the author, which did not get into the bookstores. It is reprinted in[28], pp. 77–100; for an English translation, see[64]

[44]Doležel L (1972) Jan Amos Komenský, 2nd edn. Melantrich, Praha

[45]Zezulčík J (1995) Marie Walburga hraběnka Truchsess-Zeil. Významná žena moravského osvícenství. In: J. Zezulčík: *Zámek Kunín—perla moravské architektury*. Okresní vlastivědné muzeum v Novém Jičíně. Nový Jičín

[46]Marvanová L (1971) First impulse to Mendel's scientific education. FM 6:31–40

[47]Orel V (1996) Gregor Mendel. The first geneticist. Oxford University Press, Oxford, Translated by Stephen Finn

[48]Orel V, Vávra M (1979) Pedagogue Johann Andreas Edmond Schreiber (1769–1850) evoked in Gregor Mendel first interest in natural science. FM 14:243–250

[49]Nippert V. Personal communication

[50]This description of the house is based on the few items preserved in it. The description of the farmer's day and year as well as of the year in a village is put together from a variety of sources and from the senior author's own childhood experiences on a farm not far from Hynčice

[51]Ulrich J (1907) Gregor Joh. Mendel. Biografische Skizze. Rainer Hosch, Neutitschein, Reprint from the Illustrierter Neutitscheiner Volkskalender für das Jahr 1908

[52]Forbes E (ed) (1967) Thayer's life of Beethoven. Princeton University Press, Princeton, NJ

[53]The description and interpretation of the various Catholic ceremonies and rites is based on two main sources[15,62]

[54]In a letter addressed to his parents and dated March 24, 1853 (see[28], p.114), Gregor Mendel acknowledges the receipt of a *Glückwunsch zu meinem Namensfeste* (good wishes to my name day), which he got on March 12

[55]Van der Pas PW (1972) The date of Gregor Mendel's birth. FM 7:7–12

[56]The New Testament, Luke 8:2 in[32]

[57]Okey R (2001) The Habsburg Monarchy. From Enlightenment to Eclipse. St. Martin Press, New York

[58]The New Testament Matthew 26:26 in[32]

[59]The New Testament Luke 22:19 in[32]

[60]Klein J (2000) Johann Mendel's field of dreams. Genetics 156(1):1–6

[61]The Czechs call Maundy Thursday *Zelený čtvrtek* and the Germans *der grüne. Donnerstag*, but also *Mendeltag*, presumably because it was a day on which *Mandelbrot* (almond bread) was distributed or because *menden* in middle German means to rejoice (see[19])

[62]New Catholic Encyclopedia. 15 volumes, 2nd edn. Thomson–Dale, Detroit 2003

[63]Richter O (1943) Johann Gregor Mendel wie er wirklich war. Verhandlungen des Naturforschendes Vereines, Brünn 74(2):1–262

[64]Olby RC (1985) Origin of Mendelism. Constable, London 1966, 2nd edn. University of Chicago Press, London

Johann Mendel's *Lehrjahre*: The Second Decade

4

...alles was uns begegnet lässt Spuren zurück,
alles trägt unmerklich zu unserer Bildung bei...
Johann Wolfgang Goethe:
Wilhelm Meisters Lehrjahre[1]

Long, very long ago, dense pine forests covered much of northern Europe. When the trees wept, their resinous tears petrified into amber. Later, when the Baltic Sea claimed some of the land, the forests vanished, but the petrified tears remained, the tides washing them up on the beach every now and then. People, who took them for precious stones with magical powers, collected them and made jewelry and talismans out of them. The tears, the amber, became a much sought-after commodity all over Europe, and enterprising merchants sent caravans of traders to the shores of the Baltic Sea to procure it in great quantities. The caravans traveled along established *amber routes* crossing Europe from south to north, from the Adrian Sea to the Baltic and back. One of the main amber routes took the travelers along the Odra River valley and across the Moravian Gate to the Morava River valley.[2]

It was a short segment of the amber route that Anton Mendel traveled with his son in the late summer of 1833 on their way to Lipník (Fig. 4.1). From their house, they drove through Hynčice to get to the main road connecting Odry with Hranice near the place called Železná brána (Iron Gate). It used to be an ancient tollgate, which separated two domains (Odry and Hranice) as well as two countries (Moravia and Silesia). Heading south on the main road, they soon entered Bělotín, a village that already then stretched out for more than two kilometers along the way. Exiting Bělotín, they crossed the Luha River and entered a narrow European watershed divide of two major river systems. The Luha flowed east and then turned north to join the Odra River, which took its waters to the Baltic Sea. A few kilometers down the road, they could see the Bečva River flowing in the opposite direction, first west and then south, to join the Morava River, which via the Danube sent its waters to the Black Sea. By that time they were already in Hranice, a town, which in the past was also called Bělokostelí or, to distinguish it from a town in Slovakia, Moravské Bělokostelí (*Alba Ecclesia* in Latin and *Mährisch-Weisskirchen* in German). The name Hranice derives from the town's location at the entrance into an ancient forest

J. Klein and N. Klein, *Solitude of a Humble Genius - Gregor Johann Mendel: Volume 1*,
DOI 10.1007/978-3-642-35254-6_4, © Springer-Verlag Berlin Heidelberg 2013

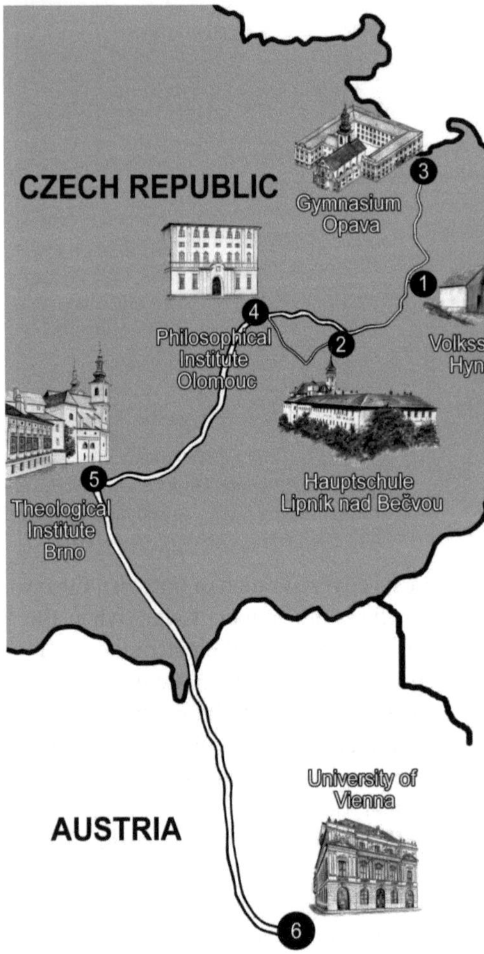

Fig. 4.1 The six stations on Johann Mendel's "amber" journey after education. The *first station*: the *Volksschule* (elementary school) in Hynčice, 1828–1833, five years. The *second station*: the Piarist *Hauptschule* in Lipník nad Bečvou, 1833–1834, one year. The *third station*: the *Gymnasium* in Opava, 1834–1840, six years. The *fourth station*: the Philosophical Institute in Olomouc, 1840–1843, two years + one year repetition. The *fifth station*: the Theological Institute in Brno, 1843/1844–1848, four years. The *sixth station*: the University of Vienna, 1851–1853, two years

that once covered the frontier region between Moravia and Silesia.[3] The origin of the name Bělokostelí, which means "White Church (Town)," is uncertain: It may have referred either to the circumstance that it belonged to the Premonstratensian monks who wore white robes or to the white stone of the walls in one of the churches.

Having Hranice and the village Drahotuše behind them, they could see in the distance on their left-hand side the towers and the roofs of a castle peeking above the treetops on a hill. It was Helfštýn, one of the two castles guarding the southern

Fig. 4.2 The Piarist monastery with the tower of the St. Francis of Assisi Church in Lipník nad Bečvou: view from the road

entrance to the Moravian Gate. The other, Drahotuše, topping the hill on the travelers' right-hand side, was by then already reduced to ruins. The sighting of Helfštýn was a sign that they were nearing Lipník nad Bečvou, their destination. Soon the church tower marking the monastery (Fig. 4.2) appeared; after some two hours of travel (Anton's horse was no racing steed), they reached the ramparts of the inner city. Instead of entering through the Hranická brána (one of the two gates by which the city could be accessed), they took the road along the southern part of the fortifications and then stopped in front of an imposing building, the Piarist monastery. There the father delivered his son into the hands of the Piarist monks and after a short rest started his trip back to Hynčice. For the first time in his life, Johann was left alone in unfamiliar surroundings.

Lipník nad Bečvou

Who were the Piarists and what were they doing in Lipník? For answers to these questions, we must go back to the early history of Christianity. In nearly every major religion, some individuals desire to disengage themselves from the "madding crowd" and seek spiritual fulfillment and inspiration in solitude. In Christianity, this trend first appeared in the third-century Egypt[4,5] As long as they were persecuted for their faith, Christians tended to adhere strictly to the creed and live their lives in accordance with Christ's teachings. Once Christianity became the state religion, however, many paid lip service to the faith, but ignored its principles in their way of life.

It was then that zealots yielded to the urge to flee from a depraved society and retire to desolate places. They became known as *hermits* (from the Greek *erēmos*, desolate) or *monks* (from Greek *monachos*, "a person, who lives alone," Greek *monos* meaning "single" or "alone"). The Egyptian and Syrian deserts provided unlimited opportunities for those wishing to lead an ascetic life in isolation, self-denial, prayer, and meditation. It was, however, not so much the need for solitude that drove the zealots into the deserts, as the desire to lead life according to the Christian principles. It did not take long for them to realize that to fulfill this desire, it was more convenient for like-minded persons to band together and form a community within the society, but buffered and protected from it physically by an enclosure, as well as spiritually by a set of rules, which all the members of the group agreed to follow. The enclosures to which the monks chose to retire came to be called *monasteries* and the rules they pledged to obey defined the *religious order* they thus formed. Since different groups of monks pledged to obey somewhat different rules, a variety of religious orders arose. To become a monk of a certain order, a person had to make a *vow*, to pledge solemnly before God to lead a certain way of life. Three components of the vow were common to the different orders: obedience, poverty, and chastity. To these the different orders added certain other pledges that differentiated them from one another and that committed them to dedicate their lives to certain specific goals.

Piarists distinguished themselves from other religious orders by their pledge to devote their lives to the education of children from poor families[4,6] The Spanish priest Joseph of Calasanza (1556–1648) founded the order after he witnessed in Rome the appalling conditions in which these children, many of them orphans as a result of epidemics and wars, lived there. In 1597, he gathered enough support to open in Rome the first free school for children. In 1617, he established a *congregation* of devotees prepared to pursue the same goal of improving the lot of these children. And in 1621, he converted the congregation officially into the *Ordo Clericorum Regularium Pauperum Matris Dei Scholarum Piarum* ("The order of the regular poor clerics of the mother of God of the pious schools"), the *Piarists* for short. (The difference between a "congregation" and an "order" is that in the former, the members of the religious community are bound by a common rule, but not necessarily by a solemn vow; to become a member of an order, a solemn vow is required.) From Rome, the order spread to other places in Italy, to Spain, and to central Europe. It came to Moravia in the wake of the Battle of the White Mountain.[6] Prior to 1620, Lipník had been the seat of the Husites and the Jednota Bratrská. In 1620, the owner of Lipník, Jiří Bruntálský of Vrbno, happened to be on the losing, anti-Habsburg, side, and the Habsburgs retaliated by confiscating all his possessions and giving the town to Franz Cardinal von Dietrichstein, seated in Olomouc. The cardinal started to re-Catholicize his possessions immediately, but did so in a nonviolent way. The Jednota had to go, of course, but he invited the Piarists (rather than, e.g., the aggressive Jesuits) to replace it. The Piarists were known for their relative tolerance in matters of religion and language. True to their reputation, the Piarists of Lipník opened their schools also to non-Catholic children and made the Czech language one of the compulsory subjects.

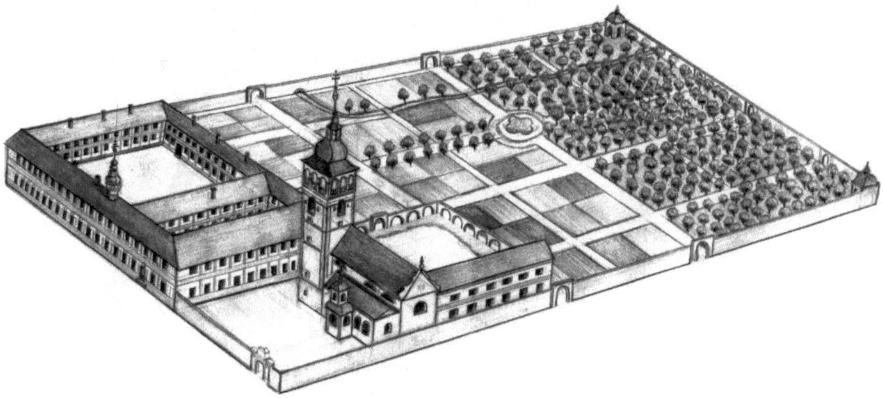

Fig. 4.3 The Piarist monastery in Lipník nad Bečvou: bird's eye view of the building complex and the garden

When the Piarists arrived in Lipník in 1634, all they had at their disposal was the meeting place and an adjacent building of the Jednota Bratrská. When they moved out in 1884, they left behind a complex of buildings enclosing a rectangular space divided into two square courtyards by a crosswise building (Fig. 4.3) The complex served as a monastery—residence of the monks—as well as a college, an institution of learning. It included the school, the novitiate (the quarters assigned to the novices—the candidates for membership in the order, placed on a two-year probation before taking the vows), an oratory (a small chapel), and the living quarters of the clerics and their pupils. The original meeting place itself became a Baroque church consecrated to Saint Francis of Assisi.[4] Part of the estate was a garden, which supplied flowers for the church, provided opportunity for manual work for both the monks and the students, served teaching purposes, and was a place of enjoyment and rest. The inhabitants of the monastery had also access to the large park of the neighboring manor house. The Piarists strived to encompass the entire scale of the educational system, from the elementary to the university level, but for a variety of reasons (shortage of students or teachers, lack of funds, and others), they had not always been able to attain this goal.

The move to Lipník must have been quite a shock to the 11-year-old Johann Mendel, excitement and curiosity mixing with anxiety and pangs of homesickness. Torn out abruptly from the warmth of home, he found himself in an environment in which everything was new to him. In Hynčice, everywhere he went, he would encounter familiar faces, people whom he knew and who knew him. Here, though, he was a stranger among strangers. In his home village, most of the people were farmers, and the village life revolved around land tillage and husbandry. In Lipník, he could see farmers, too, but they did not live there; they came only to sell their vegetables, fruits, cattle, and other produces. The burghers themselves were artisans, merchants, clerks, and clerics, acting and behaving differently from the people he had been familiar with until then. Communication, too, became a

Fig. 4.4 Lipník nad Bečvou: panoramic view from the southeast. The two tallest towers are those of the St. Francis of Assisi Church on the left with the Piarist monastery garden in front and the St. Jacob Church with the belfry in front in the middle. Some of the lower towers are associated with the bulwarks surrounding the city. In the foreground is the Bečva River

problem. In Hynčice everybody spoke German or, more precisely, a German-derived dialect, but in Lipník, where some 66 % of the inhabitants were Czechs, a language prevailed, which he, until then, had not had much opportunity to hear. Even at the monastery, where the main language was German, many of the pupils spoke Czech among themselves. The town itself must have made the greatest impression on the young Mendel. Compared to the 70 houses and 200 inhabitants of Hynčice, Lipník with its 400 houses and 5,262 inhabitants[7] seemed like a metropolis, which he was eager to explore.

And at the first opportunity he did so, undoubtedly beginning with the ramparts, which enclosed the inner city like a girdle. The double walls, six to eight meters high, two meters wide, three kilometers long (Fig. 4.4), with battlements at the top, and rondelles at the corners, must have been an awesome sight.[8] In his time, Johann could still walk the entire length of the fortification without an interruption. A decade or so later, the two gates were torn down, and the process of demolition of the city walls had begun; today only segments of them remain. From the ramparts, Johann might have been drawn by the clamor to the central square, the liveliest and noisiest area within the town walls. It was not a square, actually, but an L-shaped space, lined along its entire length with Renaissance and Baroque houses, their façades rising above broad arcades. The place of markets and fairs, the square teemed with people—sellers and buyers, local artisans, housewives, servants, and farmers from neighboring villages. The cacophony of sounds from both traders and beasts in the cattle and horse markets must have been deafening.

Here, Johann might have encountered now and then, a familiar face from Hynčice, a farmer who might have brought him news from home and taken messages to his family. Seeking a respite from the racket, Johann might have found refuge nearby in the Saint Jacob Church. Here, he was in the oldest part of the town, where a group of lime (linden, Czech *lípa*) trees (sacred to ancient Slavs) might have once stood and which might have given the town its name.[3] The church stood at this site since the fourteenth century, at least. More than the church, however, the exotic-looking tower next to it may have arisen Johann's curiosity. At some point later, he might have learned its story. In the past, church bells had been an important means of communication. Not only did they tell time, call people to church services, and announced the death of a member of the community, but they also sounded alarms when fires broke out, severe storms approached, or marauding troops emerged on the horizon. The church bells were the pride of the community. They were christened formally and nicknamed informally, and communities competed for the title of having the largest bell in the region. In 1604, Lipník won the title by acquiring Michal (named in honor of Archangel Michael), a bell that weighed 5,000 kg, stood 180 cm tall, and had a diameter of two meters. There was a problem, however: It was a bit too big for the existing church tower. When it rang, it hit the tower's walls and the whole church shook. The city fathers solved the problem by having a new, correspondingly larger tower, a belfry, built for Michal, next to the existing one. The architect, an Italian, paid homage to his native country by designing the tower in the style of a campanile, and thus creating a structure not commonly seen in Moravia. Michal was, however, not the first member of the Lipník bell community. Jacob, dating from 1464 and weighing 1,500 kg, claimed that title. Another member of the community, a featherweight Barborka (250 kg), distinguished itself by functioning as the death bell.

It is doubtful, however, that Johann had an opportunity to explore the town on his own too often. The Piarists lived their lives according to a strict schedule, in which time for sightseeing and *Stadtbummel* (stroll through the city) was not included. Presumably, the pupils had to adjust their activities to this schedule, too. They got up at five o'clock in the morning (six o'clock in winter), gathered for a morning prayer, service, and meditation at five thirty, and then had breakfast. Classes started at seven thirty and went on until ten o'clock. The lunch break extended until two in the afternoon, when the classes resumed, ending at four. After early dinner followed an evening prayer, meditation combined with searching one's conscience, and then a night's rest. Initially, Thursday was a day off for the pupils, and Saturday was reserved for recapitulation of the material learned during that week. Some of the school reforms the government introduced at the beginning of the nineteenth century forced the monks to change this schedule, however. Outside of the classroom, both the teachers and the pupils were expected to work in the garden adjacent to the monastery. By purchasing land from the town and acquiring other as a gift from their neighbor, the lord of the manor, the Piarists expanded the original small backyard into a combination of an orchard, vegetable and flower garden, as well as a park, which exceeded in size the area taken up by the monastery itself (Fig. 4.3). With a small brook running through it, but otherwise enclosed by a continuous wall, it was a peaceful place in which it was pleasant to walk on paths lined with decorative trees and bushes or to rest on a bench in the sun or shade. Johann, the budding botanist, must have spent many a happy hour in it. Founding and maintaining such a large garden required a great deal of manual labor on the part of the monks, novices, and pupils, however, and so took up much of everybody's free time. Whatever free time remained, the novices could use to go out of the monastery. Whether the boys (for girls were not admitted to the monastery school) boarding at the college were granted similar freedom of movement is rather doubtful. One may assume, therefore, that Johann Mendel spent much of the year in Lipník more or less confined to the perimeter of the monastery and that his life during the one year revolved mostly around the school he attended there. To understand the nature of this school (as well as of the one to which he transferred afterward), it may help to place it in a proper historical context.

The Empire's Schools[9]

In ancient Greece and Rome, the acquisition of knowledge had been a privilege of aristocrats, and in medieval Europe the prerogative of the clergy. As a result, the first schools in western and central Europe began to form at cathedrals and in monasteries, and the process of education had remained firmly under church control until the second half of the eighteenth century. An effort to separate schools from religious institutions and to bring them under the control of the state began only during the Enlightenment, with the Habsburg Monarchy issuing the *Allgemeine Schulordnung* decree in 1774. It took some time, however, for the decree to take effect, and even then, religion remained firmly embedded in the school curriculum.

After taking over the educational system, the government kept changing it under pressures from different interest groups, in particular the church, the emerging industrialists, and the nationalists. The church wanted to regain control of the system, the industrialists pressured for changes in the curriculum to prepare the graduates for the new professions, and the nationalists were fighting for the survival and preservation of their cultures. The government responded to these demands mostly by making minor, tentative adjustments in the system. On three occasions, however, it was forced, after years of debate, to undertake major reforms. The first of these reforms was the *Allgemeine Schulordnung* (general school regulation) cosponsored by Maria Theresia and Joseph II in 1774. The second was the *Politische Schulverfassung* (political constitution) issued by Franz I in 1805/1806. And the third was the *Schulbildung* (statutory education) law signed into effect by Franz Joseph I in 1867/1869. The most profound of these reforms, the *Allgemeine Schulordnung*, established several new school types and made them accessible to all school-age children of the empire. In effect, the reform added a new, primary level to the existing two levels of the educational system, the secondary level represented by the *Gymnasien* and the universities. The new school types were the *Volksschule,* the *Hauptschule,* and the *Normalschul*e (standard school). To these, the reforms of 1805/1806 added several *Realschulen.*

The *Volksschule* (literally "the people's school") Johann Mendel attended in Hynčice was fairly characteristic of this type of institution, with the exception that in cities boys and girls were instructed separately. In the Middle Ages, the scholastics or schoolmen called the three core subjects (reading, writing, and arithmetic) the *trivium* (literally "the place where three roads meet"), and so the schools were also referred to as *Trivialschulen.* While there was at least one *Volksschule* in every parish, the *Hauptschulen* (literary "main schools" but commonly translated as "intermediary schools") were limited to district towns only. They were either three- or four-grade schools intended to provide somewhat broader general education than the *Volksschulen,* in preparation for a transfer to the secondary level of schooling. The curriculum of the three-grade schools consisted of German grammar and composition, in addition to the trivium and religion. In the four-grade schools, the pupils were also taught drawing, geometry, geography, and natural history. The basic Latin taught originally in the higher grades of the *Hauptschule* was eliminated by the 1805/1806 reforms, as was natural history. In contrast to the *Volksschule* in which instruction of the different grades took place in a single classroom, in the *Hauptschule,* each grade had its own classroom. As in the *Volksschule*, however, in the *Hauptschule,* a single teacher, the *Klassenlehrer*, instructed his class in all the different subjects, with the exception of religion. The *Klassenlehrer* (there is no English equivalent to this term; the "form teacher" had somewhat different functions) also remained with his class through the different grades. Pupils entered the *Hauptschule* by transferring from the fourth grade (i.e., at the age of ten years) of the *Volksschule*, but transfers from the fifth grade of the *Volksschule* to the third grade of the *Hauptschule*, as in the case of Johann Mendel, were also permitted. Later (from 1869), the *Hauptschule* developed into the *Bürgerschule* (school of town's people or burghers), so designated because of its restriction to towns. The school's attendance was compulsory for children who completed successfully five years of the *Volksschule.* Children from rural areas could also attend the school if they had means of reaching it. The school had only three grades originally, but later another two grades were added. *Normalschulen* educated teachers of *Volks-* and *Hauptschulen.* They were limited to the capital cities of individual provinces and had a similar but expanded curriculum like the *Hauptschulen.* The reform of 1805/1806 changed the *Normalschulen* into *Musterhauptschulen* (model intermediary schools), which served for practical training of the future teachers. In addition to these general education schools, the 1805/1806 reform established also several types of specialized schools educating young people for professions in industry, mining, forestry, horticulture, agriculture, and the like. The schools admitted 14-year-old students who completed successfully the *Volksschule* and the *Hauptschule.* The reform also placed most of the general education schools back under the control of clergymen: the parish priest (*Volksschulen*), the dean (*Hauptschulen*), and the canon (schools in capital cities). (The dean was a priest chosen by the bishop to supervise parishes within his diocese, the district under the bishop's jurisdiction; the canon was a clergyman serving in a cathedral or a collegiate church.)

At the secondary level of the educational system, the main school type has been the *Gymnasium*. The name is a misnomer. In ancient Greece, where the word originated, it had been a designation of "a place for naked (physical) exercises." It has kept this meaning in some English-speaking countries, although there the exercises in the "gym" are no longer in the nude. It is something of a mystery why the medieval schoolmen and specifically the Jesuits adopted the word as the designation for schools preparing young people for study at a university. It was with this denotation that the word passed into the German and some Slavic languages. When Maria Theresia, at the urging of Joseph II, dissolved the Jesuits and the state took over their *Gymnasien*, the schools had five grades. The Jesuit called them *Rudimenta*, *Grammatica*, *Syntaxis*, *Humanitas*, and *Rhetorica*, according to the chief subjects in each grade's curriculum. In 1819, however, all *Gymnasien* became six-grade schools, and the grades were renamed to *Parva*, *Principia*, *Grammatica*, *Syntaxis*, *Poesia*, and *Rhetorica*. In 1849, the government added two additional grades, which originally constituted a two-year course called *Philosophy* at the university. The *Gymnasien* thus became eight-grade schools.

From a present-day perspective, the designations of the grades look weird for schools of general education. This is so because the schools were founded on humanistic ideals of the fourteenth and fifteenth century, and as a result, their main subject was Latin. The preoccupation with Latin had remained the characteristic of the *Gymnasien* throughout most of the nineteenth century, when in some of the grades 11 of the 18 weekly hours taught at the schools were devoted to the Latin language. It was for these reasons that the institution was often called *humanistische Gymnasium* and that in the Middle Ages, before the Jesuits took it over, the humanists called it *Lateinschule*. The reason why Latin enamored the humanists was that it was the language of ancient Rome, in which they found a culture much to their liking, a culture centered on a secular world here on earth, rather than in the afterlife. The works of Cicero, Livy, Tacitus, Seneca, Plautus, Virgil, Ovid, Lucretius, Catullus, and Horace revealed to them the beauty of the Latin language (contrasting starkly with the barbaric Latin of the schoolmen), the eloquence of Roman orators, the rich imagery in the works of the ancient poets, and the admirable way of thinking and acting of virtuous Romans, at least as presented by the historians. The humanists thought they might be able to instill some of these standards of excellence in the young people by having them to devote five to six years to the study of the classical Latin language and of the world in which it had been spoken.

Needless to say, it did not work out quite the way the humanists had planned it. Years of mindless learning of grammatical rules, of dissecting sentences like cadavers under the unforgiving eyes of pedantic teachers, and of memorizing long passages from Cicero's speeches, Tacitus' *Annals*, Seneca's essays, and Horace's poems had, as a rule, the opposite effect than intended. The graduates could quote the classics for the rest of their lives, as was then expected from a cultured person, and they could communicate in Latin, but most of them did not become eloquent writers or orators in a language they learned to detest. What effect the years of Latin had on their way of thinking and acting had never been subjected to a controlled study. At any rate, in the nineteenth century, the prevalence of Latin in the Austrian *Gymnasien* was patently out of step with the developments in the rest of central and western Europe. The humanists have long since realized that the Roman culture they admired so much was actually a derivative of a much more ancient culture, the understanding of which required mastering of the Greek language. This realization, together with the rising sympathy of western Europe for the Greeks, then occupied and oppressed by the Turks, had gradually, if not replaced, then at least counterpoised the adulation of Rome with the growing interest in the Hellenistic culture. Inevitably, even in the ultraconservative schooling system of the Habsburg Monarchy, these developments forced a change in the curriculum. Latin had to make room for Greek, Cicero for Demosthenes, Tacitus for Thucydides, Virgil for Homer, Horace for Pindar, and Plautus for Aristophanes. Another, far more profound development that cried out for a change in the school curriculum was the rise and unstoppable march of natural sciences in the Western civilization. Fuelled by the industrial revolution, in the nineteenth century, science had been gaining such a powerful influence that its inclusion in the curriculum of schools at all levels could no longer be delayed.

With the rise of modern science, however, began a process of polarization in the Western culture. At one pole, the *humanities* began to assemble, steeped in classical Latin and Greek languages and literature, and by extension also history, philosophy, and mathematics, all subjects believed to be cultivating true humanity in humans. The humanities also went under the name *liberal arts* or simply *arts*.[10] At the other pole gathered the *realities*, a designation which in German assumed a meaning not common in the English language. In German, *Realien* are all the subjects in schools of general education that teach factual knowledge convened under the denomination *exact sciences* or simply *sciences*. They included geography (which until then was taught as a mere postscript to history), biology, chemistry, and physics. Hence, long before the term "two cultures" was coined,[11] the concept had become embedded in the European civilization. One consequence of this polarization was the appearance of *Realschulen* with a curriculum focused on factual knowledge and hence on exact sciences, often with practical orientation. Some of these schools specialized in preparing their students for specific branches of industry and commerce. The specialized schools mentioned earlier and introduced at the beginning of the nineteenth century were in effect early *Realschulen*. The *Gymnasium*, on the other hand, remained for a long time largely unresponsive to the new developments in the society. When it finally did respond in the middle of the nineteenth century, it was in the form of splitting it into two types, the *Klassische Gymnasium*, true to its humanistic tradition, and the *Realgymnasium* with curriculum reflecting the rise of sciences. In the latter, the number of hours devoted to Latin and to other humanistic subjects was cut down, and in their stead, new subjects were introduced—living languages (French, Spanish, English), drawing, geometry, geography, natural history, chemistry, and physics. With time, most of these subjects were also introduced, in a more limited extent, in the classical *Gymnasien*. The main aim of these two school types had remained the same: to prepare the students for institutions of higher education, classical *Gymnasium* for the university and *Realgymnasium* for university-level technical institutions.

The school year in the *Gymnasien* and other schools consisted of two half-year periods (semesters) separated by short vacations around the Easter holidays. Longer vacations of approximately one and half to two months set apart sequential school years. Originally, the onset of these vacations was flexible and varied from school to school. In the Lipník Piarist schools, for example, the vacations started in the middle of September and lasted until the beginning of November. The educational reform of 1774/1755 fixed the main vacations in all schools for the months of July and August, and they remained that way until 1854, when a new decree changed them for secondary schools to a period from July 16 to August 15. Outside of the Easter and main vacations, other free days from school were restricted to various church holidays.

During the school year, teachers examined students more or less continuously. Every day a student could be tested on material presented in earlier lessons, and he would never know when his turn would come. There was always high tension in the classroom at the beginning of a lesson, as the teacher pulled out his notebook to choose a victim for the day. Some teachers seemed to have a nearly sadistic pleasure out of prolonging this tension by taking time to make the selection. Major examinations were held at the end of each semester, at which time the teachers graded each student's overall performance in the individual subjects, as well as his moral conduct and diligence. The teacher recorded the grades in the classroom catalogue and issued a *Zeugniss* to each student, a school report spelling out his performance in the various subjects. The *Matura*, *Reifeprüfung* (maturity examination), or *Abiturium* (final examination; from Latin *abire*, to go away) at the end of the last school year was introduced only by the 1849/1854 reform but has been retained to this day in secondary schools in the lands of the former Habsburg Monarchy. On this occasion, a committee of educators from within and without the school examined the students for their knowledge of the material covered in selected subjects during the entire school period. The examination consisted of written and oral parts, and the achieved grades were recorded in the certificate issued by the school.

The grading scale changed with time, but it consisted generally of five or six grades with the first being the best and the last signifying the student's failing in the subject. The grades were given either in words (in Latin, German, or Czech) or in numerals, Roman or Arabic. In words, moral conduct could be appraised (in approximate English equivalents) as praiseworthy, satisfactory, lawful, less lawful, or unlawful; diligence as diligent, appropriate, sufficient, inconstant, or slight; and performance in the different subjects as excellent (very good), commendable, good, sufficient, insufficient, or quite insufficient. In the documents preserved from Mendel's studies, the grades are given in Latin and are combined with the student's rank among all the classmates. Because Mendel was consistently at the top of his class, we find only these grades in his records: I (= *prima classis*; first in the class); I *em* (= *prima classis cum eminentia*; first in the class with distinction); *ad em* (= *ad eminentia*; next to distinction, in other words, second best); I *accedens* (second best); I *praemif. access.* (first after the "praemiant" or the prizewinner); and III *inter eminentes* (= *tertius inter eminentes*; third among the best). Many schools celebrated the end of the school year with a public ceremony at which the best students, the *Prämianten*, received, often to the accompaniment of fanfares and rolling drums, a medal (at the time of Maria Theresia with her likeness impressed on it) and a prize.

Prior to the beginning of the nineteenth century, a single teacher, the *Klassenlehrer*, conducted all the instructions in the various subjects. Then, for a short time and only in some schools, *Fachlehrer* (expert teacher) took over the teaching in Gymnasien in their specialties, but by 1818, the system returned to the exclusive use of *Klassenlehrer* again. It was not until 1849/1854 that expert teachers had been reintroduced gradually in all *Gymnasien*, this time more or less permanently. Even then, however, one of the professors (usually the one with the heaviest teaching load in the particular class) assumed the role of a form master. He handled all organizational matters concerning the class, acted as mediator between the students and the teachers' body as well as between the teachers and the students' parents.

The *Prämiant*

If the Piarist *Hauptschule* in Lipník celebrated the end of the 1833/1834 school year in the manner that some other schools did, then in the summer of 1834 the 12-year-old Johann Mendel stood on the tribune as one of the prizewinners, the *Prämianten*. Johann's achievement was quite remarkable considering the odds stacked against him. He changed from a single-classroom village school with little competition to a city school with separate classrooms of selected and hence much more competitive pupils. He had to adapt to being alone, away from home, and to an entirely new way of life, new environment, new teachers, new classmates, and new roommates. And he was one year younger than most of his classmates who were pupils of the *Hauptschule* from the first grade.[12] Despite all these handicaps, he finished the school at the very top of his class, achieving the best grades possible (*sehr gut*, very good) in all the subjects and left the school as the *erster Prämiant seiner Klasse* (the first prizewinner of his class).[13] This achievement bore witness not only to Johann's aptitude and diligence but also to the quality of the training he received at the Hynčice school. The subjects he was instructed in at the *Hauptschule* included German, in addition to the trivium and religion, and were taught by his *Klassenlehrer*, Julius Baigar.[12] Up to this point, Mendel still had no Latin, because the reforms of 1804 abolished its teaching in the *Haptschule*.[9] If the celebration did take place and Anton Mendel, who might have come to Lipník to take Johann home, attended it, he must have been mightily proud of his son as he watched him accepting a prize to the sound of trombones and rolling drums. Indeed, the occasion might have contributed to the father's decision to allow Johann to continue his studies in the *Gymnasium* in Opava.

Back home, Johann could rest his brain, but not his hands, for it was once again harvest time and his hands were needed in the fields. Once the definitive decision had fallen that Johann should continue to study, Makitta and Schreiber presumably prepared for him the application for his admission to the Opava *Gymnasium*. It was also at that time that Schreiber sent to the *Gymnasium* the birth certificate with the wrong birth date. The teachers recommended the Opava *Gymnasium* because it was the closest to Hynčice and because of its good reputation. Johann was accepted and reported to school on December 15, 1834.[14] It seems unlikely that the school year began this late since in accordance with the latest school reform, all *Gymnasien* were required to begin their instructions in October, after the summer vacation. The reason for Johann's delayed arrival could have been that it took this long to process his application. If so, he would have to, again, start at a new school with the handicap of having to catch up with a class that was already two months in session. Whatever the case might have been, in December 1834, Anton Mendel once again hitched his horses to drive his son to a new place—to Opava. Their route followed again the amber trail through the Moravian Gate, only this time in the northerly direction (Fig. 4.1). Reaching the main road north of the Iron Gate, they crossed the Odra River in the city of Odry and then continued in the northeasterly direction toward Fulnek, a town huddled around a castle on a hill in its center. From Fulnek, the road took them straight north, through the small villages Vrchy, Březová, and Lesní Albrechtice, and then a long stretch through a dense forest. Emerging from the forest, they reached Hradec nad Moravicí and a manor house standing on a site where a castle once stood that guarded the northern entry into the Moravian Gate. Those who think of this part of the country as a cultural backwater might be surprised to learn that around the time of Mendel's youth, the visitors to the manor house included Ludwig van Beethoven, Franz Liszt, Nicolò Paganini, and other famous musicians, all of whom gave concerts there. After Hradec, it was only a short distance to Opava, the city in which Johann Mendel would spend most of the next six years of his life, some 35 kilometers from his home village.

Opava[15]

The succession of stations in Mendel's life must have given him a series of cultural shocks. If the move from Hynčice to Lipník required an adjustment from small-village to small-town conditions, all the subsequent moves to Opava, Olomouc, Brno, and Vienna exacted an adaptation to progressively more complex urban environments. Compared to Lipník, Opava was already then a metropolis (Fig. 4.5). Not only was it double the size of Lipník both in area (827 houses) and population (14,300 inhabitants),[7] but as the capital of a district and of Czech Silesia, it also was the seat of a number of institutions, and the provider of many services. There were district and provincial courts, governmental offices, a chamber of commerce, several banks, credit and savings institutions, custom offices, numerous clerical institutions, churches, monasteries, a military garrison, a theater, a museum, archives, hospitals, charity houses for invalids and the elderly, an array of schools

Fig. 4.5 Opava: panoramic view from the northwest. (1) The parish Church of the Ascension of Virgin Mary. (2) Hláska tower at the Horní (*Upper*) Square. (3) The Church of St. Vojtěch (Adalbert) at the Dolní (*Lower*) Square, the site of the Jesuit College and of the *Gymnasium*. The hills in the background are the promontories of the Jeseníky Mountains, which Mendel had to cross on his travels between Hynčice and Opava; the city itself lies at the southern margin of the Polish flatland. The arrow points out the location of the Hradec castle. Based on the watercolor by Franz Kalivoda from the first half of the nineteenth century

for general and specialized education, and rapidly growing textile, machine, and other industries. Since 1788, Opava also had its own newspaper.

Opava owed its origin to a ford, at which the amber route crossed the Opava River. It was located in present-day Kateřinky, a suburb of Opava. The city itself arose on the river's right bank, at a site now taken up by the *Dolní náměstí* (Lower Square). Later, the city's center shifted to the *Horní náměstí* (Upper Square), dominated by a 72-meter-high watchtower, *Hláska*, so named because it was used to watch for and report (*hlásit* in Czech) fires. In spite of it, however, the city burned down several times in its long history. (Its recorded history begins in 1224, but its unrecorded existence must have extended deep into the Celtic period.) Like Lipník and most other European cities, Opava was surrounded by fortifications, but these were gradually demolished during the years 1800–1839. By the time Johann Mendel arrived in the city, only one of the original three gates (*Ratibořská brána*) was still standing; the other two (*Jaktařská* and *Hradecká brána*), as well as most of the ramparts, had been taken down already.

In the Middle Ages, several religious orders established their presence in Opava: Claretians, Dominicans, Minorites, Franciscans, Johanites, and Jesuits, all had built their convents, monasteries, churches, colleges, and other institutions in the city. Later the Husite and Protestant movements curbed the expansion of these orders,

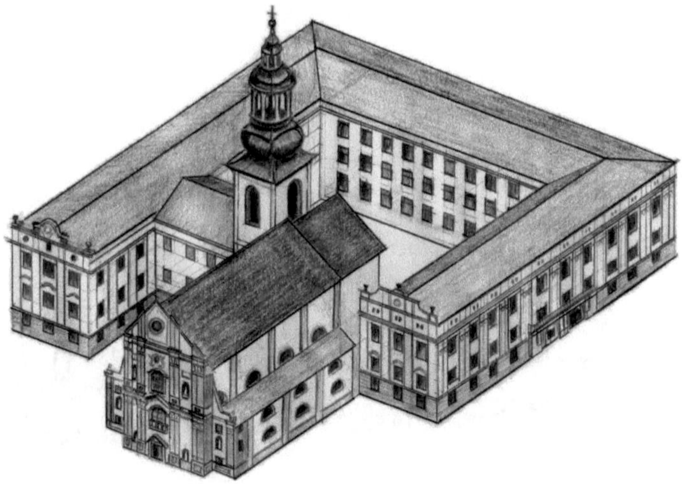

Fig. 4.6 Opava's former Jesuit College with the Church of St. Vojtěch (Adalbert) in the *middle*. The *Gymnasium*, the school that Johann Mendel attended from 1834 to 1840, was located in the wing on the *left*

but most of them regained their former strength in the post-White Mountain period. Joseph II dissolved some of the orders and repossessed their buildings. The buildings owned by the Minorites were converted into governmental offices, a library, a military hospital, and archives; those of the Dominicans into storage places; and those of the Franciscans into a general hospital. From the Society of Jesus, dissolved in 1773, the government requisitioned its college, in which the Jesuits ran since 1630 a *Gymnasium*, with a library of some 35,000 tomes.[16] The college, rebuilt by the Jesuits from 1711 through 1723 and subsequently remodeled by its new owners, the Silesian estates, stood behind the Church of Saint Vojtěch (Adalbert) at the Lower Square (Fig. 4.6). How stately the Baroque building was is indicated by the fact that for the time of the congress of the Holy Alliance, from September to December 1820, it served as a temporary residence of the Austrian Emperor Francis I.[17]

It was in this building that Johann Mendel went to school for six years, from 1834 to 1840, as the commemorative plaque at its entrance now informs passersby. Alas, sprightly teenagers' voices no longer resound through the building's staircases and corridors, for the *Gymnasium*, which still functions as the main secondary school in Opava, has since moved into a new building nearby. In the former *Gymnasium* now reigns a great silence appropriate for a place devoted to storing records of past ages: The building now harbors the province's archives. In the new building, too, now sound different voices than in Mendel's time. Then it was German, Latin, and Greek that was spoken there; now it is Czech and tortured German or English when foreign language classes are in session. The change in the spoken language from German to Czech reflected a transformation in the ethnic composition of Opava's residents. While in the medieval times, the city's populace was about equally split between the two ethnic groups, by the time Mendel began his studies there, the proportion of Germanic people in the city rose to about 90 %. From the second half of the nineteenth century, the Czechs struggled to assert themselves and succeeded finally to obtain permission to open a Czech *Gymnasium* parallel to the German one. An abrupt change in the composition occurred after World War II with the expulsion of the Sudeten Germans.

The head of the Opava *Gymnasium* at the time Mendel studied there was the priest Ferdinand Schaumann, professor of religion and a member of the Order of Saint Augustine.[18] He came to Opava from the same monastery in Brno that Mendel would enter a few years later. Contrary to claims one encounters in articles about Mendel, however, Schaumann did not teach Mendel and apparently had no influence on his decision to become a priest. Mendel's religion professors were two other priests, Bartolomäus Müller and his substitute, Ludwig Tidl.[19] Mendel's other two teachers in Opava were Thomas Zauhar, *professor grammaticalis*, and Martin Beck, *humanitatis professor*. The former was the *Klassenlehrer* in Mendel's first four (grammatical) years at the school, the latter in the last two (humanities) years. Zauhar must have been active also outside of the school, because in 1842, the city made him its honorary citizen.[16] By a curious coincidence, another student of the *Gymnasium*, Karel Křížkovský, was also to become an Augustinian friar and Mendel's colleague in Brno.[20] Although Křížkovský and Mendel overlapped partially at the *Gymnasium*, they were not classmates (Křížkovský was two years older than Mendel) and may have known each other in Opava only superficially.

As in other *Gymnasien* of the Austrian Monarchy, the school's curriculum in Opava consisted of six years of Latin, geography/history, arithmetic/algebra, and religion, and of six years of Greek. There was no formal instruction in any of the natural sciences—biology, chemistry, and physics. Indeed, Mendel had not received any formal instruction in these subjects in either his primary or his secondary schooling. At the village school in Hynčice, he got only tidbits of instruction in natural history in a somewhat clandestine manner, against the will of the authorities. In Lipník, he might have been served further crumbs during his work in the garden, assuming that one of the teachers there was able to provide them. And in Opava the only two places for pursuing his awakening interest in nature were the library and the museum, both in the same building as the *Gymnasium*. The library was a legacy of the Jesuits. Although it contained mostly theological and philosophical works, there were also books on natural history in it. The museum was founded in 1814 by three enthusiasts, Faustin Ens, a professor at the *Gymnasium*; Franz Ritter Mükusch von Buchberg, a retired military officer; and Johann Joseph Schlösser, Opava's mayor.[21] Faustin Ens (1782–1858)[22] originated from Rottweil, which is now in southern Germany, but was then under the sovereignty of the Habsburg Monarchy. At the age of 17 years, he joined the fight against Napoleon, then entered the Benedictine order, but left it after the novitiate. He became a private teacher in Opava and from 1812 to 1844 professor at the Opava *Gymnasium*. He taught first mathematics and natural history and later geography and history. Ens published several books, among others a history of

Opava. In 1848, he retired to Bregenz, Austria, and was active in a museum there. Ens and von Buchberg were ardent naturalists, who made natural science collections the main theme of the museum, the first of its kind in the Czech lands. The museum retained its association with the school until 1895, when it was moved to another building erected in the park across the present-day *Gymnasium*. It seems likely that curiosity about nature drove Mendel to both the library and the museum and that he encountered Ens in both places. Zauhar might have alerted Ens already to the bright new student, and Ens might have approached Mendel to seek his help in the preparation and maintenance of the exhibits. If so, the two might have spent much time together in the museum, and the student might have learned some natural history, especially botany, from the professor in this informal way. We should keep in mind, however, that Mendel had not received any systematic instruction in natural science; it may help us to understand his later failure to pass a teachers examination.

Mendel's *Abgangszeugniss*, a certificate attesting the successful completion of his studies at the *Gymnasium* and summarizing his overall performance in the 12 semesters, reveals him once again as a *Musterschüller*, a student whom teachers like to give as an example to the less diligent classmates. If all students in the class were like Mendel, there would not have been any need for a six-mark scale; a single mark in three variants would suffice: *primus*, *primus eminens*, and *ad eminens*. These are the only marks Mendel's *Abgangszeugniss* shows. In three of the 12 semesters, his overall rating was *III inter eminenti*, *I accedens*, and *I praemif.*

Where Mendel lived in those six years in Opava is not known. The original Jesuit *Gymnasium* was part of a college and as such probably provided lodging for out of town students. After the dissolution of the order, the government abolished the college status of the institution, forcing the nonresident students to seek private lodging, which could be either of a full-pension or half-pension type. The former included bed and board, the latter only bed (i.e., no meals). In Mendel's case, his parents could afford neither the full-pension lodging for their son nor the tuition the school charged other students. The school waved the tuition, and he had to satisfy himself with a half pension and had to live on the small amount of cash provided by his parents and supplemented by food packages sent from home whenever haulers happened to have a trip to Opava.[23] (In his biographical novel about Mendel, Werner Heinenen[24] cites a letter Johann purportedly wrote to his parents from Opava; in it he identifies one of the haulers as Wladislaus Kratky. In reality, however, the letter is apocryphal, as Heinen himself admits in his final footnote.) The packages might have contained a big loaf of home-baked bread, a chunk of butter, a jar of purée made from boiled plums, eggs, smoked meat, or fruit. It was tough going for Johann Mendel at Opava. After paying the rent, he had little cash left in his pocket to supplement his frugal meals with food from a grocer, let alone a restaurant. His rapidly growing body, like that of any healthy teenager, demanded an adequate supply of energy, but his diet failed to meet this demand, and so he probably felt hungry much of the time. Since he could not afford any of the treats which the city had to offer and which his classmates from well-off families indulged in, he probably avoided their company. He may not have had friends not because he didn't want to but because his poverty isolated him from other students. Alone, without confidants, he must have struggled the more intensely with the physical and emotional vexations any adolescent must cope with. To numb the awakening demands of his body, he worked very hard. The school curriculum required a great deal of memorizing and mental drilling, which must have taxed his strength, especially since it had not been compensated by any physical exercise. Johann, the conscientious son that he was, must have felt that he had to give his utmost in his school performance. He was well aware of the sacrifice his family had to make in order to send him off to Opava, and he did not want to let them down. And so, he may have lived under a continuing stress to remain at the top of his class in order to prove to himself and his family that he was worthy of their sacrifice. Poverty, hunger, loneliness, homesickness, overexertion, anxiety, stress, and lack of physical exercise, all these accompaniments of his life at Opava eroded Johann's health. In the first four years of his studies, though, he still managed to recover his strength during summer vacations. For two months each year, he became a farm boy again, having enough to eat, leaving all his existential worries behind in Opava, working and sweating in the fields, and feeling that he was contributing to the welfare of the family rather than sapping it. Browned by the sun, exhausted physically, but relaxed in spirit, he then returned in the fall to Opava with the firm resolve to tough out another grueling school year.

Then, however, in 1838, in his fourth year in Opava, catastrophe struck. In his so-called autobiography,[25] Mendel speaks of *mehrere schnell aufeinander folgende Unglücksfälle* (several misfortunes following rapidly one after the other), but posterity knows of only one of them, apparently the gravest one. Sometime in January or February of that year, Johann's father had an accident in which he was seriously injured.[23] While felling trees for the lord as part of his *corvé* obligations, Anton Mendel's chest was crushed when a tree trunk rolled over it. It seems that Johann was not informed about the accident until he came home for his summer vacation. By then it became clear that Anton Mendel was going to survive but that he might never recover fully. Thus, the question of his successor and of Johann's future began to loom over the two months in which he would have otherwise recovered his strength. At the end of his vacation, the family decided that he should return to Opava to finish his studies but that he would have to support himself, for under these circumstances he could not expect any financial help from home. On the way back to Opava, Johann might have thought that his dream of getting a full education might have been coming to an end. He realized that if his father would not be able to resume fully the work on the farm, Johann would have to take over and become a farmer after all. Strangely, after the accident, Anton Mendel was more resolute than ever to have his son complete his studies. He apparently believed in his full recovery, but chiefly he wanted Johann to escape the cursed *robota* and have a better life than his own. Johann, however, returned to Opava with a great worry on his mind; how would his mother and sisters cope in Hynčice?

Earlier, perhaps anticipating some of these developments, Johann had taken steps toward self-sufficiency. While continuing his studies at the *Gymnasium*, he enrolled in a course for *Schulkandidaten und Privatlehrer* (school candidates and private teachers) at the *Hauptschule* in Opava, to obtain the permit necessary for tutoring other students.[25] On the examination at the end of the course, he scored the highest mark of "very good" in all subjects but one: Ironically, in religion he was merely "good." Highly recommended by the school, he was able to find enough students to keep his head above water financially for the last two years at the *Gymnasium*. In the spring of 1839, however, the exertion, stress, and privation caught up with him, and just before Pentecost he collapsed, suffering from an unspecified illness,[18] presumably a nervous breakdown. His condition was so serious that Anton Mendel had to pick him up and bring him home. Johann stayed at the farm for the rest of the school year, but in September, after the summer vacation, he returned to Opava and was allowed to enter the final, sixth, grade without having to take a special examination for the period of his absence from the school. Since in 1839 White Sunday fell on May 19, Johann stayed at home for three and half months, from the middle of May until the beginning of September. As usual, his stay at home did him good; he recuperated quickly and was able to help with the harvest, at the time when help was needed most on the farm. Back in Opava, he completed his studies at the *Gymnasium* with flying colors.

With the *Abgangszeugniss* in his pocket, the 18-year-old Johann Mendel was confronted with the problem of what to do next. The certificate from the course for *Schulkandidaten und Privatlehrer* made him eligible to become a teacher at a primary school,[9] if he could find a vacancy at a school at which a teacher had just retired or died. He would also have to be well connected to secure such an appointment. Not only did he not have these connections, but also he was not enthusiastic about the prospect of spending the rest of his life drilling children in the multiplication table. By this time, Johann apparently began to feel the pull of natural sciences even though he had very limited exposure to them. He might have been dreaming of becoming a professor at a *Gymnasium* and a naturalist like Faustin Ens. To qualify for such a position, however, he would have to continue his studies. Minimally, he would have to take two years of "philosophy" and possibly a few semesters at a university to pass the qualifying examinations. Encouraged by the success he had in Opava in finding tutorial jobs, he decided—after consultations with his family and possibly his professors in Opava—to enroll at the Philosophical Institute in Olomouc, a city some 50 kilometers southwest of Hynčice.

Olomouc

Whether it was again his father who drove him to Olomouc or whether he hitched a ride with one of the long-distance haulers, we do not know. In either case, Johann found himself traveling the amber route again, this time crossing the entire watershed of the Odra and Morava rivers (Fig. 4.1). The first part of the crossing was familiar to him because it was the same road that took him seven years earlier to Lipník. On the second leg of the journey, the traveler traversed the northern tip of the Haná region, first in the northwesterly direction through the villages of Dolní Újezd and Horní Újezd, and then, from the small town of Velká Bystřice, south west, along the Bystřice ("the rapid one") River to Olomouc. In Haná, the rich deposits of loess created the most fertile soil in all of Moravia, and the hardworking, good-natured people of the land, the Hanáci, knew how to get the most out of it.[26] Even though his trip must have taken place close to harvest home, from the wide expanses of stubble and the few remaining "mandels," he could tell that it had been a bountiful year. In the villages, the spacious, well-kept, richly decorated farm houses attested to the prosperity of their owners, who were distinctive in their dress, manner of speech, behavior, customs, and life philosophy. The ostentatious weddings in Haná were proverbial and a Hanák's image of heaven was familiar throughout the Czech lands. A Hanák in paradise was said to lie on his belly at a pond filled with buttermilk (Johann Mendel's favorite drink) on which lumps of freshly churned butter float. When he cannot drink any more, he rolls on his back and opens his mouth to receive clods of cheese rolling down from the Sýreček Mountain looming over the pond, *sýreček* or *tvarůžek* being a special aromatic white cheese, for which the region around Olomouc was famous. But Johann might have viewed all these signs of prosperity with apprehension: Is he going to have enough to eat in Olomouc or are two more hungry years awaiting him in the heart of the bountiful Haná?

After Velká Bystřice, on a plain as flat as a pancake, the travelers could soon discern the Svatý Kopeček (the Holy Hill) as it emerged from the haze, like a little lamb that had strayed away from the herd of the Jeseníky Mountains in the background. The white walls and the two steeples of the Church of Virgin Mary's Visitation on the top of the hill shining through the bluish haze beckoning a Christian traveler to make a pilgrimage to the Mother of God. Soon afterward, the skyline of Olomouc itself began to appear on the horizon, first the spires of the five largest churches and then the entire panorama (Fig. 4.7). In 1840, the year Mendel entered it, Olomouc was a town of some 700 houses and nearly 12,000 inhabitants, including over 5000 military personnel stationed in the local garrison.[7] Like Opava, Olomouc arose from a settlement near a river crossing, here the river being Morava.[27] Archeology provides evidence that since the Paleolithic age, humans have lived on the three swells on which now stand the three oldest churches of Olomouc, Saints Michael, Peter, and Václav. It has been, however, only since the Neolithic era that permanent settlements of farmers began to arise at the ford, at which traders traveling this leg of the amber trail used to cross the Morava River. Later the Celts, followed by various Germanic tribes, followed by the Slavs, had taken possession of the site. Even the Romans, who had otherwise stopped their expansion into northern Europe at the Danube, had ventured temporarily all the way to this trading post. The claim that they called the place *Mons Julii* or *Juliomontium*, the Mount of Julius (Caesar), and that the name "Olomouc" is a corruption of this Latin designation is, however, a figment of humanist scholars' unrestrained imagination. The Přemyslids, when they unified Moravia and attached it as a margrave to Bohemia, recognized the strategic and economic importance of the place and erected a castle there in the eleventh century and made it their seat. The castle stood where the Church of Saint Václav stands today, but all that remains of it are a few Romanesque windows. With time several settlements sprang about the castle

Fig. 4.7 Olomouc: panoramic view of the city in Mendel's time. The Philosophical Institute, in which he studied, was close to the Church of Our Lady of the Snows

and later congealed into a community, which acquired the status of a royal town sometime between 1239 and 1246. Long before this date, in 1063, Olomouc became the seat of a bishop, the overseer of a church district or *diocese*. The bishop placed his throne, his *cathedra*, first in the Church of Saint Peter, but later moved it to the Church of Saint Václav, which thus became a *cathedral*. The bishop himself did not reside in the church, of course, but in a splendid palace nearby, though he also had an even more opulent residence in the city of Kroměříž, where most of his lands were located. In 1777, the Olomouc bishopric was upgraded to archbishopric, the seat of the archbishop, the overseer of several dioceses. The earlier developments made Olomouc the capital of Moravia, a position it retained until 1640, when during the Thirty Years War, the Swedes began an eight-year-long occupation of the city. In the wake of the occupation, all administrators fled to Brno, established their offices there, and so made Brno the capital of Moravia. Afterward, Olomouc not only failed to regain the title but for many decades could not recover from the devastation the Swedes wrought onto it: Most of the buildings were destroyed, and its population was decimated to a mere 2000 people of the original 30,000.

As might be expected of a politically salient city, Olomouc figured prominently in the history of the region. Not counting the usual remittent visitations, inflicted on all larger human communities (conflagrations, epidemics, wars, pogroms, as well as assaults by Mongolians, Turks, and other marauding hordes), Olomouc entered the history books as the place where the last Přemyslid king, Václav III, was murdered (in 1306 by an unidentified assailant); where the King of Hungary, Matthias Corvinus, was proclaimed the King of Moravia and Silesia (in 1469 in the Cathedral of Saint Václav); where the peace treaty between Vladislav II and Matthias Corvinus was signed in 1479; where Marquis de Lafayette, French statesman, general, and active participant in the American Revolution, was held prisoner from 1794 to 1797; where the Austrian imperial court sought refuge during the revolution of 1848, when Ferdinand I abdicated, and Franz Joseph I ascended the throne; and, finally, in 1850, where an agreement was signed between Austria and Prussia concerning the position of these two states in the German Confederation.

In 1840, the year in which Mendel arrived in Olomouc, its inner city was still corseted by lofty ramparts, erected on Maria Theresia's orders nearly 100 years earlier. It could be accessed only through one of four gates, a situation which already then was impeding traffic and becoming an obstacle to the growth of industry in the city. Yet the government in Vienna kept denying the city's petitions for permission to tear down the fortifications. To understand the government's reluctance to issue the permit, one must go back in history, a century before Mendel's arrival in the city. In 1740, the reigning emperor, Karl VI died without an issue: There was no male Habsburg eligible to ascend to the throne. Anticipating this development, Karl made arrangements (the so-called *Pragmatic Sanctions*) with the heads of the chief European powers, which were supposed to guarantee the ascendancy of his daughter, Maria Theresia, to the royal seat. In vain, for no sooner was his body interned in the Habsburg crypt than the armies of Prussia, Bavaria, and Saxony were on the move to partition the Austrian Empire among themselves. The War of the Austrian Succession had begun. The Prussians had not even bothered to declare war on Austria; they simply marched into Silesia, proclaiming it their possession. They even went a bit further south and occupied Olomouc as well. They left the city only after two years, when Maria Theresia and Frederick II signed a peace treaty in which the former managed to salvage a small part of Silesia, henceforth called Austrian or Czech Silesia. Realizing that the Prussians might return and march through the Moravian Gate all the way down to Vienna, Maria Theresia decided to turn Olomouc into a fortress that would stand in the way of any future invaders from the north. And this she did. During the years 1742–1756, Olomouc became the "Fortress of Moravia," one of the strongest and largest citadels in the empire. The test of its strength came in 1758 when, as she anticipated, Frederick's army returned to besiege Olomouc. For seven weeks, the Prussians tried to take the city, but failed. A century later, however, the situation had changed, and the massive ramparts no longer served their purpose. It was again the Prussian army which proved this point when, in 1867, during another war, they simply bypassed Olomouc on their march to Vienna. Toward the end of the nineteenth century, Vienna finally gave permission to demolish most of the ramparts and replace them with parks. Of the four gates, only one was left standing as a historical monument, the Theresia Gate commemorating Maria Theresia's visit to the city in 1752.

Mendel spent most of his two and half years in Olomouc in the inner city, delineated on one side by the ramparts and on another side by the Mlýnský potok (Mill Brook) flowing into the Morava River. It was in the inner city where the university and most of the historical buildings and monuments were located. There

Fig. 4.8 The *Philosophische Lehranstalt* (Philosophical Institute) of Olomouc, at which Johann Mendel studied from 1840 to 1843. It was then building No. 14 in *Obere Fronleichnamgasse* (Upper Corpus Christi Street), which is now called Křížkovského Street

was much to see and admire,[28] including several squares enclosed by palaces and burgher houses and decorated by sculpture-adorned fountains or columns. The largest was the Upper Square with the town hall tower, astronomical clock, and the Holy Trinity Column. Scattered throughout the inner city were several churches, built or rebuilt at different times and in different styles, with Baroque prevailing. And then there were, of course, the ramparts, which with their decorative gates, as well as casemates and dungeons incarcerating political prisoners, provided an imposing, if somewhat intimidating, sight. The *Philosophische Lehranstalt* (Philosophical Institute) was located in the *Obere Fronleichnamgasse* (Upper Corpus Christi Street) No. 14,[29] in a splendid four-story palace (Fig. 4.8). Today, the street bears the name of Křížkovský, who arrived in Olomouc one year earlier (1839) than Mendel to study at the same Institute.[20] At the time Mendel and Křížkovský attended it, the Institute was part of the university, but a decade later, it was incorporated in the *Gymnasium* as its seventh and eighth grade. The Olomouc University was founded by the Jesuits,[30] who arrived in the city in 1566. The Collegium (college) they opened in the same year acquired a university status in 1573 and so became the second oldest university in the Czech lands, the oldest being Charles University in Prague, founded in 1348. Following the dissolution of the Jesuit order in 1773, the Olomouc University was taken over by the government and five years later was moved to Brno. It was moved back to Olomouc in 1782, but only as a Lyceum, a form of a secondary school on par with a *Gymnasium*. In 1827, it was elevated back to the university status under the name *Kaiserliche und königliche* (or k.k. for short) *Franzens-Universität* (Imperial and Royal Francis University), which it bore during Mendel's stay in Olomouc. In 1850, however, the

Fig. 4.9 The house No.171 (now No.18) at the *Dolní náměstí* (*Lower Square*), in which Johann Mendel rented a room during his studies in Olomouc. The city has still not commemorated this fact

government began dismantling it in punishment for its involvement in the 1848 revolution. The liquidation was completed in 1860, when only the theological faculty and the large university library remained in Olomouc. The university was not revived until after World War II, when it assumed the name of the Moravian-born Czech historian and statesman František Palacký. While in Olomouc, Mendel might have witnessed in 1841 the arrival of the city's first train, and so the completion of a railroad connection to Brno and Vienna.[31] Four years later, the railroad also connected Olomouc with Prague. And around the same time, a railroad connected Brno with Bohumín in northern Silesia (via Přerov), which made it possible to travel from Brno to Mankovice (and hence Hynčice) by train on the so-called Ferdinandka (the Northern Railroad of the Emperor Ferdinand).

Mendel's first worry upon his arrival in Olomouc was to find a cheap lodging, since the Philosophical Institute, like the *Gymnasium* in Opava, did not provide any. He succeeded in renting a shabby room in the house No. 171 (now No. 18) at the *Unteren Platz*[32] (Lower Square), a short walking distance from the Institute (Fig. 4.9). Next, he set out to secure a source of income in the form of tutoring laggard students, the way he did in Opava. Taking into consideration his circumstances, the Institute waved the tuition that it required other students to pay, but for the rent and for his sustenance, he had to earn money somewhere. To his great consternation, finding students for tutoring proved to be very difficult. Being a newcomer to the city, he had no contacts, no friends, nobody who could recommend him to the parents looking for a tutor for their children, and nobody who could bring him in contact with such families. In contrast to Opava, a significant proportion of the inhabitants of Olomouc were speaking the Czech language, which he still had not mastered. Ultimately, he managed to earn enough to stay alive, but barely so. Undernourishment combined with constant worry about making ends meet soon began to affect his health. Křížkovský, who found himself in the same situation, could not take it and, after few months in Olomouc, quit the school, returned to Opava, took the course for elementary school teachers, and taught for a while at the village school not far from Opava.[20] Mendel continued to struggle through the first semester, all the way to the examinations. He passed the examinations in mathematics and Latin with excellent marks and then broke down before he completed the rest of them.[33] He fell ill and returned to Hynčice, which meant that he forfeited the entire first year because, unlike the *Gymnasium* in Opava, the Institute did not allow him to take the examination later. A half year of hardship, self-denial, and privation came to nothing! Was he aware that this might happen and so went home with the intention of not returning, like Křížkovský before him? Or did he expect a postponement of the examination and was shocked when it was denied? One might even ask: Was the illness a ruse, like students sometimes use when they need more time to prepare themselves for an examination? Or was Mendel really so sick that he could no longer care about the consequences? We cannot answer these questions because we have no way of knowing how sick he really was and what transpired in his head at that time. There are, however a few facts that restrain a temptation to speculate wildly. First, there would be at least two more of such mysterious episodes of illness that coincided with examinations, one at the end of his third year in Olomouc and the fourth several years later in Vienna. At least the last of these could not be attributed to physical privation, and there is no record indicating that Mendel consulted a physician during any of the episodes. Second, when he returned to Olomouc some four months later (not after one year, as he claimed in his autobiography,[25] he lodged at the same address as in the first year. Apparently he kept his room, expecting to return for the next semester. Third, in the examination document kept by the Institute, Mendel's failure to complete the first school year is commented on laconically thus: *Während der Prüfungen Krankheitshalber ausgetreten* (withdrew during examinations owing to illness) as if the school officials did not expect him to return. When he did return in the fall, he had to enroll anew; he was registered under a different number (42 instead of 49 which he had originally) and was classified as *Repetent* (i.e., a student repeating a school year), which meant that he now had to pay tuition.[33] Fourth, when he again reported sick at the end of the second school year, he *was* allowed to take the examinations a few weeks later than the other students. All this gives an impression that initially the faculty might have suspected Mendel of simulating his illness to gain time for preparation, but later they realized that he was a serious, sincere, gifted, if rather nervous, perhaps even somewhat mentally labile, student, who deserved special treatment. If that was indeed their assessment of young Mendel, it may not have been far from the truth.

In the meantime in Hynčice, great changes were in the making.[18] Johann's father never regained his former strength, remaining an invalid for the rest of his life. Johann's sister Veronika got married, and her husband, Alois Sturm, agreed to take over the Mendel farm, which Anton could no longer manage. On August 7, 1841, the seller, the buyer, and eight witnesses signed the sale contract. A year later, Johann's parents retired, and the new owner took possession of the farmstead No. 58. Traditionally, when a farmer retired, he and his wife remained on the farm, but moved to their *Ausgeding* or *výměnek* (retirement quarter), often just a single room with a pantry, in the back of the house. They were expected to lead a separate life from that of the buyer's family; on a small pension, the new owner agreed to pay them for as long as they lived. In practice, however, if there were no animosities between the two parties, the retired couple helped the young couple in various ways, if by no other means than taking care of the children. It seems that Anton and Rosina did this for a while but later, for unknown reasons, moved to another village. Theresia, presumably, stayed on the farm for 11 additional years, until she got married in 1852. According to the sale contract, the inventory of the farm at the time of the ownership change included two horses, four cows, one heifer, one bull calf, and assorted poultry. The total value of the property was assessed at 400 guldens.[34] The contract specified not only the pension the seller was to receive but also Sturm's obligations toward Theresia and Johann. For Theresia, he was to provide the dowry at the time of her marriage. To Johann, as we know already (see Chap. 3), Sturm committed himself to pay ten guldens yearly for the duration of his studies and then 100 gulden at the time, when Johann entered a profession or became a priest.[35] In the latter case, Sturm would also be obliged to finance Johann's *Primiz*, the festive first mass following his ordaining ceremony. Should Johann, for some reason, have become incapable of leading an independent livelihood, he would have been entitled to free quarters in the *Ausgeding* and one *Metze* (0.48 acres) of arable land on each of the farm's fields.

The ten guldens was a pittance, which could not support Johann in Olomouc, but when the family circle discussed the situation, Theresia decided to forgo part of her dowry in Johann's favor, so that he could complete his studies.[23] In 1841, when she made the decision, Theresia was 11 years old. Did she really know what she was doing? Did she realize that her act might later decrease her assets at the marriage market, as it was then practiced? The family undoubtedly did their best to dissuade her from her decision, but she insisted on helping her beloved brother. And so an arrangement was made by which Alois Sturm was to increase correspondingly his support of Johann's studies. With this prospect, Johann returned to Olomouc in the fall of 1841, where, as we know already, he had to start from the beginning again.

Being a part of the university, the Philosophical Institute used the professors of the philosophical faculty to teach the different subjects according to their expertise.[33] It was the first time that Mendel heard lectures delivered by a specialist rather than by an omniscient *Klassenlehrer*. The Institute divided the study into four semesters spread over two school years, with each semester concluded by examinations. It offered both obligatory and elective courses. All students had to take religion and philology (Latin) in both years, as well as theoretical philosophy, mathematics, and natural history in the first year, and practical philosophy (ethics), physics, and world history in the second year. The elective courses were pedagogy (the art and science of teaching), Austrian history, diplomacy and heraldry, numismatics, history of philosophy, classical literature, Greek philology, esthetics, and agriculture. Philosophy, physics, and mathematics were taught in Latin; all remaining courses were conducted in German. The students had on average 16 hours of obligatory lectures per week in the first year and 20 hours in the second year. Of these, mathematics took seven hours per week in the first year and physics eight hours per week in the second year.[36] At the end of the semester, students were examined in all the obligatory subjects and those elective subjects they signed up for. In each subject, the students received separate marks for "application" (*Verwendung*, presumably diligence) and actual knowledge. In addition, they also received a mark for their overall performance in all the subjects combined. In the first, aborted semester of the school year 1840/1841, Mendel received the top mark "*e*" (*eminens*) for diligence in all obligatory subjects and for knowledge in mathematics and philology; because of his illness, he was not examined in the other subjects. In the four semesters of the school years 1841/1842 and 1842/1843, he received top marks in all the obligatory subjects, except philosophy, in which he had to satisfy himself with the second best mark "1" (*prim.*). During the first two semesters, he chose classical literature as his elective but for some reason was not examined in it. In the third and fourth semesters, he received top marks in elective pedagogy. He had natural history only in the first, aborted semester, presumably for four hours per week. In the school year 1841/1842, he was supposed to have had natural history again, but did not, apparently because of the professor's illness.[34]

The director of the Institute was an aristocrat, the canon Eduard Ritter von Unckrechtsberg (1797–1870), with whom Mendel appears to have had little contact.[32] His teachers were Dr. Michael Franz von Canaval (classical literature, philology, and esthetics), Prof. Dr. Tomáš Eichler (religion), Prof. Dr. Josef Wittgens (philosophy), Dr. Jan Helcelet (natural history, agriculture), Prof. Dr. Johann Fux (mathematics), and Prof. Dr. Friedrich Franz (physics). It is doubtful that Mendel learned much of natural history from Helcelet, who was apparently gravely ill. (Later, when Helcelet transferred to Brno and became ill again, Mendel substituted for him at the *Oberrealschule*.) The best of the faculty were Fux and Franz. Johann Fux (or Fuchs, 1785–1848) was an Austrian educated by the Piarists, whom he joined in 1801, but left a few years later. After teaching at different schools, he came to Olomouc in 1817 and remained there until his death at the age of 63. He authored the textbook *Vorlesungen über reine Mathematik* (Lectures on Pure Mathematics), which was approved for use in *Gymnasien* of the Habsburg Empire.[37] Friedrich Franz (1796–1860) was a Czech, born in Vysoké Veselí in northern Bohemia. He became canon, at the collegiate church of the Premonstratensian order in Nová Říše in southern Moravia.[38] With a doctorate from Prague, he went on to teach physics and applied mathematics at the Philosophical Institute in Brno from 1823 to 1842. During these 19 years, he lived in the Augustinian Abbey at Staré Brno and so came to know the friars there quite well, including the Abbot Franz Cyrill Napp. It was then a common courtesy among the monasteries to extend, space allowing, this form of hospitality to monks of different orders. In Brno, Franz became interested in daguerreotype, an early form of developing photographic plates coated with silver made sensitive to light by exposure to iodide fumes. Invented by Louis J. M. Daguerre in1836, the method became very popular in the middle of the nineteenth century, and Franz had been its chief proponent in Moravia. In 1842, Franz moved to Olomouc and became Mendel's physics teacher in the school year 1842/1843. In June 1843, he received a letter from an unidentified member of the Saint Thomas Abbey (it was not from the abbot, as some biographers of Mendel claim) asking him to recommend a suitable candidate from among his students for admission into the Augustinian order. Franz informed his students asking them to contact him should any of them be interested. Two students did, one of them being Mendel. As physics was the major subject in the second school year at the Institute, Franz came to know Mendel quite well. Of the two, he recommended Mendel, having this to say about him: Mendel "achieved top marks nearly continuously in both school years at the Philosophical Institute and is of a very solid character...In my area of expertise he is probably the best. He knows some Czech, not enough though, but is prepared to make an effort to master the language during the years of theological study."[42] Napp, knowing Professor Franz well and trusting fully his judgment, accepted Johann Mendel on the strength of this recommendation.

Inevitability and Serendipity: Part 2

Abbot Napp knew, of course, that the school year at the Philosophical Institute ended at the beginning of summer and that this was the time when young men finishing their studies there were making decisions about their future, and hence the best time of the year to inquire about possible candidates for a career in the abbey. In this regard, the timing of the inquiry was inevitable. All the other circumstances of the inquiry, however, were knots of serendipity. The coincidence of years, persons, and places were purely accidental. Had Mendel finished at the Institute one year earlier, as he should have, Franz would have just missed him. Had Franz found a different place to stay during his 19 years in Brno, the inquiry about a suitable candidate might not have gone to Olomouc, but to other places. Had Napp not been the abbot of the Saint Thomas Abbey, it would have been rather unlikely that Mendel would have ended up there, but even if he had, he would not have been allowed to carry out his experiments. And so on, any of the great number of alternative combinations of events would not have resulted in the development that made history. The coincidences of time, persons, and place that brought Mendel to the Saint Thomas Abbey at Staré Brno now appear almost like a *deus ex machina* resolution of a plot in a bad theater piece. But they did happen; they were reality, not a play.

The actual conjunction of occurrences led Mendel to the second most profound decision of his life. The first one was *not to become* a farmer, and the second to *become* a priest. How did he reach this latter decision? Catholic writers[39] would like us to believe that Mendel responded to a sacred calling that he felt from his childhood, presumably under the influence of Johann Schreiber. There is, however, no evidence known to us that might support this claim. On the contrary, there are at least four facts that speak against it. First, the enlightened Schreiber was the most unlikely priest to inflame a religious zeal in a child. He might have incited and kindled Mendel's interest in natural history, but not in priesthood. Second, in all his studies at the primary and secondary school level, Mendel's performance was outstanding in all the subjects except one—religion. Both in the elementary schools in Hynčice and in the *Hauptschule* in Opava, his performance in religion was merely "good." This result certainly does not indicate religious zealousness on Mendel's part. Third, as we shall learn later, Mendel had problems with discharging some of the basic duties of a Catholic priest, such as comforting the sick and administering the last unction to the dying. These were certainly not signs of having achieved in priesthood the yearning of his life. Fourth, and most important, there is Mendel's own testimony. In his autobiography, he writes: *Der ehrfurchtsvoll Gefertigter fühlte das es ihm nicht möglich sei, solche Anstrengungen noch weiter zu ertragen. Er sah sich daher gezwungen, in einen Stand zu treten, der ihn von den bitteren Nahrugssorgen befreite. Seine Verhältnisse entschieden seine Standeswahl.*[25] ("The respectful undersigned felt that it was not possible for him to endure such exertions any further. He thus found himself *forced* to enter a profession that would free him from bitter worries about making a living. His circumstances determined his choice of a profession." The "exertions" here refer to the hungry years Mendel spent in Olomouc.) These are clear and frank words. He penned them years after he became a priest, and so they cannot be interpreted as a mere poise. Neither can they be construed to mean, as some biographers[29] argue, that by *Standeswahl,* Mendel meant his entry into the abbey rather than his choice of priesthood. When entering the abbey, Mendel knew that, if he wanted to stay there, he would have to become a priest; so it is difficult to see the difference, and anyway, from the context it is apparent that by *Standeswahl,* Mendel meant his profession, that is, *priesthood,* and not a transient passing through the abbey's gates. Like Křížkovský before him, Mendel could no longer take any more hungry years, and he realized that the road to a university-level education and perhaps a professorial position led through priesthood. The fact that nearly all the professors he knew were priests made that point very clearly. The decision to become a priest was not, however, a sudden one. From the sales contract, it is clear that Mendel and his family had considered this possibility at least since 1841, but most likely even earlier. Perhaps, he hoped secretly that he might find some other way than priesthood to attain his goal, but the Olomouc experience made it obvious to him that this would not be the case. After the completion of his studies at the Philosophical Institute, he could no longer expect further help from Hynčice, and he lacked the

Fig. 4.10 Mendel to his parents. The letter then continues: ...*They will find their consumer for sure. I also look forward to something else. You would guess what it is if I were to tell you that it starts with the letter "B": Buttermilk!* The letter is undated but Mendel probably wrote it shortly after he joined the abbey at Staré Brno. A photocopy of the original letter written in German is deposited at Brno's Mendelianum. Its printed version is to be found in[13], p. 113; the translation is ours

strength to support his further study all by himself. And so he decided to become a priest, but first he returned to Hynčice for vacations and to gorge on pickles and buttermilk (Fig. 4.10).

References and Notes

[1]"... everything we encounter leaves tracks behind; everything contributes imperceptibly to our education..." J. W. Goethe: *Sämtliche Werke, Briefe, Tagebücher und Gespräche.* Vol. 4, book 7, p. 798. Deutscher klassische Verlag 1985

[2](a) Květ R (1998) Trasa jantarové stezsky na území Moravy v pravěku i historické době. Vlastivědný slovník moravský (Brno) 50; 382–387. (b) Květ R (2002) Staré stezky v České republice. Moravské zemské muzeum, Brno

[3](a) Lutterer I, Majtán M, Šrámek R (1982) Zeměpisná jména Československa. Slovník vybraných zeměpisných jmen s výkladem jejich původu a historického vývoje. Mladá fronta, Praha

[4]New Catholic Encyclopedia (2003) 15 volumes, 2nd edn. Thomson—Dale, Detroit

[5]Lawrence CH (2001) Medieval monasticism: form of religious life in Western Europe in the middle ages, 3rd edn. Longman Group, London

[6](a) Zemek M, Bombera J, Filip A (1992) Piaristé v Čechách, na Moravě a ve Slezsku 1631–1950. Prievidza 1992. (b) Bombera J (1950) Dějiny piaristické koleje v Lipníku nad Bečvou

1634–1884. Lipník nad Bečvou. (c) Musilová B. Historie piaristů v Lipníku nad Bečvou. http://mesto-lipnik.cz. (d) Bartůšek V (personal communication)

[7]Hosák L (1967) Historický místopis Moravy a Slezska v letech 1848–1960. Profil, Ostrava

[8](a) Dvořáček P, Žůrek Z (2002) Lipník nad Bečvou. Klíč k Moravské bráně. Město Lipník nad Bečvou (b) Hajduk A (1907) Geschichte der Stadt Leipnik; ein Buch für jeden Leipniker. Nach original-Urkunden und anderen zuverlässigen Quellen. L Gundel, Leipnik

[9](a) Kádner O (1931) Školství v republice Československé. In: Dědina V (ed.) Československá vlastivěda, vol. 10: Osvěta. Bohumil Janda, Praha, pp. 8–222. (b) Timp H (1868) Die Problematik "Klassenlehrer oder Fachlehrer" in der Gymnasialreformen von 1792 bis 1849. Pedagogischer Verlag Eugen Kettel, Wien

[10]Barzun J (2000) From dawn to decadence, 500 years of Western cultural life, 1500 to the present. Harper Collins, New York

[11]Snow C (1959) The two cultures and the scientific revolution. Cambridge University Press, London

[12]Ulrich J (1907) Gregor Joh Mendel Biografische Skizze. Rainer Hosch, Neutitschein, Reprint from the Illustrierter Neutitscheiner Volkskalender für das Jahr 1908

[13]Křiženecký J (1965) Gregor Johann Mendel 1822–1884. Texte und Quellen zu seinem Wirken und Leben. J.Ambrosius Barth, Leipzig

[14]This date of admission is given on the document issued by the *Gymnasium* upon the completion of Mendel's studies (see[40])

[15](a) Ens F (1835) Geschichte der Stadt Troppau. Carl Gerold, Wien. (b) Müller K, Žáček R (eds.) (2006) Opava. Nakladatelství Lidové noviny, Praha

[16]Schwerdfeger J (1930) Die Geschichte des deutschen Staatsgymnnasiums in Troppau. Zum 300-jährigen Jubileum der Anstalt. Graphische Kunstanstalt Heinz u. Co., Troppau

[17]In 1815, the rulers of Russia (Tsar Alexander), Austria (Emperor Franz I), and Prussia (King Frederick William III) formed a Holy Alliance to suppress the revolutionary movement in Europe. All the other European rulers, except the king of England, signed the agreement

[18]Iltis H (1924) Gregor Johann Mendel. Leben, Werk und Wirkung. Julius Springer, Berlin. An English translation by E. and C. Paul was published under the title *Life of Mendel* by George Allen & Unwin, London 1932

[19]Signatures of these professors appear on the document issued by the *Gymnasium* upon Mendel's completion of his studies in 1840 (see[40])

[20]Elchler K (1904) Životopis a skladby P. Křížkovského. Knihotiskárna benediktinů rajhradských, Rajhrad

[21]Orlík J (1964) Počátky gymnazijního muzea v Opavě. In: J. Duda (ed.) 150 let slezského muzea: 1814–1964. Krajské nakladatelství, Ostrava. See also chapter "Zakladatel a kustod gymnazijního muzea" in reference[22] with references therein

[22]Myška M (2003) Faustin Ens. Životní příběh slezského intelektuála doby předbřeznové. Matice slezská, Opava

[23]Schindler A (1902) Gedenkrede auf Prälat Gregor Joh. Mendel anlässlich der Gedenkenthülung in Heinzendorf am 20. Juli 1902. Private print of the author, which did not get into the bookstores. It is reprinted in[29], pp. 77–100; for an English translation, see[41]

[24]Heinenen W (1944) Der junge Genius Johann Gregor Mendel. Ein biographisher Roman, 2nd edn. Fels Verlag Dr Wilhelm Spael KG, Essen

[25]Mendel's "Autobiographie" (1850) In: Křiženecký (1968)[13], p. 74–77. For an English translation, see A. Iltis: Gregor Mendel's autobiography. J Heredity 45:231—234, 1954. Purkyně University at Brno published a bibliophilic edition of the autobiography under the title Gregorii Mendel Autobiographia Iuvenilis. Ad centesimum Quinquagesium J. G. Mendel natalen. Universitas Purkyniana Brunensis 1972

[26]Die österrechisch-ungarische Monarchie in Wort und Bild. Vol.17: Mähren und Schlesien. K. k. Hof- und Staatsdruckerei, Eduard Beyer, Wien 1897

[27]Doležil, H. Politické a kulturní dějiny král. hlav. města Olomouce. Parts 1–3. Matice školská, Olomouc 1903–1905

[28]Nešpor V (1998) Dějiny města Olomouce, 2nd edn. Votobia, Olomouc

[29]Sajner J (1974) Johann Gregor Mendel. Leben und Werk. Ein Bildbuch, 2nd edn. Augustinus, Würzburg

[30](a) Navrátil J (ed.) (1973) Kapitoly z dějin olomoucké university 1573–1973. Profil, Přerov 1973. (b) Nešpor V (1947) Dějiny university olomoucké. Nakladatelství Národního výboru Olomouce, Olomouc

[31]Hora-Hořejš P (1997) Toulky českou minulostí. Vol. 6, Příběhy a postavy českého národního obrození. Barone and Via Facti, Praha, pp. 168–170

[32]Richter O (1943) Johann Gregor Mendel wie er wirklich war. Verhandlungen des Naturforschendes Vereines, Brünn 74(2):1–262

[33]Weilig F (1993) Johann Gregor Mendel. Der Mensch und Forscher. Forscher in der Kontroverse. Medizinische Genetik 5:35–49, 208–222, 274–288, 379–393, 1993; 6:35–50, 241–252, 1994

[34]Orel V (1996) Gregor Mendel. The first geneticist. (Translated by Stephen Finn). Oxford University Press, Oxford

[35]Simunek MH, Thümler U, Sekerák J (eds) The Letters on G. J. Mendel. Correspondence of William Bateson, Hugo Iltis, and Erich von Tschermak-Seysenegg with Alois and Ferdinand Schindler. Studies in the history of sciences and humanities vol. 28. Prague 2011

[36]Weiling F (1991) Historical study: Johann Gregor Mendel 1822–1884. Am J Med Genet 40:1–25

[37]Weiling F. Fux (vereinzelt Fuchs), Johannes. *Archiv für Geschichte der Naturwissenschaften* 10: 341, 1983

[38]Weiling, F. Franz, Friedrich (1983) Archiv für Geschichte der Naturwissenschaften 7; 341

[39]Ryan PE (1941) Gregory Mendel. Abbot and discoverer of the laws of heredity I –XXXI. *The Messenger*, March 1938–November 1941. This series of articles was published anonymously in the journal of the Roman Catholic Diocese of Convigton, Kentucky, on the request of Bishop Howard. The bishop sent R. L. de Waegenaere to Brno to collect material for the article and then instructed Father Ryan, his secretary, to write the articles. (This information comes from a letter Professor Edward O. Dodson, University of Ottawa, Canada, wrote to Dr. Vítězslav Orel, Mendelianum, Brno. The letter is deposited at the Mendelianum.)

[40]Orel V, Marvanová L, Sajner J (eds) (1965) Iconographia Mendeliana pictorial publication issued for the centenary of the publishing of the discovery of the principles of heredity by Gregor Johann Mendel in Brno. Moravian Museum in Brno, Brno

[41]Olby RC (1985) Origin of Mendelism. Constable, London 1966, 2nd edn. University of Chicago Press, London

The Third Decade: The Vow

5

Primum, propter quod in unum estis congregati, ut unianimes habitetis in domo et sit uobis anima una et cor unum in deum.

Regula Sancti Augustini[1]

In articles about Mendel, it is often stated that Brno (Fig. 5.1) was a provincial town. This statement is correct in the sense that in Mendel's time Moravia was a province of the Austro-Hungarian Empire. This, however, is presumably not what the authors of the articles mean by the word "provincial." They do not use "provincial" in reference to a place, but to a society "limited in outlook" or "lacking the polish of urban society."[2] But was Brno really such a town, or is this judgment rather an expression of the authors' own provincialism and unfamiliarity with this part of Europe? To answer this question, let us take a brief look at Brno's geographical location, history, and cultural standing in Mendel's time.

The City on Two Rivers

The modern city of Brno lies in the foothills of the Czech-Moravian Heights, residing slightly north of the confluence of two rivers, Svratka and Svitava (Fig. 5.1). Both rivers now flow through it: Svratka in the west for a length of nearly 30 kilometers and Svitava in the east for about one half of that length. Around the inner city, the hills Petrov, Špilberk, Červený kopec (Red Hill), and Žlutý kopec (Yellow Hill) form a semicircle, which opens southward into the Southern Moravian Lowlands, a fertile agricultural region that connects with the Danube Basin.

For some 400,000 years, members of the human race have lived, intermittently, in the valley cut out by the two rivers. Stránská skála, not far from Brno, has yielded evidence of the presence of *Homo erectus* and of Cro-Magnon people over the entire region.[3] Celts, Germanic people, and Slavs, in that order, were the last three great waves of humanity that swept through the area, and each could lay claim to naming the city. The Celtic *brynn* ("hill town"), the Germanic *Brunno* ("well town"), and

J. Klein and N. Klein, *Solitude of a Humble Genius - Gregor Johann Mendel: Volume 1*,
DOI 10.1007/978-3-642-35254-6_5,

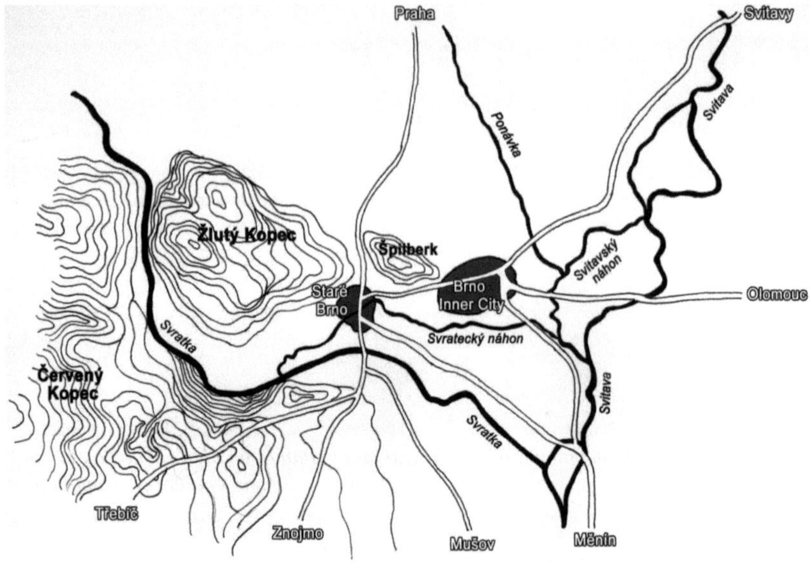

Fig. 5.1 The location of Brno and Staré Brno at the confluence of the rivers *Svratka* and *Svitava*: The situation in the fourteenth century. The smaller stream *Ponávka* still flows through Brno, though with an altered course. The two races (*Svratecký náhon* and *Svitavský náhon*), which arose from side branches of the two rivers, are now obliterated. In the past they supplied water for mills built on them. Note that roads running from north to south and from west to east crossed in Brno

the Slavic *brn* ("muddy place") have all been suggested as the roots of the Czech *Brno* and the German *Brünn*.[4] Slavs arrived in the area in the fifth century and established settlements at the ford where ancient trading routes crossed the Svratka River. Subject to spring flooding, the area was murky and sludgy, so calling it a "muddy place" would have made sense.

At the beginning of the eleventh century, after the disintegration of the Great Moravian Empire and the annexation of Moravia by the lords of the rising principality of Bohemia, the Přemyslids built a castle there. Where exactly the castle stood is uncertain. One possibility is a site between the ford and today's Mendel Square. By that time, however, the center of activity in that region had already begun to shift from the ford to the area around the Petrov Hill, so another possibility is that it stood on that mount. Alternatively, there may have been two castles: an older one near the ford and a younger one on the Petrov Hill. The hill is 247 meters above sea level and three of its sides fall precipitously to the plane below, while the fourth side connects via an isthmus with the Špilberk Hill. Presumably, the castle stood on top of the precipitous part and was enclosed by inner fortifications. Eventually, a settlement arose on the isthmus outside of the castle, which was surrounded by outer fortifications. The original settlement close to the ford was too far away to be included in the castle's outer fortified area, and so it developed separately from the *Město* (City) *Brno* and came to be called *Staré* (Old) *Brno* (Fig. 5.2). The inner and the outer parts of the Petrov castle had separate churches—St. Peter and St.

Michael, respectively. Both churches were originally Romanesque rotundas, but were later replaced by more imposing structures, the (neo-) Gothic St. Peter-St. Pavel (Paul) Cathedral and the Baroque St. Michael Church.

Further expansion of the settlement northeast of the Brno castle gave rise to the heart of modern Brno. The inner and outer castle fortifications were then torn down, and new bulwarks were erected surrounding what we shall from now on refer to as the *inner city* (Fig. 5.3). By the fourteenth century, the inner city encompassed an area of 47 hectares of land, on which stood over 500 houses, inhabited by some 7,500 residents. Villages were scattered all around the city, with one of them being the original Staré Brno. Over time, the villages turned into suburbs and then into different quarters of the Brno City. The original inner city had two large and several smaller squares. The former were the Upper and the Lower squares, which are now the *Zelný trh* (Cabbage Market) and the *Náměstí svobody* (Liberty Square), respectively; the latter included fish, horse, hay, coal, and fowl markets (Fig. 5.3; for a map with finer resolution, see Fig. S5.1).

There were altogether eight churches in the inner city and several additional ones in the surrounding villages. Associated with some of the churches were monasteries of the different religious orders: Dominican with St. Michael, Minorite with St. John, and Augustinian nuns with the Assumption of Virgin Mary churches.[5] Several other orders established monasteries in Staré Brno and other villages outside of the fortifications. The inner city consisted of four quarters—the Brněnská, Měnínská, Běhounská,[6] and Veselá čtvrt'—each accessible by a separate town gate of the same denomination; the Měnínská quarter had, however, an additional gate called Židovská (Jewish) brána. As immigrants from different parts of Europe began to settle down in the city, they tended to stick together according to their ethnicity. Germanic and Flemish people established settlements in the Běhounská and Veselá quarters and used the St. Michael Church as their spiritual center. The French-speaking Walloons congregated around the St. Nicholas Church that stood in the middle of today's Liberty Square. After the Germanization of the Walloons, the church lost its purpose and was ultimately taken down. The Jews colonized an area around their synagogue, which stood at a site now taken up by the Church of Mary Magdalene on today's Masarykova Avenue. The colony comprised a part of the Měnínská quarter. The Czechs, meanwhile, were scattered throughout the city, but their heaviest concentration was in the Brněnská and Měnínská quarters.

The fortifications served the city well, since the sieges of Brno by the Husites in 1428 and 1430, the Swedes in 1643 and 1645, and the Prussians in 1742 all failed. In 1805, however, the Austrian army abandoned the city to Napoleon's forces, which then took it without a fight. On December 2 of that year, Brno's burghers, standing on the bulwarks of the Petrov Hill, witnessed Napoleon's victory in the battle of the three emperors near the village of Slavkov (Austerlitz). By that time it had become obvious to everybody except the government officials in Vienna that the fortifications had become a military anachronism. The demolition of the walls was a slow process, which extended over half a century and was not completed until 1860. A castle built (or more likely rebuilt) on the Špilberk[7] Hill in the thirteenth

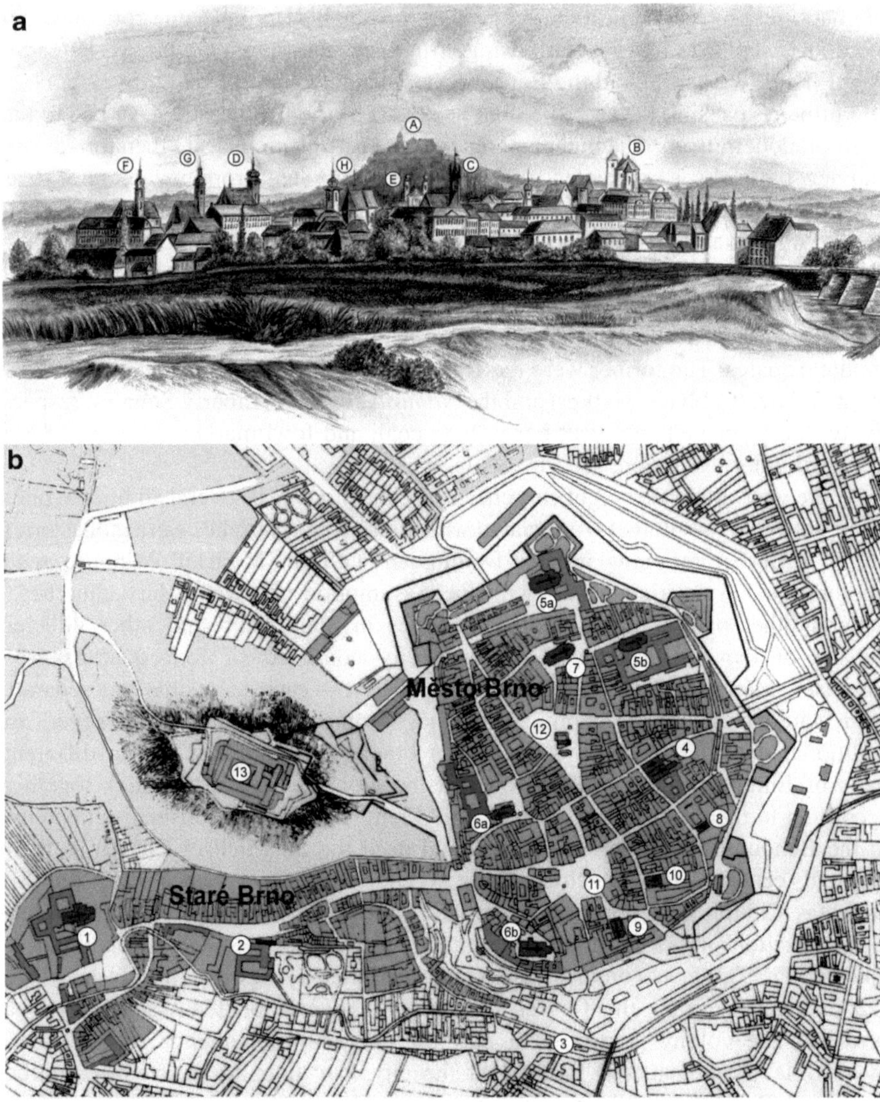

Fig. 5.2 (continued)

century and later turned into a citadel was also rendered militarily useless by Napoleon's demolition squad. After the departure of the French army, however, it was turned into a prison for both common criminals and political activists. Among the latter were several famous names, including the Italian carbonari ("charcoal burners") Silvio Pellico and Piero Maroncelli, who in their memoirs drew Europe's attention to the inhumane conditions under which the prisoners were kept there.[8]

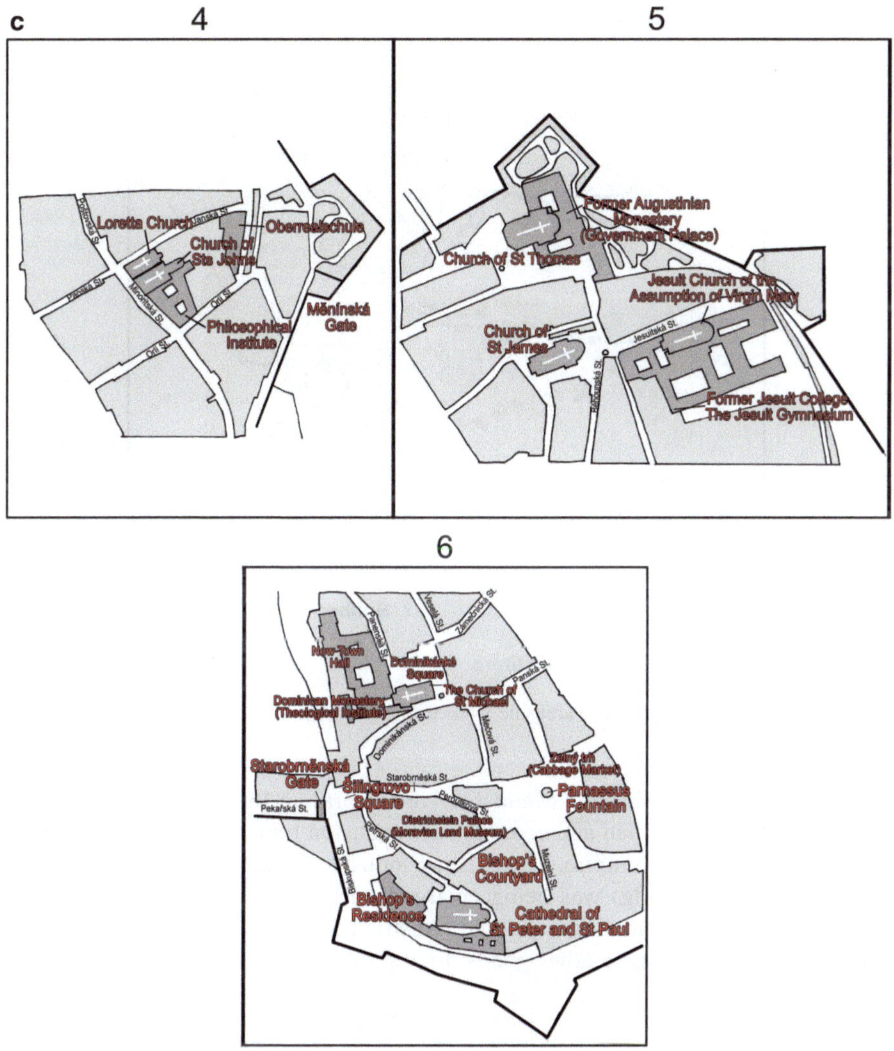

Fig. 5.2 (**a**) Brno, as it might have appeared to Mendel's eyes when he approached this fifth station on his "amber" journey after education. (A) The menacing silhouette of a hill topped with a citadel seen in the background is Špilberk. (B) St. Peter's Cathedral on Petrov Hill. (C) The Old Town Hall under construction. (D) The Church of St. James. (E) The Church of St. Michael. (F) The St. Thomas Church of the Augustinians. (G) The Jesuit Church of the Assumption of St. Mary. (H) The Minorite Church of St. John the Baptist and St. John the Evangelist. (**b**) Ground plan of Old (*Staré*) and City (*Město*) Brno in the middle of the nineteenth century. The plan places the churches shown in Fig. 3A to their respective locations in the city. It shows the city just before disposing of the ramparts and expanding into a new appearance like an imago bursting out from a pupa. The churches with their towers were more than signposts in the city; they marked the distribution of people with different religious inclinations, ethnicity, and nationality within the city. The annotated locations are these: (1) The Church of the Ascension of the Virgin Mary and the St. Thomas Abbey associated with it at Staré Brno. (2) The Chapel of St. Anna and the General

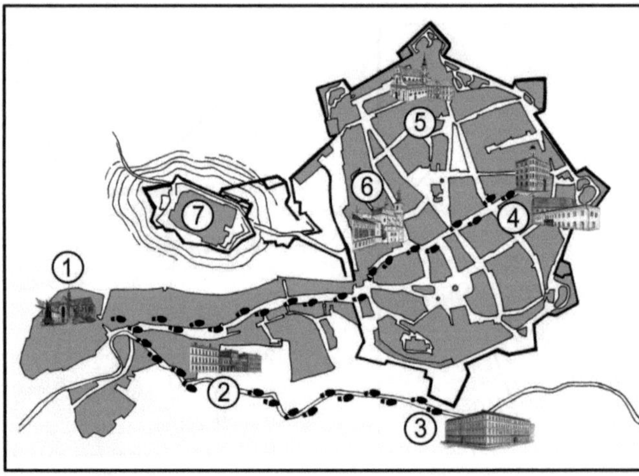

1 St Thomas Abbey
2 Hospital
3 Technical School
4 New Oberrealschule and Philosophical Institute
5 Original St Thomas Abbey
6 Theological Institute
7 Špilberk

Fig. 5.3 Mendel's Brno: the two areas he walked to most often

By the time the 21-year-old Mendel arrived in Brno in 1843, the city had been in the midst of a grand urban and economical expansion transforming it into a major industrial center of the empire. The existing industry (chiefly wool, but also cotton and linen manufacturing) was being enlarged, mechanized, and modernized; not without reason was Brno nicknamed the "Austrian Manchester" in those days. New industries manufacturing leather goods, food products, chemicals, machinery, and

Fig. 5.2 (Continued) Hospital associated with it at Staré Brno. (3) The old Technical School at the Trnitá and Tovární streets outside of the ramparts enclosing the inner city. (4) The complex of buildings including three churches (Loretto, St. John the Baptist, and St. John the Evangelist) associated with the Minorite Monastery as well as two schools (the Philosophical Institute and the *Oberrealschule*). (5a) The St. Thomas Church originally associated with the St. Thomas Abbey at Moravian Square in the inner city. (5b) The Jesuit Church of the Ascension of the Virgin Mary, associated with Jesuit College and the *Gymnasium*. (6a) The St. Michael Church and the Dominican Monastery with the Theological Institute (seminary) at the Dominikánské Square (*Rybný trh* or Fish Market). (6b) The Church of St. Peter at the Petrov Hill, the Bishop's Residence, and Bishop's Court. (7) The St. Jacob (James) Church. (8) The Church of St. Josef. (9) The Church of Holy Cross associated with the Capuchin Monastery at the Kapucínské Square (the *Uhelný trh* or Coal Market). (10) The Church of St. Mary Magdalene at the Židovská (Jewish) Gate. (11) The Reduta Theater at the Zelný trh (Cabbage Market). (12) Velké náměstí (Large Square). (13) Špilberk. (**c**) Detail of the areas 4, 5, and 6 in Fig. 5.2b

Fig. 5.4 The view of Brno from the south after the erection of the viaduct for the railroad connecting the city with Vienna. The connection was activated in 1843. In the foreground is the Svratka River, in the background on the left the Špilberk Hill with the citadel. In the middle is the Petrov Hill with the unfinished cathedral (Based on E. Gurka's drawing from 1839)

arms were being founded, along with iron and steel mills. Trade was increasing and with it also transportation, communication, banking, and services of all kinds. Brno was changing, if not from day to day, then certainly from year to year. Four years earlier, the arrival in Brno of the first passenger train pulled by a steam-powered locomotive was a cause for great celebration (Fig. 5.4). In 1843, trains arrived from and left for Vienna daily, making the 110-kilometer long trip in slightly more than four hours. In the decade before Mendel's arrival, the population of the inner city doubled, and in the following years the town continued to grow by the addition of suburbs. In 1850 alone, the incorporation of 29 suburbs spiked the city's population size to nearly 50,000. The city's streets had been illuminated by gas lamps; the course of the rivers flowing through the city had been straightened; the ramparts had been converted into roads and parks; and new houses, administrative buildings, and factories began mushrooming everywhere.

Along with the physical and material changes, Brno also grew culturally. It had several learned societies (as well as two Masonic lodges), which not only gave their members the opportunity to hear and debate reports about the latest advances in natural sciences, meteorology, agriculture, and pomology, as well as in historical research and philosophy, but also served as a means of popularizing sciences and humanities. Some of the societies published newsletters and magazines serving the same purpose. Research was conducted at the secondary-level schools and in some monasteries. Brno had a museum since 1818, a provincial archive since 1839, and since 1761 a permanent theater (Reduta), in which the public could enjoy plays by Goethe, Schiller, Lessing, Shakespeare, and Moliere, as well as operas by Piccini, Gretry, Gluck, Mozart, Beethoven, von Weber, Rossini, Bellini, Auber, and Donizetti. The city was frequently included in the itineraries of pianists, violinists, and other famous soloists such as Jan Nepomuk Hummel, Louis Spohr, Franz Liszt, Anton Rubinstein, Ferdinand Laub, and Clara Schumann. The Brno of Mendel's time also produced an outstanding homegrown musical culture typified by three names: František Sušil, Pavel Křížkovský, and Leoš Janáček. Since 1833, Brno has also been the site of international industrial expositions, a tradition that continues to this day. Hence, far from the sleepiness that is usually associated with a provincial town, Brno, in Mendel's time, was an industrialized, busy, and bustling city. And one of the main centers of the city's cultural life was the Augustinian Abbey.

The Augustinians[9]

The name of the religious order that Mendel joined in 1843 might seem to implicate St. Augustine of Hippo (354–430 CE) as the order's founder. In reality, however, the link of the Augustinians to their namesake is rather loose. According to the traditional account, students, friends, and followers congregated around the bishop in the garden at Hippo (today's Annaba, Algeria) and formed a loose community not bound by any vows, but guided by their master's teachings. Similar communities sprung up in other places in North Africa, their trademark being the renouncement of private possessions. Later in the fifth century, when Vandals invaded North Africa, some of these communities relocated to central and northern Italy, where they thrived for centuries. To stave off the proliferation and diversification of these numerous opposing groups, Pope Innocent IV (1241–1254) and his successor Alexander IV (1254–1261) unified them all into a single religious order which they then used as a political tool against the Dominicans and Franciscans who had grown too powerful. The basis for the grand unification was the *Rule of Saint Augustine*, which provided a directive under which the Augustinians were to live.[1] The origin of the rule is as nebulous as the origin of the order itself. At least three documents go under this name, of which, however, only one is agreed by all to have been penned by St. Augustine.[10] The origin of the other two documents is controversial, although both seem to be based on one authentic letter (Epistle 211) St. Augustine wrote to a community of nuns in the Hippo garden, a community headed by his sister. When she died, the members of the community could not agree on who their new mother superior should be. St. Augustine addressed this problem and added his thoughts on the conduct of life in a religious congregation in general. The name "Rule of St. Augustine" is, however, most commonly applied to a compilation of do's and don'ts for male members of a religious community. It exists in two versions: a shorter one consisting of eight chapters and a longer one comprising 45 chapters, with each chapter devoted to a specific topic. In its shorter version the document addresses the purpose of communal life, prayers, moderation and self-denial, safeguards of chastity, fraternal correction, care of the community's goods, care of the sick, forgiveness of offenses, governance and obedience, and observance of the rules.

Technically all religious orders that follow the Rule of St. Augustine are Augustinians. In this sense Augustinians also encompass Dominicans, Premonstratensians, Serrites, Trinitarians, Ursulines, Gilbertines, and others. More commonly, however, the term is applied in a narrower sense to the followers of St. Augustine who forgo their worldly possessions. But even by this delineation the term still embraces multiple orders, which fall into two categories: Augustinian Canons and Augustinian Hermits. A canon is someone living under a rule, specifically a clergyman (priest) belonging to the staff of a cathedral or other collegiate church with a body of associated clergy (chapter). The Canons Regular of St. Augustine, the Augustinian (Black) Canons, for short, or, even shorter, the Austin Friars arose in the eleventh century, when a group of ordained clergy adopted the Rule of St. Augustine to pursue a life of poverty, celibacy, and obedience, without withdrawing from the world. The designation "regular" derives from the Latin *regula*, here used in reference to the rule by which the members of the order agree to live. They are called Black Canons because they wear black robes, in contradistinction to the White Canons of Prémontré or the White Canons of St. John Lateran, who are clad in white robes.

The Order of the Hermit Friars of St. Augustine (Augustine Hermits or Austin Friars) is the official name of what is commonly understood under the term "Augustinians." The name contains two words (hermit and friar) that require an explanation. As mentioned earlier, when monasticism began to spread, it split into two kinds—eremitic and coenobitic. An eremite or hermit is a religious recluse, and the eremitic type of monasticism requires its practitioners to lead a life of almost unbroken solitude, either in a deserted place or in isolated cells of a building. Initially, all Christian monastics were hermits, but with time the practice was gradually replaced by the coenobitic kind of monasticism. The term is derived from the Greek *coen* + *bios*, "to live together," *coenobium* being the Latin name for a monastery. A coenobite is then a member of a group of monastics who live together in a communal life. Already at the time of the order's origin, however, the Augustinian Hermits did not lead life in solitude, and they did not live in monasteries either; instead, they wandered through the countryside individually or in small groups, occasionally congregating at certain places. At that time the word "hermit" might have still applied to them. Since then, however, that name has become a misnomer, because modern-day Augustinians live together in monasteries. The only leftover of eremitism is that the members of the order conduct some of their activities outside of the monastery.

The original wandering Augustinian Hermits lived from alms, as did the Dominicans, Franciscans, and Carmelites. The orders whose members begged for a living and were not bound to a particular monastery were called *mendicant* (from the Latin *mendicus*, beggar). Although the practice of begging was later abandoned in favor of working for a living, the designation "mendicant order" has been retained. A member of a mendicant order, who is forbidden to hold property, is called a *friar* (from the Latin *frater*, brother). Formally, friars differed from monks in that they were expected to lead lives of poverty and service to the community, rather than of cloistered asceticism and religious meditation. Hence, since Mendel was an Augustinian, he should be called a "friar," rather than a "monk," but since friars now live in monasteries just like the monks, the distinction between the two terms has been blurred. The Augustinian Hermits themselves did not remain a single order, as the popes in the middle of the thirteenth century had intended them to be. Various reform movements within the order led to its splitting into factions, of which three became separate orders. They are the Calced, Discalced, and Recollect Augustinians. The Discalced Augustinian Hermits are the most austere and ascetic of the three. They are quite rigid in fasting and in obeying the rules of silence, they have special rooms for penance, and they used to walk barefooted, but now wear sandals instead. They are therefore no longer "discalced," that is, "shoeless," as their name (from Latin *calceus*, shoe) implies.

In the ecclesiastical parlance the word "recollection" stands for religious contemplation, and in the designation Recollect Augustinians, it refers to the emphasis its members place on this form of spiritual activity. The regular Calced Augustinian Hermits are the most numerous of the three. They are organized into provinces headed by provincials and their counselors. Each province encompasses several monasteries, each monastery being governed by a prior (with some exceptions, as we shall learn shortly) and his deputy. The prior general with his four assistants and a secretary constitute the council or *curia*, which oversees the entire organization. The provincials, priors, and counselors are elected for three years by the provincial chapters, whereas the prior general and his curia are elected for six years by the general chapter. The order's members are priests and lay brothers—persons not authorized to perform the sacred rites of the religion. The garment of a friar consists of a black woolen habit with long wide sleeves, a long pointed hood (cowl), a cloak (scapular) hanging loosely from the shoulders, a black leather girdle belt (cincture), a hat, and shoes. The activities of the Augustinian Hermits include pastoral duties (care of souls); missionary work; participation in the practical affairs of church policy; and learning, teaching, and scientific studies.

The Abbey of St. Thomas at Staré Brno[11]

Augustinian Hermits came to Moravia in 1256, the year in which the order was officially created, and to Brno only 90 years later. In 1349, Jan Jindřich of Luxembourg (1296–1375), the son of the Bohemian King John of Luxembourg (1310–1346), became the Margrave of Moravia by the will of his brother, the Holy Roman Emperor Charles IV (1316–1378). One year later, he announced his intention to establish an Augustinian monastery at a site outside, but close to the inner city of Brno. To this end, he donated the necessary land in front of the Běhounská Gate and also several villages, which were to sustain the monastery. Subsequently the Augustinians acquired additional real estate from bequests of landlords and wealthy Brno families. In 1356, Pope Innocent VI issued the necessary permit, and in the same year the Augustinians also received the founding document from the margrave. The construction of the church began in 1353, but it would not be completed until 1397. In the meantime, the margrave died and was buried in the still unfinished church, and the monks began to pray for his soul, as stipulated in the founding document. The church, consecrated to St. Thomas, was Gothic originally, but when rebuilt in 1662–1668, it acquired its present-day Baroque appearance (Fig. 5.5). Its most precious artistic treasures were the "horizontal" Pieta and the Black Madonna. The former, possibly sculpted by the renowned architect and stonemason Peter Parler, depicts a seated Virgin Mary holding Christ's dead body on her lap in a nearly horizontal position. The latter is a Byzantine oil painting on wood depicting the Blessed Mary and the infant Jesus with dark complexions and clad in black garments. Although a legend attributes the painting to St. Lucas, one of the 12 apostles, and its emergence in Constantinople to Empress Helen, mother of Constantine the Great, who supposedly brought it to the city from the Holy Land, it may have, in fact, originated at a later date in Italy. The Holy Roman Emperor Friedrich Barbarossa acquired it somehow and then gave it, as a reward for martial assistance, to the Bohemian King Vladislav, who brought it to the Prague castle. Subsequently, Charles IV donated the icon to the Augustinians on the occasion of the consecration of their St. Thomas Church, an occasion which he and Jan Jindřich attended personally. Worshippers of the Black Madonna believe that praying to her sways her to intercede on their behalf. One such intercession is believed to have saved Brno from the Swedish siege in the Thirty Years War. Because of its alleged miraculous power, the icon has attracted droves of pilgrims to the St. Thomas Church, thus contributing to the monastery's fame and wealth. In gratitude, the Augustinians encased it in silver.

The construction of the monastery itself did not commence until 1385. But once completed, the monastery prospered thanks to the income generated by its goods and the various gifts and special favors from the margrave and the city. Later, Jan Jindřich's son and successor, Margrave Jošt (Jodocus), stepped into his father's shoes by continuing to espouse the Augustinians. Among the favors he bestowed on them was a permit he issued in 1410 which allowed them not only to open a tavern close to the monastery at the foot of the Špilberk Hill but also to serve there exclusively the wine originating from the order's own vineyards. It was not a cheap present: a tavern at this location, serving wine without a middleman involved, was a goldmine. The ever-grateful monks kept the two Luxembourgs in good memory long after the world at large forgot them. Some 350 years later, when they were rebuilding their abbey, they commissioned the sculptor to decorate the portal leading to the prelacy with statues representing their benefactors (Fig. 5.6). And so there they stand to this day, each between two columns, the father on the one side and the son on the other. Jan Jindřich holds a cartouche with the spread eagle of Moravia, and Jošt is shown with a scroll in his hand (the permit?) and crown on his head—a nod to the fact that for some ten month he was a Roman king, before he died—and with a somewhat unstable gait, as if he were just emerging from the monastery tavern.[12]

Alas, after the bounteous came the lean years, when the monastery lost most of its sources of income and its holdings were devastated. These misfortunes had befallen it during the Husite Wars, especially in 1428, during the siege of Brno, and

Fig. 5.5 (**a**) The St. Thomas Church dominating the complex of buildings—the original St. Thomas Abbey of the Augustinians in Brno City. (**b**) The portal (main entrance) to the abbey. (**c**) Detail of the portal showing the statue of Jošt Lucemburský. Margrave of Moravia from 1375 to 1411, King of Germany from 1410 to 1411, and the benefactor of Brno's Augustinians

Fig. 5.6 Eliška Rejčka (1288–1335), Czech Queen from 1300 to 1305, then again from 1306 to 1307, and Polish Queen from 1303 to 1305. She founded the Cistercian convent in Staré Brno, which the Augustinians later took over. There does not seem to exist any authentic portrait of her, although in the illuminated manuscripts, mostly prayer books, she is apparently depicted in the marginal embellishments. Since these miniatures show at best only outlines of her face and since it is not even certain that the illuminators ever saw the queen, our rendition must be taken as an imaginary portrait. Although she spent her last years in her convent, popularly known as the "klášter Králové" (The Queen's Convent) and although she dressed like a nun, she did not become one

then again during the Thirty Years War. Since the monastery was outside of the city's main fortifications, it became an easy prey to the besieging armies. (Although in 1486, the city extended the fortifications to also enclose the monastery, this measure did not fully alleviate the problem since during the siege of 1645, the Swedes damaged the St. Thomas Church so badly that it had to be rebuilt almost entirely.) Ultimately, however, the monastery always bounced back from these visitations and its renown continued to grow. The popes acknowledged this reality, by granting the convent distinctive privileges. First, in 1721, the elected head of the monastery became *prior perpetuus*, prior appointed for life. And then, in 1752, Pope Benedict XIV elevated the St. Thomas monastery to an abbey, headed not by a prior, but by an abbot, with a prior as his deputy. The abbot's position was, of course, also permanent. This decision had several practical consequences. As an abbey, the monastery was no longer part of a province and hence not under the jurisdiction of a provincial (nor of a bishop). Instead, the abbey assumed a position equivalent to that of an independent province answerable directly to the prior general. The St. Thomas Abbey was the only one in Europe to hold such a position. Another privilege of the abbot was the right to use pontifical regalia (otherwise used only by bishops and the pope) as a symbol of his office. They were the crosier (a staff resembling a shepherd's crook), the miter (deeply cleft ornamented cap with peaks in front and back), and the pectoral cross (a cross worn on the breast). The elevation enhanced the social and political standing of the abbey and of the abbot in the city and the entire province. Already since the fourteenth century, when the monastery became a major landowner, the prior had a permanent seat and voted in the provincial diet. But since the elevation, the abbot had been showered by other honors and had frequently been elected to become a member of various financial and educational institutions. The reason why the Vatican gave the St. Thomas monastery such special treatment was undoubtedly the Black Madonna. In the eyes of the church, the possession of the icon made the monastery singular, gave it a distinctive clout, and so was worthy of extraordinary privileges.

The distinction afforded to the monastery put the St. Thomas abbots under pressure to add certain opulence to its buildings, including the church, so as to make them representative of their status. Therefore, they spent considerable sums of money on renovations and refurbishing, especially in the years 1734–1742. The effort backfired, however, when the attractive buildings caught the attention of Joseph II, who was set on closing most of the monasteries in Brno and its environs. Only monasteries involved in services to community, such as teaching, were to stay open. The Augustinians escaped the axe by committing two of their members to teaching mathematics and Biblical studies at the philosophical and theological institutes in Brno. On the emperor's orders, however, the provincial government, together with the Moravian Estates, confiscated their beautiful, recently renovated buildings and used them for administrative offices. The same directive commanded the friars to relocate to the convent of Cistercian nuns in Staré Brno, one of the orders the emperor dissolved. And so the original abbey became Místodržitelský palác (the Governor's Palace), and government officials replaced the friars. The building retained this function until the disintegration of the Austrian–Hungarian Monarchy in 1918. Subsequently, it served various other functions; most recently the former prelacy has been turned into the Moravian Gallery displaying paintings and sculptures from the fourteenth to the nineteenth centuries.

At the time of the move, Staré Brno was still a village outside of Brno City, a village with its own history.[13] If there ever was a castle in Staré Brno, no trace of it has been found. The oldest structures for which archeological evidence exists are two small churches: one consecrated to St. Procopius and the other to the Virgin Mary. Presumably, separate settlements arose around them, which later fused into one. The older of the two, the St. Procopius Chapel, might have stood on the market place close to the presumptive castle and hence to the ford. The younger Virgin Mary Chapel stood, surrounded by a few houses, directly at the foot of the Žlutý Hill. Built at the end of the tenth century as a Romanesque rotunda, it consisted of a semicircular apse opening directly into the circular nave, which was covered with a flat ceiling, topped with a semidome (concha). In the second half of the twelfth century, the rotunda was torn down and replaced by a larger church with an oblong nave and a rectangular presbytery. It too did not last very long, however. In the fourteenth century, a new, much larger, and grander church arose right next to it, and this third version was fated to last to this day. The rise of the new church is forever tied to the name of an extraordinary woman, whom the Czechs call Eliška Rejčka (Fig. 5.6). She was a Polish princess baptized Rickenza after her Swedish mother, but to the Poles she was "Ryksa." She was wedded, at the tender age of 15 years,[14a] to the widowed Bohemian King Václav II (1283–1305), the son of Přemysl Otakar II, and assumed the name Alžběta; but for the Czechs she became Eliška Rejčka. When Václav II died of tuberculosis, the 17-year-old widow was—in the interest of state politics—wedded again to a Bohemian king, Rudolph of Habsburg, who forced himself on the Czechs after the assassination of Václav III in 1306 and the extinction of the Přemyslid line of descent. Rejčka's second marriage was even shorter than the first one, since Rudolph died suddenly in 1307. Twice a queen and twice a widow, Rejčka had no desire to become once again a pawn in the game of Central European politics. She befriended Jindřich of Lipé, the most powerful noble of the kingdom, and under his protection retired to her estates, first in Hradec Králové in northeastern Bohemia and then in Brno. In the meantime, the question of succession to the Czech throne had been resolved in favor of John of Luxembourg, who married Alžběta, the second daughter of Václav II. Although, or perhaps because, no love had been lost between the two Alžbětas, King John was favorably inclined toward Rejčka, and so when she asked him for permission to found a Cistercian[15] convent in Staré Brno, he not only gave his consent but also contributed an estate around the Church of Virgin Mary for that purpose. The convent's official name was to be *Aula sanctae Mariae* (The Great Hall of St. Mary), but the citizens of Staré Brno preferred to call it *klášter královѐ* (The Queen's Convent),[14b] and in the documents from that time, it often appears under the name *Aula Regina* (The Great Hall of the Queen).

The third version of the Church of Virgin Mary, now renamed to Church of Assumption of Virgin Mary, was to be the first building of the Aula. Although the construction of the church began shortly after the issuance of the convent's founding document in 1323, it was not completed until more than half a century later. Initially, the nuns had to content themselves with humble one-story buildings for their living quarters, erected in the form of a quadrangle adjacent to the church at its northwestern aspect. Later, however, as the order grew by further contributions from wealthy donors, larger and more spacious buildings were erected on the convent's land. And when the main road leading from Brno City to Staré Brno and beyond stood in the way of further expansion, the nuns procured permission to relocate it further away from the convent. Rejčka herself moved to the monastery in 1332, and when she died three years later, she was buried there,[16] next to her friend and lover Jindřich of Lipé, who died six years before her. Just like the Augustinian monastery, the Cistercian convent had its ups and downs, and since it was at some distance from the fortified city, its downs were deep: over the centuries, it had been burned down and rebuilt, only to be burned down again several times.

The second half of the eighteenth century was one of the "down" periods in the history of the convent. Joseph II adjudged the Cistercian convent as not providing sufficient service to the community and in 1772 ordered to close it down. The Augustinians' move into the convent the following year was, as one historian put it aptly, "like exchanging a palace for a hut, moreover a hut falling apart."[17] The Augustinians decided to renovate it thoroughly, to tear down entire sections, to erect new ones in their stead, and to repair the rest. To these reconstruction costs and moving expenses came at least one other large expenditure. As mentioned earlier, Joseph II spared the St. Thomas Abbey from closure under the condition that the Augustinians provide two professors to teach mathematics and philosophy at the Philosophical Institute. Since the order did not have qualified individuals among its brethren to meet this obligation, it had to pay two Piarists to do the teaching on its behalf. These combined disbursements and also the war during which the monastery was forced to take care of the Prussian and French occupiers, as well as the state bankruptcy that followed the war, brought the order into a financial crisis from which it took decades to recover.

Through the renovations, the monastery acquired the appearance it essentially has to this day. Although, because of their relocation, the Augustinians became associated with The Church of Assumption of Virgin Mary (rather than St. Thomas), they kept the name "Abbey of St. Thomas," but distinguished themselves from their former affiliation by the addition of the qualifier "in Staré Brno." An alternative name "Augustinian monastery in Staré Brno" is also used. Since Mendel spent 41 of his 62 years in this abbey, let us now have a closer look at its physical setting—the scene in which his adult life played itself out. The abbey's present-day appearance betrays its long, checkered history of constructions and reconstructions necessitated by disasters, decay, and need for modernization. Through the changes the convent grew in size and complexity. Historians distinguish four phases in its architectonic development: the first two encompassing the Cistercian and the remaining two the Augustinian eras (Fig. 5.7). All that remains from the earliest Gothic phase is the church, which has essentially been preserved in the form in which it was built in the fourteenth century, with a few walls attached to it, remaining from the original *Aula*. In the second, Baroque phase, covering the seventeenth and eighteenth centuries until 1783, the Cistercians gave the convent its present basic outline. In the third phase, from 1783 to the beginning of the nineteenth century, the Augustinians added to the abbey one wing, which they then extended in the twentieth century—the fourth phase of development. Because of its dual, though nonoverlapping proprietorship, the maze of buildings had, in reality, become two convents in one.

Apart from the church, the essential functional elements of a monastery are a *prelacy* (offices and residencies of the monastery's leaders, the abbot, and the prior, in the case of an abbey); a *refectory* (a room, in which the conventuals meet to dine and socialize) with the associated kitchen facilities; a *chapter hall* (the place where the monastery's full members meet to deliberate rising matters that concern them all); the *dormitories* or other lodging accommodations of the conventuals; and *facilities*, as well as accommodations of the supporting staff. All of these elements, excepting the church, are found in the Staré Brno abbey in two editions (Fig. 5.7), of which the Augustinians used always only one for its originally intended purpose. A common structural element of a monastery is a quadrangle, in which four wings enclose a courtyard. There are four quadrangles in the Staré Brno abbey, three of which are complete and closed, whereas the fourth one is incomplete and open (Fig. 5.7). The wings (sides) of a quadrangle are of similar design, consisting of a row of rooms and a corridor running alongside of them. Some of the older wings are or were one-story buildings, but most are now two-story affairs. The roofs of all the

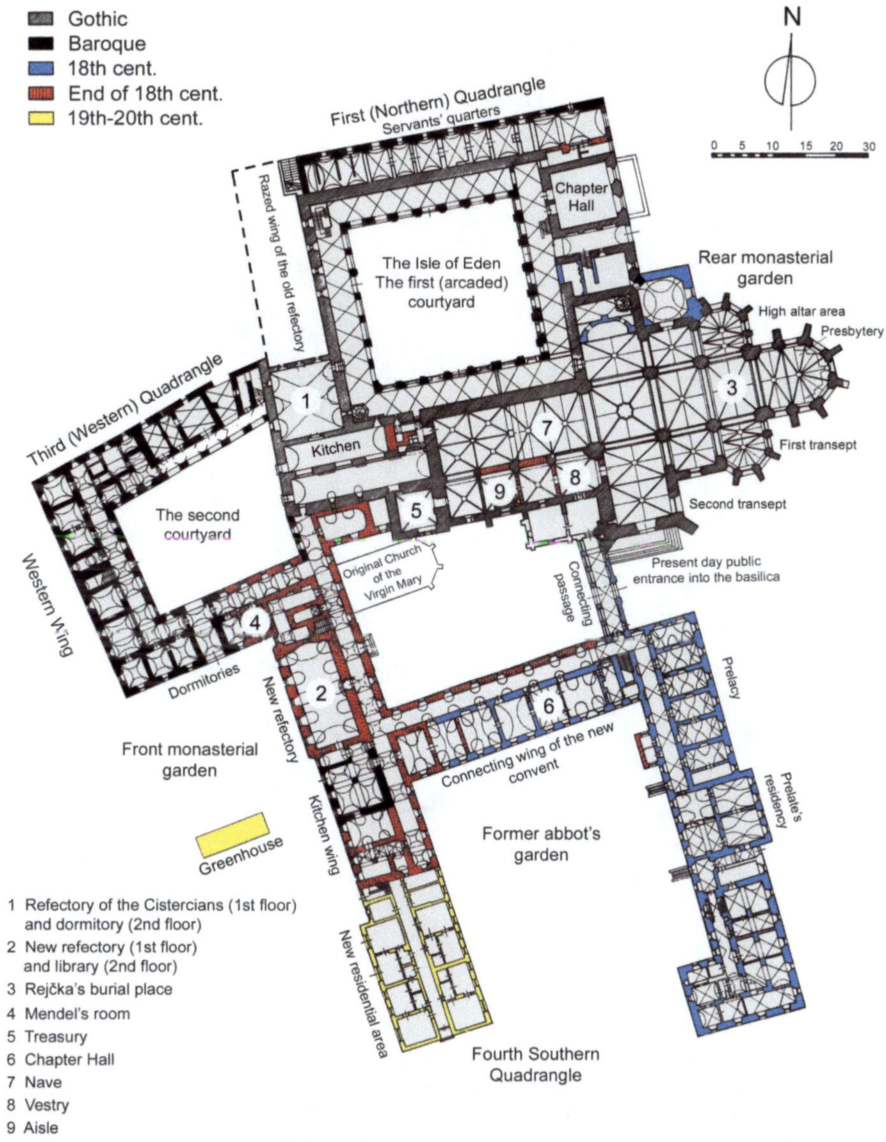

Fig. 5.7 The ground plan of the St. Thomas Abbey in Staré Brno (Based on reference 11)

buildings, except the church, are covered with red tiles. An aerial view of the entire complex is shown in Fig. 5.8; a brief description of the church and the quadrangles follows.

The church is now a *basilica minor*, a denomination referring to the special ecclesiastical privileges it enjoys. Its exterior is essentially that of the fourteenth-century late Gothic structure. Its interior has been Baroqueized, in our opinion to its detriment: The added bombastic splendor of the altars with their elaborate

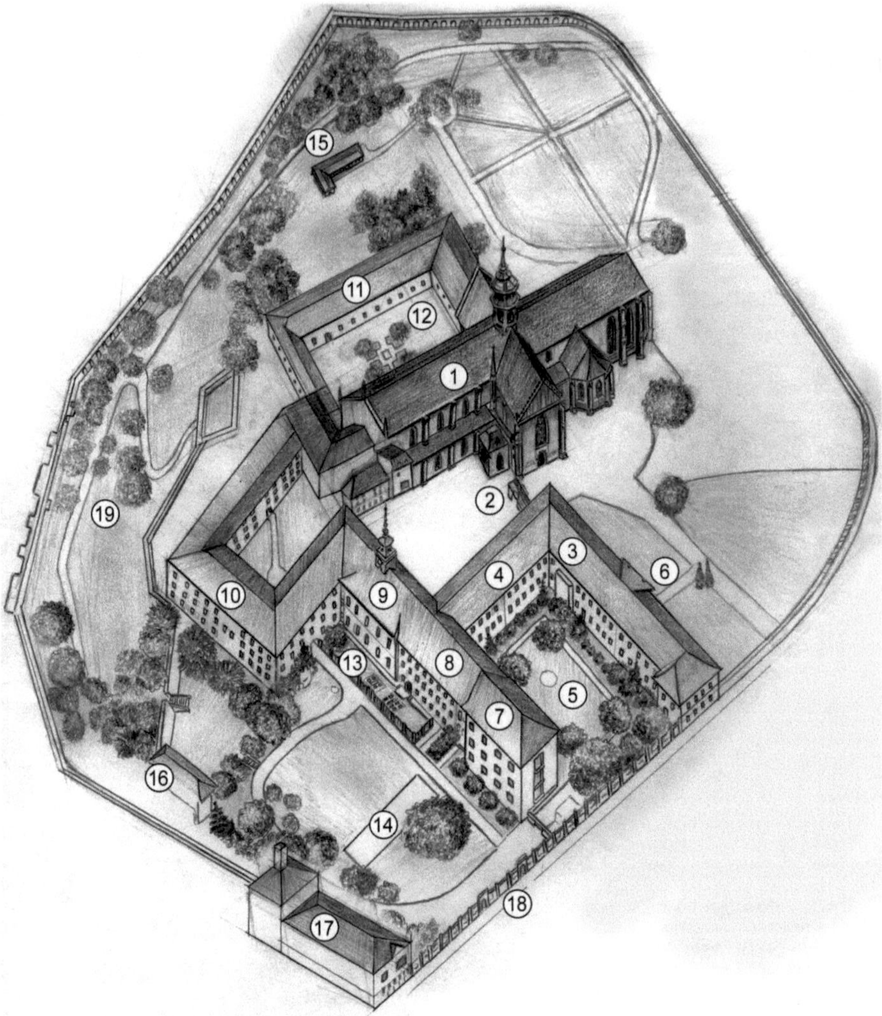

Fig. 5.8 A bird's eye view of the St. Thomas Abbey in Staré Brno. (1) The Church of the Ascension of the Virgin Mary. (2) The passage connecting the abbey with the church. (3) The prelacy. (4) Connecting wing containing the chapter hall. (5) Abbot's garden. (6) Entrance to the prelate's residency. (7) Civic addition (not part of the abbey). (8) Kitchen wing. (9) New refectory on the first floor and the library on the second. (10) Friars' living quarters. (11) Servants' quarters, administrative building, storages, stables, etc. (12) The Isle of Eden. (13) Mendel's experimental garden. (14) The greenhouse. (15) The beehouse. (16) The summerhouse. (17) The beer brewery. (18) Walls enclosing the abbey. (19) Orchards and gardens

ornamentation does not harmonize with the simple beauty of the slender arcade columns, graceful arches, and rib vaults. The church is built from dark-red brick, with stone used sparingly for decorative purposes only. Its basic ground plan, like that of most Gothic churches, is that of a cross and consists of the east–west oriented main vessel, transected with two transepts, the first one short and the second long

(Fig. 5.7). The second transept divides the church into a western part, the nave, and an eastern part, the choir, ending with a trefoil consisting of the presbytery (the part of the church reserved for the clergy) and two side chapels. The church has altogether ten altars, nine in the side chapels and the high altar, which displays the Black Madonna icon in its silver casing. The tall structure lacks an imposing tower, having instead only a modest belfry centered where the main vessel transects with the second transept. A passageway leads from the western side of the second transept's southern arm to the abbey. A public entrance to the church is also found on this side.

Leaving the church, we turn now to the abbey itself. The oldest of the four rectangles (the one which we designate arbitrarily as the "First" in Fig. 5.7) is attached through its southern wall to the northern wall of the church. It differs from the other three in that its wings are arcaded. It encloses a garden, which has been dubbed *Rajský ostrov* (The Isle of Eden). And indeed, its isolation from the bustle of the city on the outside, its freshness of greenery, and the airiness of the arcades all conspire to impress on a visitor that time has stopped there. This part was the original convent of the Cistercians, with the kitchen and refectory (now largely razed) on the first floor and the dormitories on the second floor of the western wing and the chapter hall in the eastern wing. The one-story northern wing was taken up by the lay personnel: the managers, administrators, housekeepers, gardeners, cooks, and others on nonecclesiastical duties. After the first, the Cistercians added a second quadrangle, which shared its eastern wing with the first and served as a dormitory. Before becoming abbot, Mendel occupied two rooms there on the second floor above the present-day passage from the front garden to the paved inner court. The western wing of the third quadrangle had the Augustinian refectory on the first floor and the library hall on the second. The entrance from the garden to the refectory dates, however, from the 1960s. In the southern wing of the quadrangle was the Augustinian chapter hall, as well as the guest rooms. The passageway to the church from the eastern wing of the third quadrangle and the wall of the church form most of the northern wing. In the northwestern corner of the third quadrangle once stood the original Church of Virgin Mary (the second version and before it the first version). In the connecting passageway is the main entrance to the abbey, to which once led a road lined with linden trees from the present-day Mendel Square. The Augustinian prelacy takes up the eastern and northern wings of the fourth quadrangle, the prior's rooms being located on the first floor and the abbot's on the second. The western wing of this incomplete quadrangle contained kitchen facilities on the first floor and additional library and study rooms on the second. The wing's newly added extension contains offices that are not part of the abbey. The space partially enclosed by the three wings of the fourth quadrangle used to be the abbot's garden.

In Mendel's days a visitor wishing to see the abbot would approach the abbey from the Zámecké Square by way of the linden tree alley leading to the main entrance in the church-connecting wing. At the entrance door he would be confronted by a porter who was on duty around the clock for he had his lodging connected with his outpost. The porter would inquire about the purpose of the visit, and if he found it legitimate, and if the abbot was willing to receive the caller, he would lead the visitor upstairs on a wide, carpeted staircase to one of the reception rooms. A group of visiting dignitaries would be led into the chapter hall. Both the reception rooms and the chapter hall were furnished to impress visitors. In the case of the chapter hall, the size of the room, the large windows, the polished parquet floor, the exquisite rococo and Biedermeier furniture inlaid with mother of pearl and ebony, and the collection of oil paintings displayed on the walls, all these conveyed a message that the visitors were guests of a major landlord. (A plebeian might wonder, what would St. Augustine have thought of this display of affluence, had he been the visitor.) The abbot's own living quarters and study, the guest's rooms, and the prior's residency downstairs were well furnished, but not opulent. If the visitor were to dine with the abbot, the two of them would walk downstairs and then the entire length of the west wing, in a corridor decorated with more oil paintings, to the refectory. There the assembled friars would already be waiting to greet the pair. Stewards would then begin serving the meals, which had already been prepared in the adjacent kitchen, and wait on the two large tables. Table music would redouble the culinary pleasure. If, after the meal, a distinguished guest expressed the wish to see the abbey's renowned book collection, he would be led upstairs into the main library hall above the refectory. At the entrance, he would be asked to exchange his shoes for woolen slippers. This measure was necessary to protect the precious old books from dust and the parquet floor from dirt and abrasion. What the visitor saw upon entering the room was not an ordinary library but a showcase of rare books. After the great chapter hall of the prelacy, the main library hall was the most sumptuous room in the whole monastery (Fig. 5.9). Three of its walls were covered by bookcases that stretched from the floor to the high ceiling above. On the fourth wall, opposite to the entrance, these grandiose bookcases alternated with five tall windows, which afforded a view of the garden below. The bookcases alone were precious pieces of art, handmade by one of the friars when the Augustinians still lived in their old home. Each bookcase had a cabinet at the bottom, followed by nine bookshelves and topped by a piece richly ornamented with marquetry. In the four corners of the room stood decorative columns and between them statues presumably symbolizing the main continents. Stuccowork and paintings embellished the ceiling. The room itself was not intended for study, but rather for small receptions, concerts, and other social gatherings. It contained the most valuable books, while the rest of the total of some 30,000 books was stored in the eastern part of the wing. Since this wing had three rather than two stories, the floor of the second story was at a somewhat lower level than that of the main library room. It could be accessed through a door concealed by bookcases that opened onto a short, low passageway, with a few steps leading down to the first of several, more modestly furnished rooms. It was this part of the library that the friars used for study. The books in the library came from a variety of sources.[17] The library probably began as a humble collection of books that the members of the order brought with them, supplemented by manuscripts copied by scribes in the monasterial scriptorium. But the majority of the books were donated by or purchased from private collectors. Some were also obtained from the dissolved monasteries. Most of the books in the library were theological and philosophical, but during the nineteenth century books on agriculture, husbandry, and natural sciences had also been added to the collection.

Fig. 5.9 The main library hall of the St. Thomas Abbey in Staré Brno

Walls dating from different time periods enclosed the entire complex of buildings. A high, medieval red brick wall separated the northwestern section of the property from the slopes of Špilberk and Žlutý kopec. The wall in the northeastern section was from the Baroque phase, while the walls in the southeastern and southwestern parts dated back to the nineteenth century. Gardens and orchards extended from the buildings to the walls, the largest of them being those in the back of the old chapter hall and in front of the new refectory (Fig. 5.8). Small constructions—an arbor, a summerhouse, a beehouse, a wine cellar, an orangery, tool sheds, and even a gym—were scattered throughout the garden. Large sections of the wall have since been demolished, so that the front of the prelacy wing now opens onto Mendel Square with only a wrought iron fence delineating it. On the isthmus of the Žlutý kopec, outside of the walls, stands a beer brewery originally owned by the Cistercians and then by the Augustinians. It later changed hands again[18] and is presently producing *Starobrno pivo*.

In addition to the land and the buildings of the abbey itself, the monastery also owned land and villages in the area around Brno. The inhabitants of the monastery lived off of the income of these possessions. In this regard, the monastery did not differ from any other landlords of that time. Essentially, the friars owed their comfortable lifestyle to the hard work of the peasant families, like those Mendel left behind in Hynčice. The friars, some of whom, like Mendel, came from peasant families, must have been aware of the obvious social injustice upon which their livelihood was based. It would have been interesting to know how Mendel felt about all the luxury amidst which he suddenly found himself. For compared to what he had known before, life in the monastery *was* a luxury. As he sat down for the first time to a sumptuous meal in the refectory, might he not have thought of his family toiling from dawn to dusk so as to enable others to sup like this, while they had to content themselves with milk and potatoes for dinner? How did he reconcile this injustice in his mind? How did his conscience deal with the situation in which he found himself all of a sudden? We will never know, of course, but apparently, whatever his initial qualms might have been, he eventually adjusted quite well to his new life.

The Novice Gregor Mendel

The letter Professor Franz wrote to recommend Johann Mendel for admission to the St. Thomas Abbey was dated July 14, 1843. Less than two months later, on September 7, the abbey, apparently waving the otherwise standard requirement for interviewing the candidate,[19] informed Mendel officially that he had been admitted.[20] Three documents were then required for the actual admission: The first of these was the parental consent,[21] a letter addressed to the abbey, stating that Anton and Rosina Mendel had no objection to their son's choice of a clerical profession. Johann wrote the letter and his parents signed it on September 19. For the second document, Johann had to walk to Odry, for a medical examination by the municipal physician Dr. Schwarz.[22] The doctor found Johann perfectly healthy, apparently fully recovered from the earlier episodes of illness in Opava and Olomouc. The third required document was a letter releasing Johann from the jurisdiction of the archbishop in Olomouc, thus allowing him to be placed under the ecclesiastical jurisdiction of the bishop and the abbot in Brno. This document was issued on September 27, 1843. After these three formalities had been taken care of, Mendel traveled, by unknown means to Brno to begin his novitiate. On October 9, 1843, at a *robing ceremony*, he received the habit of a St. Augustine friar and assumed the monasterial name Gregorius (Gregor in German, Řehoř in Czech, and Gregorios in Greek, meaning "vigilant"). From then on, he signed his name as Gregor Mendel, Řehoř Mendel, or Gregor Johann Mendel, the monasterial name assuming precedence before the baptismal name. Why he chose the name Gregor, we do not know, but surely the fact that in the history of the church, 16 popes assumed this name on their election might have had something to do with it. Of the 16, the best known is the first, St. Gregory (not to be confused with St. Gregory VII or Gregory the Great (540–604)). He was one of the church doctors responsible for ecclesiastical and monasterial reforms, including an elaboration on the plainsong chanted in Roman Catholic churches, and the Gregorian chant (the Gregorian calendar was the doing of another Gregory, however). St. Gregory's Day is on March 12, and this was when Mendel celebrated his name day (see Chap. 3). The abbey admitted three other young men along with Mendel: Wilhelm Rösner, Jan Rambousek, and Antonín Cigánek. The four of them were selected over 13 other applicants.

By putting on a friar's robe and taking a new name, Mendel did not become a friar, however. He would still have one year to decide whether he really wanted to become one. For this period he was on probation as a novice. In ancient Rome, a *novitius* was a new slave, but to Mendel the one-year novitiate might have seemed close to spending one year in an earthly paradise. True, his daily activities were strictly regulated by rigorous rules, some of which might have become tedious after a while, but they were not overbearing. And within the prescribed schedule, he had enough time to pursue his interests in natural sciences in a highly stimulating intellectual environment, without having to worry about where his next meal would come from. Another novice, Tomáš Bratránek, admitted a few years before Mendel, described, in a letter written on October 19, 1834, to his friend, Josef Šebestík, what a day in the abbey was like. He wrote: *I get up at half past five in the morning and at six go to minister the prior, who is also the novice-master. I finish at six thirty, clean up my room, breakfast bread and water or whatever heaven blesses me with, and then idle until eight. At eight o'clock I join the novice-master to pray from the breviary and come back at nine o'clock. From nine to twelve I either go to the library or read in my room. At twelve we dine and this takes full two hours, often even longer, which exasperates me, as you can imagine. From two to three is again a breviary hour and from three to seven the time belongs to me once more. I spend it reading; at seven is dinner and at eight o'clock it's all over for the day and so I read until ten o'clock*[23].

Bratránek's description reveals how two main factors, one ecclesiastical and the other secular, determined the daily schedule of a monastic. Let us take the ecclesiastical factor first. From its beginning, the church has strived to control the believers' minds by preoccupying them with thoughts of God. When the incessant prayers, in which the early monastics spent their days, proved to be impractical for most people, the church introduced the concept of *canonical hours*, requiring believers to turn their thoughts to God at certain specified hours of the day.[10] The idea of punctuating the day with prayers and rituals was not new: The monotheistic Hebrews practiced it long before Christianity emerged, as did the polytheistic Romans. The Christian Church formalized—canonized, in its terminology—the practice by prescribing precisely the kind of rites (*offices*) to be performed at each canonical hour. It divided the day into seven canonical hours called matins and lands (two separate prayers counting as one canonical hour), prime, terce, sext, none, vespers, and compline; the first being the morning prayers at dawn, followed by prayers at 6 AM, 9 AM, noon, 3 PM, at sunset, and at bedtime, respectively. The offices comprised different combinations of reciting specific psalms (sacred poems collected in the Book of Psalms of the Bible), singing hymns (songs praising God), reading short passages from the Scriptures, and saying standard prayers. To perform all these offices, a worshiper needed several books: the Scriptures, the Psalter, the lectionary (which specified the passages to be read at each hour and gave information on how to find them), the hymnal, and the prayer book. In the church, the priest would have them all, but a layperson did not have easy access to them. To solve this problem, the church introduced the *breviary* (Latin *brevis* meaning "short" or "concise"), a single book containing all a worshiper needed for the offices, and the entire *Opus Dei* (work of God, for according to St. Benedict "to pray is to work and to work is to pray"). The canonical hours were a far cry from the original idea of a "prayer without ceasing," but even they could be practiced everyday to their full extent only by the "professionals," the priests, and the monastics, while the lay believers just did not have that much time to devote to God. And even the "professionals" were often forced by circumstances to cut corners and skip some of the hours. In Bratránek's description, the reference to "ministering" (acting as an acolyte to a priest at a service) indicates that the Augustinians apparently combined the matins, lands, and prime into a morning service at the Virgin Mary basilica, the 9 AM reading from the breviary representing the terce and the sext, the noon office, being combined with the noon meal, and the none being the breviary reading from 2 to 3 PM. Normally, there also should have been an evening service, the vespers, and the compline, but Bratránek does not mention them, either out of an oversight or, more likely, because they were left to each monastic individually.

The second factor that determined a monastic's daily schedule, the secular factor, was the timetable of the meals that they took together. The bread and water breakfast may give the impression of a frugal diet, but in reality, all the other meals were anything but self-denying. The original hermits might have lived on roots and locusts, but for their nineteenth-century successors, devotion to a good cuisine was an infraction in which they happily indulged. Like every other convent, the St. Thomas Abbey had its own kitchen, which happened to be of outstanding quality, thanks primarily to its *chef de cuisine*, Mrs. Luise Ondráčková. The gastronomical standards in Moravia were generally quite high, but Mrs. Ondráčková's kitchen would have earned a five-star rating, had a Michelin Guide been evaluating Moravian monastery kitchens of that period. The monastics had four daily meals—breakfast, lunch, an afternoon snack, and dinner. The breakfast might have been rather simple, but it is doubtful that it was always as meager for all the conventuals as Bratránek describes it. All the other meals were certainly not. Here is an example of a menu on a typical day. For lunch, they had pea soup with croutons, pork chops with steamed green peas and boiled potatoes, apple strudel with chopped nuts, and tea or coffee. The afternoon snack comprised croissants, scones, cream rolls, tea, coffee with cream, and liqueur. And the dinner consisted of braised kidneys with potato-hash browns. The lunch, the main meal of the day, was a combination of a social and ecclesiastical occasion, often attended by outsiders, usually conventuals visiting Brno. As the bell of the clock on the church tower rang four times, announcing a full hour, and then, at a different pitch, 12 times, heralding noon, the assembled friars, novices, and guests stood behind their chairs in the refectory, thus greeting the entering abbot and prior. After the Amen concluding a short thanksgiving prayer (grace) spoken by the abbot, they all sat down, with the abbot at the head of the refectory table and the novice master (the prior) at the head of the novice table. As the meal was served, lively conversations developed, especially after the main course, with topics ranging from mundane administrative matters to theological and philosophical subjects. Often, readings from sacred or theological texts were also on the program, as were small concerts, of which more will be said later. The afternoon snack may have been a less organized and less formal affair, with the conventuals coming and going individually within a certain time period. Dinner was served again in the refectory at seven o' clock in the evening. After taking their meals, the monastics may have indulged in a weakness to which some of them became addicted: They lit up their cigars. It was apparently in the monastery that Mendel, too, picked up the habit, since before that time he would not have had the money to spend on cigars. Toward the end of his life, he became a heave smoker, and his nephew Alois Schindler, by that time a physician, castigated him in vain, pointing out that smoking very likely aggravated the symptoms of the various illnesses that were plaguing him.[24]

The monastics were about evenly split into those who had Czech and those who had German as their mother tongue. Although most of them were bilingual, the monastics generally held their mealtime conversations in German. Only when the Czech monastics were among themselves did they communicate in their native tongue. Because a majority of the Staré Brno parishioners did not speak German, active knowledge of Czech was required of all the Augustinian priests so that they could perform their pastoral duties in that language. We can assume, therefore, that the novice Mendel, conscientious of his duties as always, might have preferred the company of the Czech monastics, in which he could practice the language. How well he mastered the language is uncertain, however. Some biographers claim that he spoke Czech fluently, while others classify his mastery of the language as poor.[25] It seems that he was able to deliver sermons in Czech, so he must have had more than just a rudimentary knowledge of the Czech language. On the other hand, he apparently began to learn Czech earnestly only after he was past the age when learning a second language comes naturally to a person. In all probability, therefore, his Czech was heavily accented and grammatically flawed: It is very likely that he sometimes used the wrong gender with nouns so that his sermons may have sounded somewhat comical to the Czech parishioners.

During the entire novitiate he remained confined to the monastery. He could decide, however, at any time during the one year to quit, should he discover that the confinement, the rigidity of the daily schedule, the long hours spent in prayers, the demand of absolute obedience of the superiors, the sexual abstinence, and all the other restrictions prescribed by the monasterial rule were not for him. After the completion of the probation period, leaving the monastery legally would become more difficult. The novitiate was, however, also meant to give the monasterial community the opportunity to find out what kind of person the candidate was and whether he could be expected to integrate well into the brotherhood. After all, their community was small, and they were forced to spend much time together under circumstances in which it was easy to get on each other's nerves, if their characters proved to be totally incompatible. The community, therefore, reserved the right to declare the candidate unsuitable at the end of the probation period or even expel him during that interval.

Resorting to such a measure proved to be unnecessary for the four novices. Although they would not be required to formally swear an allegiance to the order until two years later, informally, they became members of the community at the end of the novitiate period. As an expression of the order's commitment to them, they were allowed to study, at the cost of the convent, to become priests. Of course, this step was also very much in the interest of the abbey, especially at the time, when three members of the order died within a short interval and priests were in short supply. And so, after one-year rest, Mendel became a student again. But before we describe these additional four years of studentship, let us have a brief look at his colleagues in the abbey. We begin with the abbot.

Abbot Cyril Franz von Napp (1792–1867): Scholar, Diplomat, and Moravian Patriot[26]

South of Mainz, in western Germany, in the region where the Rhine River forms the border between the present-day states of Rheinland Pfalz and Hessen, lies, in the former state, the small village of Dexheim. Here, an entry in the parish register of the local protestant church indicates that on August 17, 1757, a boy named Georg Ludwig was born to Jost Peter and Maria Nab. Maria was actually Jost Peter's second wife; with the first he had three children, while Georg Ludwig was the fourth child from this second marriage. One year after Georg Ludwig's birth, however, his father died and the orphan was apprenticed to a glove maker. Several years later, after having learned the trade, Georg left Dexheim, though he was probably no more than 12 years old, to wander the world. Where he went and how he lived in the next eight years, we do not know. We know only that at the age of about 20 years, he enlisted in the Austrian army and served in it for some six years as a simple soldier. Somewhere along the way, the soldier changed his name from Georg Ludwig Nab to Ludwig Napp and his confession from Protestant to Catholic. Toward the end of his enlistment period, he was stationed in Moravia, north of Brno, and there he met Johanna Friedl from the village of Bezděčí (now part of Trnávka). In January 1783 Ludwig and Johanna got married in the parish church in Jevíčko, a few kilometers south of Trnávka.

At this point the fates of Cyril Napp and Gregor Mendel began to intertwine, even before either of them was born. For in Jevíčko was an old Augustinian monastery (founded in 1353) within which Johanna or her parents must have had some contacts. Apparently, Johanna had a very high opinion of the Augustinians, and so she insisted on having her marriage consecrated by an Augustinian priest and that the newlyweds settle down in Jevíčko. Ludwig went back to his trade and opened a glove shop, which prospered sufficiently for the family with four children to live "from hand to mouth," as the future abbot was to state later. They could even afford to move into a small house and to send their son Franz, born on October 5, 1792, to schools beyond the *Volksschule* and *Hauptschule* in Jevíčko. Because of the boy's outstanding performance in the public schools, his parents hired a private instructor to teach him everything he would otherwise learn in the first two grades of the *Gymnasium*. Only then did they enroll him in the third grade of the *Gymnasium* in Olomouc. He did well in all the grades and also in the two years he spent at the Philosophical Institute. When his father died unexpectedly at the age of 52 (not 48 as the abbot believed), the widow could not support her son's further studies. At his mother's urging, Franz, after a moment of hesitation, applied for admission to the Augustinian Abbey in Staré Brno (the monastery in Jevíčko having been abolished by that time). His application, written in elegant Latin, so impressed

Benedikt Eder, the abbot of the monastery, that in November 1810 he accepted the applicant. When taking the habit, Napp assumed the name Cyril and so became Cyril Franz Napp, or just Cyril Napp. After the novitiate, he began his studies at the Theological Institute in Brno and once more performed brilliantly throughout his four years there. In addition to his studies, Cyril also taught at schools in Bohunice and Lískovec near Brno, delivered sermons regularly in Czech at the parish church, and began preparing himself for a professorship. It was apparent to everybody who came to know him that he was an unusually gifted young man with excellent prospects for the future. As could have been expected, as soon as Cyril completed his theological studies, he was ordained a priest and passed the examinations for professorship (all with top marks). Brno, Olomouc, and even Vienna showed interest in him. He chose to stay in Brno, and in March 1816, the Theological Institute, from which he just graduated, appointed him professor of the Old Testament. His plan was to obtain a doctorate in theology and embark on a scientific career. To achieve this goal, he had to go through four *Rigorosa*, four highly demanding oral examinations, spread over several years, each in a different specialty. He chose to become an expert on the exegesis of the Old Testament and ancient oriental languages, such as Aramaic, Chaldean, Syrian, and Arabic. By 1824 he had passed two of the four *Rigorosa* and was well on his way to assaulting the remaining two, when, in March, he was informed that he had been elected abbot of the St. Thomas Abbey. There he was to succeed Benedikt Eder, who had died a year earlier in 1823.

Cyril Napp was then 32 years old and his young age was apparently one of the factors favoring his election. Since the government taxed the abbey heavily after each change of leadership, the friars naturally wanted to go through the process as infrequently as possible. This, together with the realization that Napp had been developing into a public figure, one who would represent the abbey well to the outside world, must have been the prime reason behind his election. But why did Napp accept the position when he must have realized that by doing so he was sacrificing his ambitions for a scientific career? One reason was that Cyril Napp was indeed by his nature "a man of the world." Look at his pose in Fig. 5.10. It is the pose of a person who knows how to appear dignified, almost imposing, despite his short stature, early corpulence, and a face that could not be called handsome. It is the pose of someone who enjoys being in the limelight, relishes his influence, and knows his worth. Dignity, influence, and yes, one might even say power were among the potentialities which came with the position of an abbot. A historian once called Napp "the last feudal," and this characterization might not be far off the mark. To Napp, the career of a diplomat, moving in the milieu of landlords, nobles, and politicians, might have seemed an acceptable alternative to the career of a scholar. The second reason Napp took the position of abbot has to do with his background. His father was evidently of Germanic extraction. His mother was a Moravian, but judging by her maiden name, possibly also of Germanic descent, although some biographers have claimed Czech roots for her. Whatever the case might have been, Cyril Franz Napp was equally fluent in both languages, suggesting that his family might have spoken both during his childhood. Napp was sympathetic to the demands of the Czech majority to teach their children the Czech language at

Fig. 5.10 Cyril Franz Napp (1792–1867) was the abbot of the St. Thomas Abbey in Staré Brno from 1824 to 1867 (Based on a photograph)

schools and to have the same rights as those that were granted to the German-speaking minority. Napp also befriended several renowned Czech nationalists—Josef Dobrovský, Pavel Šafařík, František Palacký, and Josef Jungmann (for whom he celebrated a Requiem mass on his death). Finally, he chose as his monasteric name that of one of the two saints (Cyril and Methodius) who, although not Slavs, brought the Slavic form of Christianity into the Czech lands. None of this made Napp a Czech nationalist, however. Napp would undoubtedly have protested against being labeled a "Czech patriot," but he would have been content with the epithet "Moravian patriot," for his love of the land had no bounds. It was this emotion that was behind Napp's tireless engagements in various governing bodies for the benefit of Moravia. He must have realized that in the function of an abbot, he would have many more opportunities to serve his land than he would have had otherwise. This realization must have become an important factor in his acceptance of the position of abbot.

There were two main institutions in which Napp could play out his efforts to influence the cultural and political development in Moravia: the Provincial Diet and the Agricultural Society, both located in Brno. The former was one of the two masters of the province, the other being the central imperial government in Vienna. The Provincial Diet (*Landtag*) was an assembly of representatives of the four Moravian Estates (*Mährische Stände*): the clergy (*Prälaten*), the landlords (*Herren*), the nobility (*Ritter*), and the commoners (*Bürger*; note that the largest section of the population, the peasantry, had no representation in the Diet). Since the two masters had pursued different interests, there had been continuous strife between them. The imperial government with its centrist and absolutistic tendencies had been eager to limit the powers of the Diet and to rule the province unopposed. The estates, on the other hand, had been weary of any attempts to limit their rights and were continually on guard to defend their historical privileges. The actual executive body of the Diet was the *Landesausschuss*, a committee elected by the members of the Diet from their midst for a period of six years. Each estate had three representatives in the committee. The *Landesausschuss* then elected from its midst a hetman (*Landeshauptmann*), who presided over the Diet.

As an abbot, Napp automatically became a member of the Diet on account of his high ecclesiastical rank (i.e., *prelate*, the title which came with the abbotical office) and because of his standing as a major landowner. For this reason, the election of a new abbot had to take place in the presence of two government representatives who functioned as observers, making sure that the friars conducted the election according to the rules. The hetman had to report the election of a new abbot and his admission into the Diet to the emperor. The *Landtag* officially admitted Napp at its October 18, 1824, meeting. Although, as mentioned earlier, the members of the Diet were concerned about their historical rights and privileges; they were mostly ignorant of history and the legal aspects of the Diet's relationship with the crown. By contrast, Napp, the conscientious person that he was, took the time to familiarize himself with both the history of the relationships between the estates and the emperor and with the documents granting the various privileges. As soon as the members of the Diet realized that they had an expert on these questions in their midst, they began seeking his advice. It did not take long before he became a candidate for the *Landesausschuss*. Before he could be elected, however, Napp had to become an *Auskultant*, a member of the *Landesausschuss* without voting power. And so, on March 17, 1826, the *Landesausschuss* admitted him to their deliberations in this function. Elected into the *Landesausschuss* three years later, on October 16, 1829, Napp became a full member on November 16 of that year and remained so for six years. His service proved to be so invaluable that in 1835 and then again in 1841, he was encouraged to run for a second and a third term, and both times he had no difficulty being reelected. Normally, regulations required the candidate to serve three years as an *Auskultant* even between two terms, but in Napp's case, the emperor waved the requirement both times. As a member of the *Landesausschuss*, Napp influenced cultural and political developments in Moravia and Silesia, specifically the foundation of new schools (e.g., the Technical Institute in Brno), new chairs (e.g., the chair for Czech language and literature in Olomouc), and charitable institutions (e.g., institutions for people with impaired vision or hearing at various places in Moravia). Additionally, the emperor appointed Napp director of secondary school education in Moravia and Silesia (*k.k.Gymnasial-Studiendirektor für Mähren und Schlesien*) and a member of the committee for the finalization of the Moravian land register (*Mährisch-schlesische Verwaltung für Katasterangelegenheiten*). In the former function he inspected the *Gymnasien* in the province and prepared a detailed report evaluating the quality of teaching in each of them and recommending improvements. The task of the land register committee was to determine the actual distribution of agricultural land among the individual landowners and thus provide a basis for a fairer allocation of taxes.

An imperial decree established the Agricultural Society in 1770 under the name *Kaiserliche-königliche mährisch-schlesische Gesellschaft zur Beförderung des Ackerbaues, der Natur- und Landeskunde in Brünn* (Imperial-Royal Moravian-Silesian Society for the Advancement of Agriculture, Natural History, and Regional and Cultural Studies in Brno). Even after abbreviating the first two words to "k. k." as was then customary, the title was still too long, so the institution was informally referred to as the *Ackerbaugesellschaft* (the Agricultural Society). Its stated aim was to promote agriculture, forestry, natural history, and regional/cultural studies through scientific research, dissemination of knowledge, and counseling of governmental agencies. It was run by elected officials: a president, vice-president, secretary, and a twelve-member executive board. The elected president had to be approved by the emperor. Originally, the society covered all the areas fitting under the umbrella of its title, but later it began to split into daughter societies or sections, with each section specializing in one of these subjects. By 1861, there were 18 sections, each administered more or less independently, most of them publishing their own proceedings (*Mitteilungen*) and having their own libraries. The main sections were agricultural, forestry, natural sciences, pomological, viticultural, horticultural, and apicultural. The membership in the society and its sections grew steadily from 318 in 1821 to more than 8,000 in 1867. The society played an important part in establishing agricultural schools (six in Moravia and one in Silesia by 1870), running the Franzens-Museum in Brno, appointing commissioners for state examinations in the different subjects, and various other activities.

Napp was active in both the parental organization and some of its sections. He became a member of the Agricultural Society in 1825, member of the executive board two years later, acting president in 1849, vice-president a few months later, and finally, in 1865, president, which, incidentally, was a position that previously had always been occupied by an aristocrat. He was also the president of the pomological section, and in 1847 the government made him a member of the Agricultural Congress (*Landwirtschaftlicher Kongress*) in Vienna, which was an advisory body to the Ministry of Agriculture. In these functions, he advocated the introduction of new species and varieties of crops, as well as new varieties of fruit trees and vines; encouraged the adoption of new methods of crop cultivation and improvement; organized annual meetings of the societies and publication of reports on progress in the various fields; instigated competitions with prizes for the solution of important agricultural problems; and collaborated with other members on experiments designed to test new practices of crop growing. He authored or coauthored several articles in which he described the results of these experiments and observations. His publications attest to his expertise in areas as diverse as vine growing and sheep breeding. For his many contributions, successive emperors decorated him. In 1836, he became *Ritter des kaiserlichen Leopold-Ordens* (The Knight of the Emperor Leopold's Order); in 1850, *Ritter des kaiserlichen Franz-Josef-Ordens* (The Knight of the Emperor Franz-Josef's Order); and in 1859, *Kommander der österreichischen kaiserlichen Ordens der eisernen Krone* (Commander of the Austrian Imperial Order of the Iron Crown). These distinctions made Napp a nobleman, giving him the right to place the preposition *von* before his last name. Out of modesty, Napp himself never used the predicate, but in the post-1860 monthly reports of the pomological section, the editors listed him as "von Napp." As much as Napp might have been pleased with these distinctions, he was probably most gratified by the *Doctor honoris causa* degree in theology that the University of Olomouc bestowed on him in 1832.

Because of these accomplishments, Napp became a well-known personality in the province of his time. Nevertheless, history, which views the past through the lens of an inverted telescope, would have undoubtedly forgotten all about him by now, were it not for what he made out of the abbey. Two other abbots preceded him at the helm of the monastery since the move to Staré Brno: Vincentius Polzer from 1794 to 1809 and Benedikt Eder from 1809 to 1823. They did their best to pull the abbey out from the financial and moral low into which it sank because of its relocation, but the task seemed to have been beyond their capabilities. A full recovery required a man with great managerial skills, excellent connections, and vision. Napp was such a man. He achieved the abbey's economical recovery primarily by increasing the productivity of the monasterial estates. He actually took over the management, preparing himself for the task by studying the literature on the subject and by seeking advice from the experts among his friends, especially Franz Diebl, professor of agriculture at Brno's Philosophical Institute. By following their recommendations, Napp introduced crop rotation, cultivation of leguminous fodder plants, and rational methods of sheep breeding. He also conducted field trials with new crops, established fruit-tree nurseries for improved varieties, and experimented with new methods of pest control. He described his experiences in the publications of the Agricultural Society so that others might profit from them. This way he managed to bring the abbey's wealth to new heights while at the same time completing the reconstruction of the monasterial buildings initiated by his predecessors.

But all of this effort was a mere preparation of material grounds on which to realize his vision of making the abbey one of the foremost cultural centers of the province. To this end he actively sought candidates and carefully screened all applicants for admission into the abbey; promoted those who showed special talents, be they for philosophy, literature, music, or natural sciences; enabled these talents to obtain the education necessary for their full development; encouraged members of the monasterial community to become involved in teaching and other cultural activities outside of the monastery in order to foster contacts with the outside world; and opened the gates of the abbey to a stream of distinguished visitors to whom he extended hospitality, culinary pleasures, and a refined and enlightened intellectual atmosphere of the monasterial community. The entries in the Guest Book kept in the abbey's library bear witness to the variety and caliber of the visitors: in places they read like pages from a Who's Who in Moravia/Silesia. Church dignitaries alternate with Czech nationalists, philosophers, linguists, historians, scientists, theologians, *Gymnasium* and university professors, journalists, physicians, merchants, aristocrats, high-ranking provincial clerks, and army officers. The abbey also welcomed a group of actors from Brno's theater, monks that were passing through, and students. Under the date November 27, 1842, there is a curious, if somewhat enigmatic, entry (presumably made not by the guest himself but by the librarian); it reads *Rabbi Hersch Dänemark*, with the added note *der wohl nicht schreiben kann, aber kuriose Kunste treibt* (who apparently cannot write but pursues curious arts).

What kind of person was Napp? As with other highly placed personalities, there were several sides to his character. In the company of aristocrats and feudal lords, he was respectful, but not servile, friendly but not obtrusive; he was one of them. To his subordinates, he appeared cold because of his insistence on etiquette and his keeping a distance from them. He was a strict disciplinarian, even a stickler, one might say, when it came to the observance of monasterial rules. (It is therefore a great irony that the bishop of Brno criticized him, as we will learn later, for his lax attitude toward discipline.) A memo he wrote illustrates this side of his character. It is addressed to the prior, that is, his deputy, and concerns Mendel.[20] In translation from German it reads thus:

> *To the reverend prior Baptist Vorthey.*
> *It has come to my attention that the ordained priest P. Gregor attends lectures in fourth year theology without wearing the order's cowl. Although he is a priest already, P. Gregor is nevertheless still a student and for this reason and for the reason of retaining uniformity in his outer appearance with the student-priests of the order, the reverend prior should inform P. Gregor that he has to wear the order's cowl when he attends lectures, just like any other clergyman of the order.*
> *October 18, 1847* *Napp*

At least four points are worth noting about this memo, for they allow us a peek through the keyhole into the internal affairs of the abbey. First, surely Napp could have taken Mendel aside after lunch and told him in two sentences that he should not walk around without the cowl, but instead he chose to reprimand him formally and indirectly, via the prior. By doing so he not only added weight to the reprimand but also avoided encroaching on the authority of the prior, who was in charge of not only the novices but also the students. Additionally, the memo might also have been a veiled reprimand to the prior himself, for he should have been the one to notice the violation and taken the proper steps to correct it. Second, the memo makes clear how strict Napp was in matters of discipline. Mendel almost certainly did not dispose of the cowl out of vanity or to show off his status. He simply found this part of the habit a nuisance, especially during his walks to and from the Theological Institute, when wind might have flapped it over his head and into his face. But Napp showed no benevolence toward comfort that violated the rules. Third, the opening sentence of the memo suggests that Napp learned about the transgression indirectly, that is, somebody must have told him about it. If so, it would indicate that the friars were informing on each other. Such behavior would be in accordance with monasterial code of conduct, but viewed from the outside of the abbey might not be considered honorable. And fourth, the memo is an example of adults being disciplined like little children in the monastery, a practice some monasterial candidates might have found demeaning. Whether Napp's insistence on etiquette went as far as requiring his own mother to address him as "Your Grace," as claimed by some of his contemporaries,[20] cannot be verified.

Quite a different side of Napp's character is revealed by his actions on behalf of his subordinates. Later, we will describe several examples of Napp's assistance to his fellow friars when they needed help. The examples reveal Napp as a kindhearted

Fig. 5.11 Napp's Eleven: The full members of the St. Thomas Abbey in Staré Brno in Mendel's time (Based on a photograph)

human being, compassionate with subjects under his charge, empathic with their frailties, and extending a helping hand to them even if it sometimes meant that his superiors would not approve of his actions.

The St. Thomas Quartet and the Other Augustinians

In Mendel's time, the community of friars at the St. Thomas Abbey had about a dozen members in residence and a few more at various outposts (Fig. 5.11). About one half of the friars in residence devoted most of their time to pastoral duties at their affiliated church and in the parish, particularly in the local hospital. The parish had some 12,000 believers, all of whom technically belonged to the Church of the Assumption of Virgin Mary. Obviously not all of them attended church services regularly, nor did they all require priestly assistance, but those who did kept the five or six priests assigned to parish work busy. The remaining friars taught at local educational institutions and in their spare time pursued their scientific and other learned interests. In 1843, a short time before Mendel and the three other candidates began their novitiate, two friars (Aurelius Thaler and Gelasius Platzer) died, so that at the end of that year the abbey had 18 members: 14 friars and four novices (Fig. 5.12).[27] Of these, four friars lived permanently outside of the abbey: Franz Xaver Wiesner was the professor of Biblical studies at the Theological Faculty of the University of Olomouc, while Philipp Gabriel, Anton Franz Alt,

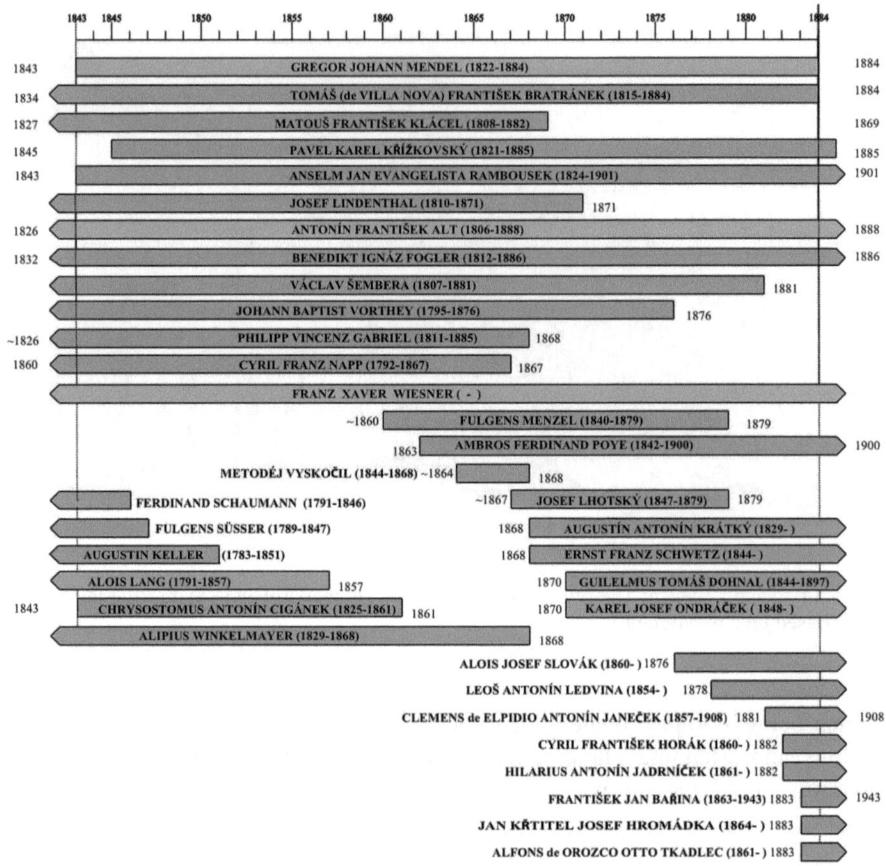

Fig. 5.12 Friars of the St. Thomas Abbey in Staré Brno fully or partially overlapping with Mendel's tenure there. Dates in parentheses give years of birth and death of each individual; dates on the margins of each bar give dates of entry and of exit the abbey (where known to us) (Based on data found in ref. 23 and 27; prepared by Dr. Akie Sato)

and Ferdinand Schaumann were directors of *Gymnasien* at Český Těšín (in Silesia), Bratislava (in today's Slovakia), and Opava, respectively. And one friar, Tomáš Bratránek, lived temporarily in Lvov, where he taught at the university. Of the nine in-residence friars, two (Abbot Cyril Napp and Prior Baptist Vorthey) spent most of their time on administrative duties, three (Augustin Keller, Václav Šembera, and Josef Lindenthal) on parish work, and two on teaching (Benedikt Fogler taught French and Italian at the *Gymnasium* and *Realschule* in Brno, and Matouš Klácel read philosophy at Brno's Philosophical Institute); while the remaining two (Fulgentius Süsser, who preceded Klácel at the Philosophical Institute, and the parish priest Alois Lang) were retired and in poor health. In the following few years, the situation changed somewhat in that three friars died (Süsser in 1847 and Keller and Lang in 1851); Klácel was dismissed (in 1844) from his teaching post and turned to research in the abbey; and the abbey acquired an additional novice (Pavel Křížkovský in 1845). Of the three earlier novices, Wilhelm Rössner left the abbey, Anton Cigánek died in 1861, presumably the victim of parish work in the hospital, and Anselm Rambousek taught Czech language at the *Gymnasium*, in addition to carrying out pastoral work.

The education level of the friars was quite high. They all went through secondary- and university-level schooling; the latter represented by the Theological Institute and other specialized university courses in some cases. They all spoke several languages—German, Latin, Greek, and one or two additional living (Czech, Italian, French, English) or dead (Hebrew, Arabic, Syrian) languages. They were all honed in mathematics and philosophy and some were experts on other select subjects. Most had some background in music and participated (some actively, but most of them passively) in Brno's cultural life. Some of the friars were well traveled, at least within the realm of the Austro-Hungarian Monarchy, and some were well-read. Because of all this, one can assume that anyone of them could have participated in high-standard, intellectual conversations on a wide variety of topics. Nonetheless, posterity records do not reveal much more than the names and dates of births and deaths of all these men, with the exception of five: Napp, Klácel, Bratránek, Křížkovský, and Mendel. These five project high above the abbey's intellectual skyline. That they met at the same time and the same place must be credited to Napp, who assembled the other four, recognized and nurtured their talents, and provided conditions for their development within the convent's limits. The four, in their turn, influenced one another, and so we refer to them as the "St. Thomas Quartet" (Fig. 5.13). They were friends, although the intensity of their interrelationships varied from person to person and with time. They had things in common but differed in many others. They all came from meager circumstances: Klácel was the son of a shoemaker, Bratránek of a clerk to a landlord, Křížkovský an illegitimate son of a servant woman, and Mendel of a farmer struggling to keep his head above water. The last two had particularly harsh experiences behind them when they joined the abbey. (Even in the abbey, Křížkovský remained the target of ill-suited jokes about his illegitimacy.) They all came from different regions: Klácel from Česká Třebová in Bohemia, in the valley of the Třebovka River, near the border with Moravia; Bratránek from Jedovnice near Brno; Křížkovský from Holasovice, a village at the site of a prehistoric Slavic settlement, some ten kilometers west of Opava; and Mendel, as we know already, from Hynčice in Silesia. They differed in their ethnic origin: Klácel's family was Czech (although his father Germanized the family name to Klatzl, the son adhered consistently to its Czech spelling); Bratránek's mixed (father, Czech; mother, Austrian); Křížkovský's Czech, but Germanized; and Mendel's Germanic. Klácel and Bratránek learned Czech at home and German at school, whereas Křížkovský and Mendel began to seriously learn Czech only in the abbey. Klácel was the only Czech nationalist among the four; Křížkovský became sympathetic to the Czech nationalist movement under the influence of Klácel and upon exposure to Moravian folksongs, whereas Bratránek and Mendel had remained neutral in the Czech-German conflict that flared up in the Brno of their time. Mendel was the youngest of the four; Klácel and Křížkovský were one year older and

Fig. 5.13 The Saint Thomas Quartet. From left to right: Gregor Mendel, Pavel Křížkovský, Tomáš Bratránek, and Matouš Klácel (Based on a sketch by Alois Zenker)[28]

Bratránek seven years older than Mendel. But all four died within a range of three years of one another: Klácel in 1882 at the age of 74 years; Bratránek and Mendel in 1884 at the age of 69 and 62 years, respectively; and Křížkovský in 1885 at the age of 65 years. All four had a similar primary and secondary school education: Klácel, Piarist *Gymnasium* and Philosophical Institute in Litomyšl; Bratránek, *Gymnasium* and Philosophical Institute in Brno; and Křížkovský and Mendel, *Gymnasium* in Opava and Philosophical Institute in Olomouc. After joining the abbey (Klácel in 1827, Bratránek in 1834, Křížkovský in 1845, and Mendel in 1843, at the age of 19, 19, 25, and 21 years, respectively), all four attended the Theological Institute in Brno. All four also joined the abbey for the same reason: because they could not find any other way to continue their studies and pursue their intellectual interests.

They all showed early signs of giftedness, without which sons of poor parents had no chance of getting a *Gymnasium*-level education. Klácel could read fluently after only a few weeks in elementary school; Bratránek impressed his elementary school teacher with his verbal capabilities; Křížkovský taught himself to play the clarinet and read music; and young Mendel soon multiplied better than Makitta. Of course standing out in a crowd is one thing, but having the right person to notice you is quite another. A teacher may or may not recognize talent, but even if he or she does, he or she may not be in a position to nurture it. Often, an encounter with another person is essential in order to give direction to the development of a student's potential. In Mendel's case, Makitta recognized his intelligence, but it was Schreiber who galvanized his budding interest in natural history.

In young Křížkovský's case, his musical aptitude was all too apparent to his uncles, themselves accomplished village musicians, who had to thrash the boy frequently for stealing into taverns to hear their band play. It was, however, Jiří Janáček, the father of the composer Leoš Janáček, who took it upon himself to foster the boy's talent, after Křížkovský's mother, at a loss with what to do about her child's obsession, brought her son to him. Though the 16-year-old Jiří Janáček was only five years older than Křížkovský, he already held the position of an assistant teacher at the school in Neplachovice, a village next to Křížkovský's birthplace. Janáček got Křížkovský involved in choir singing at the church in Neplachovice and later also in Opava and thus launched the career that ultimately brought Křížkovský to Staré Brno. Bratránek's Muse, on the other hand, was Ottilie von Goethe (née Pogwisch) and her sons Walter and Wolfgang (Ottilie being the wife of Johann Wolfgang von Goethe's only surviving son August). The acquaintance with the von Goethe family not only opened the door for Bratránek to other aristocratic and patrician houses in Vienna and elsewhere but also provided him with unrestricted access to Goethe's archive in Weimar. It was this access that made Bratránek an internationally recognized expert on Goethe's works and the editor of Goethe's correspondence with the founder of the National Museum in Prague, Kašpar Graf von Šternberg, and with the brothers Alexander and Wilhelm von Humboldt. Crucial to Klácel's intellectual development was his acquaintance with his teacher, the philosopher Bonifác Buzek (1788–1839), at the Philosophical Institute in Litomyšl, the birthplace of the Czech composer Bedřich Smetana.

Since no member of this quartet was a religious zealot who had chosen the monastic mode of life because of a calling, each member had to come to terms with his chosen way of life somehow. Mendel and Křížkovský seemed to have accepted it as the price they had to pay for escaping the poverty into which they were born and for getting a chance to pursue their chosen careers. If the monastic life was an onus to them, they bore it without complaint. The two philosophers, perhaps because they never experienced starvation in their lives, were of a different disposition. Klácel in particular had a hard time incorporating his rebellious nature into the straitjacket of the monastic rule and in the end decided to withdraw from the abbey. Bratránek put up with it for as long as he could live outside the abbey, but once he faced the prospect of going back to Staré Brno, he too bailed out, although ultimately he did return into the community of his monastic brothers to die.

Philosophy and Linguistics at the St. Thomas Abbey

Let us now take a closer look at the three friars who, after Napp, must have had the greatest influence on Mendel's intellectual development in the third decade of his life. Let us begin with the two philosophers first. It was probably on Buzek's recommendation that František Klácel (1808–1882; Fig. 5.14) applied for

Fig. 5.14 "The Saint Thomas quartet"

admission into the St. Thomas Abbey in Old Brno. He was admitted in 1827, assumed the monastic name Matouš (Mathew), studied at the Theological Institute in Brno from 1829 to 1833, and in 1833 was ordained a priest.[29] Napp then fulfilled Klácel's wish to study at the University of Olomouc for a doctoral degree in philosophy. However, in 1835, shortly before the completion of his studies (after his third *Rigorosum*), Napp asked Klácel to return to Brno and assume the position of a professor of philosophy at the Philosophical Institute. He replaced Fulgentius Süsser (1789–1847), an elder friar from the St. Thomas Abbey, who gave up the post due to his declining health. Up to this point, life seemed to have been unfolding well for Klácel. His dream of becoming a philosophy professor like Buzek had come true, he enjoyed teaching, and the students, responding to his youthful enthusiasm, liked him. Then, however, things began to go awry. Brno's Bishop Schaffgotsch, under whose jurisdiction the Philosophical Institute operated, knew from experience that whenever students liked their professor, there was a reason to be watchful. Sure enough, his Excellency's spies soon began informing him that Klácel was a Czech nationalist, a freethinker, and a rebel, who was teaching Hegelianism and pantheism in his philosophy classes. The bishop may not have been sure of what it was exactly that was wrong with Hegelianism, but he was aware that it was considered a dirty word in the clerical circles, and he knew, of course, that pantheists did not believe that God was a person, but instead claimed that the whole universe, with all its laws, forces, and manifestations, was God. So, he let his displeasure be known to Klácel, ordering him to stop corrupting innocent minds with poisonous ideas. Klácel, however, in his enthusiasm did not heed the ban, and so, in 1844, the bishop dismissed him from the post. Like Buzek before him, Klácel became a marked man, who would not be allowed to teach philosophy again.

Napp tried to ease the impact of the blow by sending Klácel away from Brno. On the invitation of Baron Antonín Veith, Klácel went to the baron's manor house in Liběchov, on the right bank of the Labe River, some 40 kilometers north of Prague. Veith was a Czech nationalist who made his residence the meeting place for important personalities of the time. At Liběchov, a small, isolated sleepy place, Klácel kept himself busy by organizing the baron's large library, an activity in which he had considerable experience, since in his early years at the St. Thomas Abbey he functioned as a librarian for some time. But he soon realized that he had become a fugitive. Once his presence at Liběchov became known, the archbishop of Prague put pressure on Veith to send Klácel packing. Klácel then returned to Brno to devote himself to philosophical studies. But in 1848 a sudden unexpected change in the political situation rescued him from the state of depression into which he had slowly been sinking since his dismissal from the Philosophical Institute. Even though the change, about which we will have more to say later, proved to be temporary, while it lasted, it raised the hopes of the whole nation. Klácel traveled to Prague, where he became politically active in the Czech nationalistic movement. Soon, however, the Austrian army crushed the Prague uprising, and Klácel was forced to go into hiding in Česká Třebová, before he could safely return to Brno. Embittered, frustrated, and depressed, he sought solace in research and writing and escape in the private tutoring of girls from the wealthy families of Brno's burghers. The former activity, however, could not console him, because there was little hope that its fruits would ever be allowed to reach readers outside of the monastery. The latter was a rather risqué activity, since the Augustinian Rule contains a detailed prescription for a monk's behavior in female company, and any kind of intimacy is certainly not among what is allowed. It seems, however, that Klácel had pretty much ignored not only the rule but also certain social norms of the period.

In the late 1850s, Klácel found himself in a situation that he could control no longer. He was legally bound to an organization, the Roman Catholic Church, which obliged him to participate in ceremonies and rituals that no longer meant anything to him. Ultimately, he even stopped professing its central creed: *Credo in unum Deum, Patrem omnipotentem* (I believe in one God, the Father Almighty), for he indeed forsook a personal God for a god in the form of the universe. Furthermore, the institution required him to follow a certain way of life that was becoming more and more difficult for him to adhere to. He felt like he was living in a cage, and so he began to weigh the various options of escaping. Already in the early 1850s he began to favor one of these options. In 1852 he confided to one of his lady friends (Veronika Vrbíková, see Vol. 2 Chap. 1) his plan to emigrate to the United States,[29a] and one year later he wrote to Bratránek about his desire to organize a religious mission in Texas.[30] But how would he arrange it? He was sure that the bishop would not grant him the necessary permission: Should he leave without permission? He vacillated, until in 1867 events more or less forced him to come to a decision. In that year Napp, his protector, died, and in the next year the chapter elected Mendel to become Napp's successor. This development made him make up his mind, and on June 28, 1869, he left Brno secretly, sneaking out like a thief, to embark on a long journey, by train across Europe and then by boat across the Atlantic. He arrived in

New York on July 21 and two days later fired off a letter to Mendel informing him about his arrival in the United States. The circumstances of Klácel's departure from Brno are somewhat obscure, particularly in regard to the parting of Klácel and Mendel, each of whom later called the other his friend. It seems, however, that Mendel was among the few who were in on Klácel's plans. It may even have been Mendel himself that signed the travel document Klácel needed for the journey. If so, the remarks these two friends later made about each other are rather puzzling. Mendel's comment about Klácel was recorded by his nephew Alois Schindler in a letter to Hugo Iltis, dated January 26, 1923: *One Sunday I went to see my uncle the Abbot. I noticed a photo album lying open there. I leafed through it and found at the end a photograph of a priest in the Augustinian habit, placed loosely inside. There was plenty of room left in the album, so I tried to put the photo into it. But my uncle told me to leave it, that the photograph did not belong in the album, since the priest had broken his vows. But he used to be a friend of mine, he added.*[31] (One wonders whatever happened to that album. Might it still be somewhere in the monastery or has it fallen victim to the cleaning up mania of Mendel's successor?)

What Mendel had said about Klácel is formally correct: Klácel left the abbey without permission and so broke the vows. From the standpoint of the Catholic Church, he committed a crime—he was a deserter and a traitor— and as such he had no place in the album. (Indeed, in the catalogue listing friars of the Old Brno Abbey during the period of Mendel's abbacy, Klácel's name is not included)[23,27] he was a *persona non grata*. Yet, Mendel kept the picture in the album, even if in an easily removable form, which could only mean that he had remained fond of his departed friend. Klácel, on the other hand, had this to say about Mendel in his autobiographical sketch: *Prelate Napp died at an advanced age—elected was a young professor, learned and free-thinking, my friend, but a hypocrite, who knew how to please everybody, since he didn't care about a right side or left, but only about the "golden calf," which he hid under the fleece of the Eastern Lamb, and swore onto anything the more willingly, the less he believed in it.*[31]

Very unkind words, these, about someone, whom the writer still calls his friend; one would not wish to know what Klácel might have said about his enemies. Even if taken as having been written in a state of bitter disillusionment, the brutality of his words contrasts strikingly with the civility of Mendel's remarks about Klácel. They reveal more about Klácel himself than about Mendel. What they do reveal about Mendel's character we leave for later analysis.

In the same autobiographical sketch, Klácel gave this reason for his defection from the abbey: *Seemingly, I was living quite well, but my soul was in ruins and my spirit, which for some years longed for America, felt lonely, not only in the monastery, but also in the Czech crown lands, in the Holy Church, in the whole of Europe.*[32] He went to the United States as a prophet of a new philosophy, his philosophy, to which he hoped to convert the US citizens of Czech descent. Like other immigrants in the country, Czech descendants tended to settle down together in small communities, primarily in central Texas, Iowa, Illinois, and Wisconsin, and tried to keep not only their original language but also their customs, traditions, and other aspects of their culture. Klácel intended to reach his compatriots through the printed word and word of mouth. Under the name Ladimír Klácel, he began to indoctrinate them through the medium of the *Slovan Amerikánský* (American Slav). It was a daily newspaper published by a certain Jan Bárta in Iowa City, the same man who invited Klácel to come to the United States and become the paper's editor. When the daily folded for lack of readers, Klácel tried to start several other periodicals on his own, but they all failed for the same reason. He then made an effort to spread his word through lectures, but again with only limited success. Equally unsuccessful were his attempts to organize communities that lived by his philosophy. He did not seem to be able to realize that he was dealing with simple people, who were unreceptive to his utopist ideas. They were mostly pragmatic people, already deeply imbued with the spirit of their new homeland, with entrepreneurship, veneration of private property, and worship of the dollar. So all his preaching about socialism and communism, about a society in which individuals should live in harmony like parts of a single body, about establishing a new social order based on science, and about the brotherhood of all people on the mystical principle of pantheistic cosmicity was obviously falling on deaf ears. His disillusionment was immense. Eventually, Klácel's savings ran out and he began to starve. The combination of his disillusionment and the physical distress caused by starvation left him a broken man. He lived on charity, wandering from town to town, looking for a place where he could rest his old bones. In Belle Plaine, Iowa, a local butcher, a certain Frank Zaleský, stemming from the Litomyšl region like Klácel, felt sorry for him and took him into his home. There he died a few months later. A characteristically Czech postscript followed: Shortly after his demise, the citizens of Belle Plaine erected a monument in his memory (as did the American-Czech community in Chicago). The Czechs in Brno took more than a century to commemorate Klácel in a similar way and then only in the form of a statue that looks like a hula-hooping stack of papers.[33]

While Klácel had tried to work out a compromise between living in the abbey and working outside, Tomáš De Villa Nova František Bratránek (1815–1884, Fig. 5.14) arranged his affairs so as to spend as little time in the abbey as possible.[34] The only time he actually lived in the abbey were the years of his novitiate, together with the years he spent studying at Brno's Theological Institute (a period from 1834 to 1839); then, the period from 1843 to 1851; and finally, from 1881 until his death in 1884. Although on paper he was a member of the abbey for 50 years, altogether he actually lived in the monastery for only 16 years. How did he achieve this? As soon as he was ordained a priest in 1839, he left the abbey, with Napp's blessing, to study philosophy at the University of Vienna for two years, from 1839 to1841. Almost immediately after taking his doctorate, he left for the Ukrainian City of Lvov (which the Austrian occupiers then called Lemberg, in the province then called Galicia), where he became an assistant to Professor Ignác Jan Hanuš (1812–1869) at the Philosophical Faculty of the university. When the bishop dismissed Klácel from the Philosophical Institute in 1844, Napp called Bratránek back to Brno to take over Klácel's post. While teaching philosophy (and at one point also natural history), he only lodged and dined at the abbey, spending most of the day at the Philosophical Institute, instead. Also, for vacations he left the abbey to travel through Austria, Switzerland, northern Italy, and the Rhine Valley, visiting notable representatives of literary and scientific life wherever he went. Through the contacts that he made during his trips, he had himself appointed an extraordinary (in 1851) and then (in 1853) ordinary professor of German literature at the Jagiellon University in Kraków (which was situated in the part of Poland then annexed by Austria). He occupied this post until his retirement in 1881, serving at one time as a dean of the Philosophical Faculty and at another as the president (*Rektor*) of the university. Bratránek managed to keep himself so busy that he only spent his vacations in Brno.

From this account, one might think that Bratránek, in contrast to Klácel, must have been contented with the course of his life. Although a friar, he lived much of the time outside of the abbey and so escaped the dreary ecclesiastical routine he loathed. Facing the prospect of having to return to the abbey after his retirement from the university, he took the penultimate step to extricate himself from monastic life—*secularization*. The authorities granted him his wish to live permanently in the world (*saeculum* in Latin) outside the abbey and the order, while still remaining, in essence, a priest. Unlike Klácel, he had avoided taking sides on controversial issues and so stayed out of trouble (undoubtedly one of the reasons he quit the teaching post at the Philosophical Institute was that, after the Klácel affair, the ground had been too hot for him). He had behind him a distinguished career as a teacher, philosopher, and researcher in the field of German literature. Numerous decorations, medals, and honorary memberships in prestigious societies attested to his high standing among his peers, as did his acquaintances with the foremost representatives of cultural life in the monarchy. What else could he have wished for?

Had he, in advanced age, not written an autobiographical sketch (similar to Klácel's),[35] there would be no reason to doubt that the productive life of a tranquil person had come to an end when he died. From his autobiography, however, a different view of Bratránek's life emerges. He comes across as a man who believed that he had miscarried his life by always trotting in the middle of the road, not daring to deviate from the course lest he run into trouble, a man who felt like he was neither Czech nor German, neither servant nor aristocrat, neither priest nor layman, and not a human being, only a monk. Despite all the contacts and acquaintances, he realized that he had no true friends and nobody to turn to. An insurmountable aversion to life and work overwhelmed him at the end of his life. Despite his secularization, he returned to the St. Thomas Abbey to die there in 1884, the same year as Mendel.

Music at the St. Thomas Abbey

Having lived off of the toils of others in this world, wealthy nobles sometimes took out insurance for a good life in the afterworld. A popular form of such insurance used to be the establishment of a foundation with the stipulation that its beneficiaries would pray for the benefactor's soul. Brno's noblewoman Sibylla

Polyxena Francesca von Montani née, countess of Thurn and Walsassin, a scion of an old lineage that originated in Lombardy, northern Italy, but spread subsequently through other lands of the Habsburg Empire, undoubtedly had eternity on her mind when she established what became known as the Thurn Foundation at the St. Thomas Abbey.[36] At that time, the Augustinians still resided at their original location near what is now Moravské náměstí (Moravian Square), and St. Thomas was still their affiliated church. The stated purpose of the foundation was the development of church music at the St. Thomas Church.

As in many other religions, in the Catholic version of Christianity, music played an important part in its rituals and ceremonies. Not only did the congregation as a whole sing hymns to the accompaniment of an organ, but their souls were also lifted by the sound of music performed by amateur and professional singers, soloists, and a choir. In Mendel's time, every respectable parish church employed a *Regenschori*, a choir master or director, who was responsible for preparing, rehearsing, and conducting performances of church music on Sundays and religious holidays. The word choir had a dual meaning back then: It referred to the organized group of singers performing in the church, but it also signified the elevated part of the church (usually above the main entrance) from which the group sang. Since the Renaissance period, composers have written masses, Requiems, Te Deums, litanies, and other pieces based on the sacred texts of the Catholic Church. It had been the responsibility of the *Regenschori* to choose compositions appropriate for the different occasions and present them to the congregation. The Thurn Foundation, however, supported not only the *Regenschori* (who was generally a member of the monastery anyway) but also a veritable music school—a musical conservatory. Indeed, this institution was loosely modeled on the first conservatories, which arose in the Renaissance period in Naples, Venice, and other Italian cities. The name "conservatory" was originally chosen not because it implied the "conservation" of music, but because *conservatorio* was then a term used in reference to a workhouse (from Latin *servire*, to serve) or an orphanage in which teachers instructed the foundlings in music. In France after the revolution of 1798, conservatoire stood for an institution in which talented citizens could receive musical training. Ultimately, the Thurn Foundation did develop from an organ school (in 1819) into a true conservatory. The foundation supported six to 13 boys from underprivileged families and a staff of teachers that instructed the pupils in musical theory, singing, and playing various musical instruments. The boys received free lodging at the monastery, meals, candlelight, and medical care. The foundation also admitted additional students who could afford to pay the tuition of 100 florins a year. Candidates in both categories had to be nine to 12 years old and had to be able to play several instruments. Because of the navy-blue uniforms they wore, people called them affectionately *modráčci* in Czech and *Blaumeisen* in German ("blue tits"). They had a grueling daily schedule: a wake-up call at 5:00 AM, prayer and study until 6:45, morning mass with singing at 7:00, then breakfast and school (elementary or lower middle) until lunch, a short walk in the afternoon, and then study and practice until 6:00 PM, dinner, brief free time, an evening prayer, and bedtime. This routine could be disrupted, however, by their performances in churches, at concerts, in the theater, at balls, in welcoming ceremonies for visiting dignitaries, and in funerals for distinguished deceased. While singing *Salve Regina* and litanies, they were supposed to think of the countess. Each student was trained to sing either solo or in the choir, play several instruments in their "harmony" (the orchestra of 20–30 players), and also to conduct. They performed in small groups, as a whole ensemble, or integrated into large choirs and orchestras of professionals and amateurs. It seems that Brno's cultural life would have ground to a stop without them. From their midst sprang accomplished musicians, virtuosi, soloists, and composers, including Moravia's greatest—Leoš Janáček. The success story of the Thurn Foundation peaked in the second half of the nineteenth century and did so thanks primarily to two persons—Napp and Křížkovský. Napp loved music. He and Křížkovský played in the string quartet of Count Michal Bukuvka. He went to concerts whenever his busy schedule allowed him to do so, and on his insistence, the "blue tits" performed "table music" at noon meals in the refectory. But most importantly, Napp recognized Křížkovský's potential, appointed him *Regenschori*, and created conditions in which his creativity and organizational talent could develop.

Pavel Karel Křížkovský[37] (1820–1885, Fig. 5.14), then still signing his name as Karl Krischkowsky, came to Brno in the fall of 1843, at nearly the same time as Mendel. Unlike Mendel, however, he still had to complete his secondary education before he could embark on the path leading to priesthood, toward which he was leaning by then. Therefore, he enrolled in Brno's Philosophical Institute, where two Augustinians from the St. Thomas Abbey were teaching at that time: Matouš Klácel and Filip Gabriel, the former taught philosophy and the latter mathematics. Gabriel, however, was also the *Regenschori* at Staré Brno's parish church, and through him Křížkovský established contact with the abbey and the Thurn Foundation. At the same time, he also continued his own musical education. The dominant figure in Brno's musical life back then was Gottfried Rieger (1764–1855), music teacher, conductor, and composer.[38] Rieger, like Křížkovský, came from an impoverished family, his father being a cottager and village musician in Opavice, a Silesian hamlet, which the parties in the Austro-Prussian war split by a frontier. Rieger's father managed to have his 13-year-old son accepted into the service of a local count, first as a footman and later as a member of the *Hauskapelle* (house orchestra), where he learned to play several instruments. In 1787, the young musician came to Brno, and there he soon earned himself the reputation of an excellent music teacher. Ultimately, he got himself appointed music director of the theater, for which he wrote *Singspiele* (musicals). Except for four years, he remained in Brno for the rest of his life. In 1828 he founded Brno's first music school, which over the years gave Moravia many professional musicians. The music he wrote (19 masses, several cantatas, and some orchestral pieces, including a symphony) fell into oblivion, but his pedagogical and organizational activities made Brno the musical capital of the province. Whether Křížkovský actually studied with Rieger (who was by then nearly 80 years old) is uncertain, but the two must have met at concerts and at the choir of Staré Brno's parish church where Rieger occasionally helped out. An indication that Křížkovský knew Rieger more closely than by reputation alone is that he commemorated the latter's passing with a performance of Cherubini's Requiem.

A profound influence on Křížkovský's artistic development was the priest František Sušil (1804–1868), a poet, translator, collector of Moravian folksongs, and ardent Moravian patriot.[39] Born in the village Nový Rousinov near Vyškov in southern Moravia, he graduated from the Piarist *Gymnasium* in Kroměříž and then attended the philosophical and theological institutes in Brno. After taking orders and doing pastoral work for some ten years in different places around Moravia, he was appointed professor of the New Testament at Brno's Theological Institute. Already as a student, while on vacation in his native village, he began collecting folksongs and kept at it for some 30 years, gradually covering the whole of Moravia and Silesia, as well as western Slovakia and the Slavic enclaves in northern Austria. With the help of local priests and teachers, he would gather willing peasants in the local pub or visit them at their homes, listen to their singing, and then record on paper both the words and the melody of each song. He published, at his own expense, his collection of 2,361 songs in three volumes, which appeared in 1833, 1840, and 1860. The collection became the source of inspiration for Czech composers from Křížkovský, through Janáček, to Bohuslav Martinů. Shy and modest to a fault, Sušil was revered by those who came in contact with him, but probably by none more than Křížkovský and Janáček.

To Křížkovský, Sušil's collection was a revelation and a tremendous impulse for his creative activity. Although he already began composing before he came to Brno, under the influence of the city's cultural atmosphere, Sušil's enthusiasm, and the encounter with the pristine melodic beauty of the Moravian folksongs, he began developing his own compositional style. Sušil, however, also had a different kind of influence on Křížkovský. As a fervent Czech nationalist, he did not miss an opportunity to bring an uncommitted or lukewarm person over to the cause. Collecting Moravian folksongs was one expression of his nationalism; another was organizing meetings and other social events that promoted Czech culture. He was a member of a Czech patriotic circle, which included Matouš Klácel, the physician, and since 1850 professor of natural sciences and agriculture at Brno's new Technical Institute, Jan Helcelet (1812–1876), and the politician Ignat Wurm. Gradually, Křížkovský came to know all the partakers in the nationalist movement and began to warm up to the ideals that they stood for. He returned to the original Czech spelling of his name, worked on improving his ability to speak the Czech language, read Czech books and magazines, and composed music to Czech texts. Little by little, mainly under Klácel's influence, a decision ripened in his mind to apply for admission to the St. Thomas Abbey.

And this he did, in the fall of 1845. Napp must have been apprised to the presence of a talented young musician in Brno, and so Křížkovský's application sailed smoothly through the admissions process. Because Napp desperately needed a good *Regenschori* for the Staré Brno parish church, he also accelerated, as much as the rules allowed, the procedure that would bring the novice to this position. He rushed Křížkovský through the novitiate (1845–1846), the simple vows (1846), and the theological studies (1846–1850) and had him take the Holy Order of priesthood even before he finished his studies (in 1848). That same year he appointed him the choirmaster of the parish church and the director of the Thurn Foundation. With Napp's blessing, Křížkovský plunged into Brno's musical life. In the Reduta Theater he staged Sophocles' Antigone to the music composed by Felix Mendelsohn-Bartholdy. In concert halls he introduced the cantatas, oratorios, orchestral works, and chamber music of Michael and Joseph Haydn (most memorably the latter's The Creation), Mozart, Beethoven, Cherubini, Spohr, as well as the works of several Czech composers. He played the viola in the string quartet in a series of concerts spread over Brno's musical season. He became one of the founders of the *Männersingerverein* (Men's Choral Society) and choirmaster of the *Beseda brněnská*.[40] His performances of sacral music at Staré Brno made the parish church the pilgrimage site of musical connoisseurs. And on top of all these, Křížkovský composed his own music. He started with church music, but when he discovered the Moravian folksongs, he began harmonizing them at first, adapting them next, and at last composing in his own style using their texts and under their influence. Some of his best works for male and mixed choruses are from this last category. They include the jewels of Czech choral music, the *Utonulá* (The Drowned Maiden), *Dar za lásku* (The Love's Gift), *Odvedeného prosba* (The Recruit's Prayer), *Výprask* (Threshing), and *Odpadlý od srdca* (The Faithless Heart).

Through his activities and the performances of his compositions, Křížkovský's renown began to grow throughout Brno and Moravia. Even Prague, which had the reputation of looking down on Moravian culture, received *Utonulá* and his other works warmly. His career as a composer of secular music seemed assured, and he himself experienced a period of a great creativity. He planned to write a monumental hymn celebrating the Czech nation, a cantata commemorating the five hundredth anniversary of Charles IV death, and an oratorium solemnizing St. Procopius. A theater in Prague even approached him with a proposal to compose an opera based on Czech text. All this, however, came to naught with one stroke of the bishop's feather. The bishop followed Křížkovský's creative ascent in secular music and his conversion to Czech nationalism with displeasure. It was outside of his power to stop the latter, but well within it to put a brake on the former. He explicitly forbade him to write or perform any secular music outside of the monastery. Křížkovský had no choice other than to obey. He retreated from the outside world to the seclusion of the abbey and its church. He continued to compose church music, but with few exceptions, it lacked the originality and luster of his profane compositions. Deeply wounded, he sought atonement in nurturing the talents of the "blue tits" and in reforming church music.

As for the former, one talent in particular gave him much gratification as he watched it to unfold—that of Leoš Janáček.[41] Křížkovský never forgot the kindness and help Jiří Janáček, Leoš's father, extended to him in Neplachovice. So, when the father brought the ten-year-old Leoš to him for an audition, he not only admitted the boy to become one of the "blue tits" but also took him under his wing when he realized how exceptionally talented he was. And so it came to pass that at one time the St. Thomas Abbey in Staré Brno harbored two of the greatest geniuses Silesia/

Moravia had produced, both still unrecognized: one who a few years later would supply the answer to the puzzle of heredity and the other who would develop a highly original musical language and give the world *Její pastorkyňa*, *Intimate Letters*, and *In the Mist*. The two must have met on numerous occasions, but their areas of interest were so different that any closer relationship between them could hardly be expected to have developed.

The other activity in which Křížkovský sought solace was his effort to bring back into church music the purity it, in his opinion, once had. He was not alone in his striving. A whole movement developed in central and western Europe of the second half of the nineteenth century. Its proponents argued that church music had become too bombastic, too theatrical, and too worldly to fulfill its original function of elevating a churchgoer's mind to God.[42] This argument is, of course, valid. By Bach's thundering organ fugues, the vaults threaten to come down. At Mozart's church sonatas for organ and orchestra, one feels the saints might begin swinging on their pedestals. And at Verdi's Requiem, one almost expects the deceased to rise and applaud a well-executed aria. The backers of this so-called *Cecilian movement* (St. Cecilia being the patroness of church music), Křížkovský among them, wanted to get rid of all these and return instead to the simplicity of the sixteenth-century compositions of Orlando di Lasso and Giovanni Palestrina. As long as Napp, who liked showy displays, had been alive, Křížkovský tempered his urge to reform. But after the abbot's death in 1867, he began revamping musical performances at the parish church in all earnestness. The loud flourishes of trumpets were the first to go. They were followed by an all-out attack on wind music in general: Wind instruments were banished from the orchestra and their teaching was removed from the foundation's curriculum. The orchestra had gradually been downsized, until it was abolished altogether and a vocal ensemble performed a cappella in the church exclusively. And Haydn, Mozart, Beethoven, and Cherubini had to make place for the sixteenth- and seventeenth-century composers.

Křížkovský's reforms attracted the attention of the archbishop in Olomouc, where church music had been in a sore state, including music at the archbishop's own church, the Minster of St. Václav. In 1872, the archbishop's office approached Křížkovský with an offer to become the Minster's *Regenschori* and the mandate to reform and revive sacral music. Embittered by the treatment he received in Brno, Křížkovský accepted the offer and moved to Olomouc, naming the 18-year-old Janáček his successor. For Janáček this appointment could only be a temporary solution because all he would get for his effort would be free lodging and two meals per day, with an occasional remuneration for his external activities. But it allowed him to complete his studies, and, together with his appointment as a choirmaster of the *Svatopluk* musical society, it constituted an admission ticket to Brno's musical life. Alas, one year later Křížkovský was back in Brno, when the promises made to him in Olomouc remained unfulfilled, he had been in no mood to negotiate the obstacles the clergy had been placing in his path. New negotiations followed with the result that in 1874 Křížkovský returned to Olomouc, where he would then remain for nine years, until two strokes forced him to retire and to return to Brno to die there in 1885. His legacy, however, would live in Moravia in Janáček and all the other musicians he educated.

Natural Sciences at the St. Thomas Abbey

There is no record suggesting that Mendel had a taste for philosophy, linguistics, or music. He might have listened politely, amazed and amused, to Klácel's discourses on the Hegelian Absolute (see Vol. 1 Chap. 1), but he probably asked himself afterward: What does all these have to do with the real world? Similarly, he might have been entertained by Bratránek's ruminations over enigmatic places in Goethe's *Faust*, but here, too, a question probably crept into his mind: How can anybody devote a career to guessing what a poet meant by his verses? He liked to listen to Křížkovský playing the piano in the refectory, to the performances of the "blue tits" at the noontime meals, and to the singing in the church, but this was as far as his interest in music would go. For all we know, Mendel's mind might have been conditioned by an all too rustic down-to-earth approach to life to find interest in philosophy detached from physical reality, too fond of logic to let itself be seduced by fuzzy romantic imagery, and too pragmatic to spend more than a leisured while on music. For his mind was rather bent on exploring the other pole of human cultural endeavors, the pole taken up by natural sciences. It found joy in puzzling over problems arising from observations of nature and having relevance to the quality of human life. As he did not find these attributes in the activities pursued by Klácel, Bratránek, and Křížkovský, he could not muster any great interest in what they were doing. This does not mean, however, that the cultural environment his three friends created in the abbey did not influence him. On the contrary, conversations with them helped him to realize how differently his mind operated and enabled him to define his own interests. Exposure to so much idealistic philosophy protected him from becoming a philosophizing scientist of the type his contemporaries were (see Vol. 1 Chap. 1). It shaped him into a truly modern scientist, who would later restrain his deductions strictly to the data at hand and resist any temptation to indulge in unwarranted speculations. Moreover, since both Bratránek and Klácel had a good background in natural sciences, especially botany, they helped him to embark on his chosen path by sharing their knowledge and experience with him.

Interest in natural sciences, especially botany, had been a tradition in the abbey. The specific form in which this interest manifested itself reflected a general trend in natural history of the time. The eighteenth and nineteenth centuries were periods of heightened interest in biological diversity. Behind this interest were two interrelated conceptual developments: the recognition of the species as the fundamental unit of biological diversity and the standardization of biological nomenclature anchored in the species. Nature, which in the past might have appeared to philosophers as a chaotic assembly of individuals, began to emerge as an intricate, but rationally organized system standing firmly on its foundation—the species concept. The imperative of the age became: Identify, name, and classify! From this period on, a person's curiosity about nature began with the question: What's the name of that plant (animal)? Few have not asked this question, while many have made it their hobby or occupation finding an answer to it. The two centuries became the age of amateur and professional botanists, mycologists, ornithologists, lepidopterologists, herpetologists, and ologists of any of the other natural groups that piqued their fancy. In the St. Thomas Abbey, the friars' fancy tended to be piqued primarily by flowering plants and, to a lesser extent, by minerals. Some of the friars became experts on the local flora—the plants growing in Brno, its surroundings, or southern Moravia. They practiced *floristics*, the study of plant distribution in a particular region. What made the region around Brno particularly interesting was the fact that two different phytogeographical zones met in it—the boreal and pannonian. The boreal zone, named after Boreus, the Roman god of the north wind, contained species adapted to the cooler north European climate. In the pannonian zone grew species preferring the warmer, drier climate of what used to be the Roman province of Pannonia, encompassing western Hungary, eastern Austria, southern Moravia, and the northern region of the former Yugoslavia. The friars used to undertake excursions to the different parts of southern Moravia, collecting and identifying the different plant species and mapping their distribution. In his early days in the abbey, Mendel participated eagerly in the excursions. Later, however, as his body began responding to the rich Moravian cuisine of Mrs. Vondráčková, long walks to distant places and hill climbing became too strenuous for the corpulent person that he had become.[43] The friars had certain places where they especially liked to botanize. One of them was a lake near the village of Čejč, some 45 kilometers southeast of Brno, not far from the town of Hodonín and thus not far from the Slovakian border. The locality was famous for its assortment of steppe and halophilic (salt-loving) plants. The abbey owned an estate at Šardice, which was only eight kilometers away from the lake where the friars could always count on finding a good assortment of specimens. The friars pressed and dried the collected specimens and, back at the abbey, mounted them individually for the *herbarium* kept in the library. The herbarium served as a document of the species distribution, as well as a teaching tool for those learning how to recognize the species. Other means of species identification were gleaned by direct learning from an experienced botanist, consulting the books in the library, and strolling through the modest botanical garden in the abbey.

The founder of the botanical garden was actually not a member of the abbey, but an Austrian businessman Rudolf Rohrer (1805–1839), a publisher and owner of a printing press.[44] While studying economy in Vienna, Rohrer also took a course in botany and became an amateur botanist. He then combined business trips with botanical excursions, and since he traveled extensively through Austria and especially Moravia, he became an acknowledged expert on Moravian flora. In 1835, he and his friend, the economist August Mayer, manager of an estate in Heraltice near Opava, published one of the first monographs on the subject.[45] Rohrer visited frequently the St. Thomas Abbey, where he befriended several of the friars, in particular Aurelius Thaler. On one of his trips through the *Salzkammergut*, an Austrian crown land east of Salzburg, he climbed the *Grossglockner* in the *Hohe*

Tauern part of the Austrian Alps, dug out a number of the montane plant species, and carted them to Brno. The collection became the foundation from which the friars started, in 1830, a botanical garden under the windows of the refectory. The garden's first caretaker was Thaler, who also founded the monasterial herbarium. Aurelius Antonius Thaler (1796–1843) was born in Jihlava, a town in western Moravia, near the Bohemian border.[46] He joined the St. Thomas Abbey in 1818 and in 1824 was a serious candidate to succeed the deceased Abbot Benedict Eder. In a magnanimous gesture, however, Thaler relinquished his candidacy in favor of Napp.[20] He became a professor at Brno's Philosophical Institute, where he taught mathematics from 1823 and natural history from 1825. In 1833 he was forced to resign from both positions, presumably because of his predilection for having one glass too many. He died ten years later at the age of 47, three month before Mendel's admission to the abbey. He was an esteemed botanist. His herbarium was an impressive accomplishment, which served as a guide to Moravian flora for generations of naturalists (Mendel among them).[47] Together with Rohrer, he informed the public on the proceedings of the Agricultural Society and about the blooming of the different species in the botanical garden, signing his reports as "Aurel." On Sundays he also gave brief lectures on these plants to visitors. He was also the author of the "flower clock," a collection of plants, which opened and closed their blossoms at different hours of the day. He apparently carried out hybridization experiments in the garden, but their nature is no longer ascertainable. Iltis[20] relates an anecdote about Thaler's non-botanical fondness of wine and his great sense of humor: *The word has reached Prelate Napp that Thaler loves to moisten his whistle and often comes home past midnight in a not entirely sober condition. Napp determined to humiliate him waits for him clad in the full regalia of his office in the porter's lodge and toward midnight opens the door for the admission-demanding priestling. Thaler, startled at first, quickly regains composure, bows deeply and with mockingly serious "O Lord, I am not worthy entering Thy house"turns around and goes – to wash down a few more drinks.*[20,46]

After Thaler's death, it was first Klácel's and then Mendel's turn to take care of the garden. Mendel later converted it into an experimental plot. Klácel, too, may have carried experiments of some sort in the garden, but his (and Bratránek's) interest in botany was philosophically motivated, so it is unlikely that they were in any way related to Mendel's. Klácel studied carefully Darwin's *Origin of Species*, when it appeared in German translation in 1863, and later lectured on Darwin in the United States.[29d] Even though he was familiar enough with Moravian flora to introduce Mendel to it,[47] floristics was not his forté. His thoughts revolved around philosophical implications of natural sciences, and his analysis of Darwin's theory was largely concerned with its application to human society. Bratránek, too, must have had a reasonably good knowledge of the local flora, but he, like Klácel, was interested only in those aspects of botany relevant to the area of his own inquiry, which was esthetics. His major work on this subject was entitled *Beiträge zu einer Aesthetik der Pflanzenwelt* (Contributions to an Esthetics of the World of Plants),[48] which Walther von Goethe called jokingly "the green esthetics."[20] The same friars who collected plants for the herbarium also assembled rocks for the mineral collection. The choice—plants and minerals and botany and geology—is interesting. On the one hand, many a budding naturalist begins by collecting plants and minerals, simply because they are so easily accessible. On the other hand, budding naturalists also collect butterflies, beetles, and—if they have the means and the opportunity to do so—also birds and small mammals. The friars, however, did not collect anything that would require killing the creature. (Obviously, they did not consider plucking a plant an act of killing.) Although the order did not expressly forbid them to engage in such activities, its spirit inhibited this manifestation of a hunter's instinct.

Rohrer, Mayer, and Thaler belonged to a circle of taxonomists grouped around the Agricultural Society. None of them were professional botanists; all practiced botany as their hobby.[49] Other members of the circle included a lawyer (W. Tkan, 1792–1863), an accountant (F. Jellinek), a railway inspector (J. N. Bayer, 1802–1870), a Protestant pastor (Ch. F. Hochstetter, 1787–1863), and a *Gymnasium* professor (J. F. Schur, 1799–1878). Most of them were based in Brno, but some operated in other parts of Moravia. They published their reports in the society's proceedings and documented their findings in herbariums, which each of them kept. The focus of the society, however, was not floristics, or basic research in general.[50] Although the society strived to keep its members informed about advances in basic sciences, this information was selected with an eye for a potential application to agricultural praxis. The rising new generation of Moravian naturalists found this situation unsatisfactory and began exploring the possibility of splitting off from the Agricultural Society. The split was effectuated in 1861 through the establishment of the *Naturforschender Verein in Brünn* (Association of Naturalists in Brno) with its own journal, the *Verhandlungen des Naturforschenden Vereines in Brünn* (The Proceedings of the Association of Naturalists in Brno),[50] which was supported in part by a grant from the provincial government. The grant enabled the association to print copies of each issue in excess of the actual subscription and use the extra copies for exchange with other scientific societies. This way, the journal achieved worldwide distribution, and the association received foreign journals in return, which they could not afford to subscribe to. The association also began to assemble a large library and, since botanists dominated it, a central herbarium from contributions of its members. The association organized regular meetings, at which its members had the opportunity to present the results of their research. It persisted until 1943, when the German occupiers of Brno abolished it.

The taxonomists made the association their forum. As before, they included hobbyists from different walks of life, although now teachers and professors prevailed. G. Niessl (1839–1919), A. Makowsky (1833–1908), and F. Czermak (died 1911) were teachers at Brno's German Technical School. A. Tomaschek (1826–1891) and F. X. Wessely (1819–1904) were gymnasial professors.[51] I. Czizek (1839–1909), F. Haslinger (1835–1902), L. Niessner (1848–1898), J. F. Slavíček (1856–1938), A. Schwöder (1841–1934), and F. Zavřel (1834–1905) were high school teachers. Others were physicians (J. Kalmus, 1834–1873; H. Wawra, 1831–1887; and F. S. Pluskal, 1811–1901), pharmacists (C. Theimer, 1823–1870), civil servants (C. Hanaček, 1831–1904; J. Nave,

1831–1864; R. Steiger, 1823–1908), or industrialists (C. Römer, 1815–1881). In contrast to the preceding generation, the taxonomists of this generation specialized in either a particular region of Moravia or in one taxonomical group of plants. The result of their combined effort was the *Flora von Mähren und österreichisch Schlesien*[52] compiled by Adolf Oborny (1840–1924), professor at the *Realschule* in Znojmo and in Lipník and Bečvou. It was published in the Proceedings of the Association in the years 1883 through 1886. Although the names of most of these taxonomists reveal a Slavic descent, all of them communicated and published in German. Parallel to them, however, arose, in the second half of the nineteenth century, outside of the Association, a generation of Moravian botanists, who published their floras in Czech.

Through Thaler's herbarium and botanical garden, Klácel's and Bratránek's coaching, and through excursions with Tvrdy,[53] Niessl, and other botanists of the circle that would later become the Association, Mendel acquired excellent knowledge of the Moravian flora. He loved plants with the inborn passion only another passionate botanist can appreciate. He knew the heartbeat-skipping feeling a true taxonomist experiences on sighting a rare species he has heard about but not yet encountered. Yet, in spite of all this, Mendel did not become a true taxonomist. He had not penned a single report on the flora of any region or taxonomic group, not even the difficult genus *Hieracium* (hawkweed), with which he became fairly familiar. He could not brag, the way other taxonomists did, about "discovering" this or that uncommon species of the Moravian flora. But above all, he had not found the complete fulfillment of his naturalist ambitions in the floristic activity, as the other taxonomists apparently had. His curiosity about nature's diversity did begin with the imperative "identify, name, and classify," but unlike that of his peers, it did not end with it. Mendel's encounter with biological diversity made him contemplate its basis and seek an explanation for it. Later (see Chap. 9), we will argue that this contemplation of diversity may have given Mendel the first strong impetus to begin the experiments on which he would base his interpretation of heredity.

Plant and Animal Breeding in Moravia and at the St. Thomas Abbey[54]

Another potent impetus that awoke Mendel's interest in heredity came from the emergence of agricultural sciences in the province, specifically in Brno and the abbey itself. Agricultural science began to rise in Moravia about 200 years ago and right from the start it diverged into subdisciplines dealing with various aspects of farming, from soil properties, through methods of crop and livestock production, to farm management. As a son of a farmer, Mendel followed these developments with great interest and from time to time sent to his relations in Hynčice news that he thought might interest them (e.g., about the spreading of the potato blight[55] or about the prices of wheat on Brno's weekly market[56]). But he paid special attention to plant breeding, a topic that fascinated him since his childhood back home, when he assisted his father or Pater Schreiber as they grafted fruit trees and took care of them.

Since time immemorial the aim of plant and animal breeders has been to improve the domesticated forms they grow or raise. Behind this effort lies the observation that the domesticated forms, like their wild progenitors, vary in discreet characters, including those, which are useful to humans. Often, the variation decreases the value of the individuals, but occasionally it increases it. The general strategy employed by the breeders has been to spot the rare individual and turn it into a founder of a line, in which, ideally, all the individuals would possess the improved form of the character. To achieve this second step of the strategy, the obvious thing to do is to mate (breed) the exceptional individual to its closest relative, and if the character appears among the offspring, continue breeding the closely related individuals that express the character until all the progeny express it consistently. This strategy is called *artificial selection* and the mating of relatives *inbreeding*. Unfortunately, more often than not, the improved character all but disappears in the offspring, and then the breeder must gather up all of his skills and all of his perseverance to recover it, if it can be recovered at all. In Mendel's time, the breeder's greatest frustration was that he had no notion of what outcome to expect from any particular mating. He had been fumbling blindly in the dark because he had no idea what caused the seemingly erratic behavior of the character he was following. Did the character appear in response to some environmental stimulus, or did it represent an intrinsic change in the constellation of the plant or animal? Was the change permanent and constant, but inexplicably masked under certain circumstance, or was it temporary and variable? How did it relate to the reproduction of the organism? Was there any regularity in the passage of the character from one generation to the next? And what was the mechanism of this passage? The breeders may have asked these questions, but they were in no position to tackle them. To realize why this was so, we must inquire into the breeders' identity: Who were they?

Since they were not a homogeneous group, it may help to divide them into three categories. The first category contains all those breeders who carried out the actual breeding of plants or animals themselves. We can assume that large-scale breeding with the purpose of improving plant varieties and animal stocks was limited to large estates whose owners had the resources and manpower to set up experimental gardens, fields, and other necessary facilities. The estate of Countess Walpurga in Kunín and that at the Staré Brno Abbey are fine examples of such cases, but there must have been others strewn over the entire province. A few of the owners may have bred the plants or animals themselves, but most merely supervised the work. At Kunín, the countess relegated the experiments to Pater Schreiber. At the St. Thomas Abbey, Napp established a nursery garden at the Šardice estate,[47] before he was elected abbot, and later founded another one on the premises of the monastery.[57] Both nurseries were intended for testing and distributing new fruit trees and grape varieties. In his early years, Napp may have gained practical experience in plant breeding at these two stations, but later he became so swamped by administrative work that he had to leave the actual work in the nurseries to others. He entrusted the Staré Brno nursery into the care of Augustin Anton Keller (1783–1852), a procurator at the abbey and enthusiastic plant breeder, who specialized in the development of new varieties of cantaloupe melons.[58] Since Keller was at the time of Mendel's admission into the abbey in charge of the novices, it is likely that at their regular meetings their conversations often strayed from religious indoctrination to plant breeding. We can be certain that Keller found in Mendel, who had been initiated into the latter subject by his father and Pater Schreiber, an attentive listener.

In addition to the large estate owners, many parish priests (e.g., Pater Schreiber, who had a fruit-tree nursery in his parish garden at Dolní Vražné) and some medium- to small-size farm owners also practiced the art of plant and animal breeding in their gardens and on their pastures, though on a much more modest scale. The scattered distribution of these numerous practicing breeders was hindrance to the communication between them, so the spreading of information and experience was largely limited to grapevine channels. Things improved considerably with the establishment of the Agricultural Society and its different branches. The establishment was the work of the second category of breeders, epitomized by the names Napp, André, Sedláček, and a few others. None of these gentlemen had formal training in plant or animal breeding, and most had little or no practical experience in these subjects. Yet, they were highly knowledgeable in both theoretical and practical aspects of agriculture because they had extensive contacts with practicing breeders and academic theoreticians. They traveled widely, visiting places where they could gather firsthand information on agricultural topics, and they maintained contacts with many other such places through correspondence. They spoke several languages and had access to agricultural literature from all of Europe. They were sharp thinkers, excellent synthesizers of the tidbits of information they gathered from these different sources, and skillful communicators, who, through lectures and articles in the proceedings of the different societies, were able to reach a wide spectrum of audiences and readerships. They were respected and influential men, even though they were, strictly speaking, amateurs. Napp, as we have learned earlier, was an orientalist by training. Christian Karl André (1763–1831), regarded by many as an expert on sheep breeding, had, in reality, a background in economy.[59] He worked as a teacher and educator at the renowned Saltzmann Institute (see Chap. 3) and later moved from place to place, until he settled down in Brno, where he became a manager of a Protestant school. André's friend, Jan Sedláček of Harkenfeld (1760–1827), was originally a musician who became interested in horticulture and viticulture only after he was made an estate manager.[60] The three gentlemen came from different provinces: Napp from Moravia, André from Saxony, and Sedláček from Bohemia. They met in Brno, where Napp and Sedláček also died, whereas André passed away in Stuttgart, after the police, who regarded his views and actions as too liberal, forced him to leave the Habsburg Empire. The three centered their activities on the Agricultural Society in which they were major players. Napp's involvements with the societies were described earlier in this chapter. André, in addition to having drawn the program for the Agricultural Society in 1815, was a cofounder of the Pomological and

Oenological Association and of the Association of Sheep Breeders, both of which had headquarters in Brno. He also promoted contacts with corresponding foreign institutions, especially the Royal Horticultural Society in London and a similar organization in Altenburg near Leipzig. Sedláček, as the president of Brno's Pomological Association, was instrumental in founding a nursery garden below the Petrov Hill. Finally, all three were strong proponents of new methods in plant breeding, including artificial pollination and hybridization, and of the need for close ties between applied and basic sciences. They, and a few others like them, created an intellectual atmosphere in Brno and in Moravia that was unique to central, if not the entire continental, Europe.

In Mendel's time, there was no university in Brno, and in Olomouc, there was only a fragment of a university. This situation was a handicap for general education in the province, but for the developments we are describing here, the opposite might have been true. In Vienna, the nearest university city to Brno, natural sciences had assumed a dominant position over the applied agricultural sciences, and this had a detrimental effect on the development of the latter. In Brno, where natural and agricultural sciences had been emerging side by side in the absence of a university, the conditions favored interplay between the two. This brings us to the third category of "breeders," the academic breeders. The breeders in the second category not only had little practical experience, they also lacked any academic qualification for the subject they were preaching. The simple reason for this was that in their youth there was no academic institution in the province that offered courses in agricultural science. During their lifetime, however, this situation began to change, and academic breeders, the professors of agricultural sciences, began to emerge. The beginnings of this change are tied to an institution, which from its name may seem to be as far from agricultural sciences as one could get—the *Ständische Akademie* or Estate Academy in Olomouc.[61] It was called "Estate" because it was originally meant to be for one estate only—nobility, although later its admission policy was broadened to include also the estate of the burghers. It was called "Academy" because that was the designation for an incomplete university-level school back then. The idea of having a special school for the education of young men from noble families was originally conceived by the Italian diplomat and author Count Baldassare Castiglione (1478–1529). In his four-book work *Il cortegiano* (The Courtier), a treatise on courtly manners and one of the most influential books of the Renaissance, Castiglione argued that the traditional education of a perfect courtier should be supplemented by teaching him useful subjects such as living languages, arts, natural sciences, and handling of weapons. The idea caught on and institutions with a curriculum based on Castiglione's proposal began to sprout up at different places throughout Europe, either independently of or in association with universities. In Moravia, in 1725, after several failed attempts, the noble families finally succeeded in setting up an Estate Academy at Olomouc.[61] It was modeled on a similar institution founded in 1685 in Vienna, where its curriculum originally included horse riding, fencing, dancing, and French language, but later shifted its emphasis toward technical subjects such as geodesy and, later still, agriculture. The Olomouc Academy was affiliated with the university, so that when, in 1778, the university moved to Brno, the Academy went with it; and when, four years later, the university (reduced to the status of a Lyceum) returned to Olomouc, so did the Academy. At the emperor's behest, the estates included lectures on agriculture (in 1808) and on natural sciences (in 1824) in the Academy's curriculum. The lectures on these two subjects took place at the Lyceum, and the lecturers were professors at both the Academy and the Lyceum. They were Josef Wobraska (1770–1820) from 1811 to 1820, Andreas Baumgartner (1793–1865) from 1820 to 1823, Johann Karl Nestler (1783–1843) from 1824 to 1841, and Jan Helcelet (1812–1876) from 1841 to 1846. At first, the lectures were poorly attended, but when Helcelet took over, the attendance skyrocketed and they became the most popular courses at the Academy/Lyceum. They were attended not only by regular students but also by pupils from secondary schools and by nonstudents such as estate managers, clerks, and scribes. In 1846, the authorities ordered the Academy to move to Brno again, and there it developed into the Technical Institute. In 1860, the University of Olomouc, or whatever was left of it, was abolished altogether.

With the transfer of the Estate Academy from Olomouc to Brno, the latter city became the Moravian center of education in agricultural and natural sciences. In Brno, the Academy became the second institution offering courses in natural and applied sciences. The first was the Philosophical Institute, and the motivation for its creation was to recruit candidates for priesthood from peasant and middle class families. The mission of the Philosophical Institutes, in general, was to facilitate the transition from the *Gymnasium* to the Theological Institute by offering two years of advanced courses in "philosophy," which included mathematics, physics, world history, Greek and Latin philology, and religion, in addition to theoretical and practical philosophy; agricultural and natural sciences were compulsory subjects for nonpaying students only. Any city in the monarchy that had a *Gymnasium* and a Theological Institute could apply for the creation of a Philosophical Institute. Since Brno fulfilled both conditions, it petitioned the central government in Vienna for permission to open such a school on the city's perimeter. God's mills and the Habsburg administration ground slowly in those days, and so it took several years before the permission finally came in 1807. By that time the city fathers realized that they had neither the money to pay the teachers nor the building in which to place the school. They therefore entrusted the solution of these two snags to the office of the local bishop, under whose jurisdiction the government placed the school. How the office solved the first snag, we already know: The religious orders, not sacked by Joseph II, were ordered to provide the teachers from their midst. And as for the second snag, the Minorites were directed to free rooms for the classes in their monastery, which was conveniently located in the center of the city.

The Minorite building complex is on the Minoritská Street between the Janská and Orlí streets (see Figs. 5.2b, area 4, and 5.2c, panel 4). The complex includes two churches and the monastery buildings: One of the churches, the Loretto, is at the corner of Janská and Minoritská streets; the other, the Church of St. John the Baptist and St. John the Evangelist, is adjacent to it. The designation "Loretto" derives from the town Loreto in east central Italy, to which, according to a Catholic legend, angels transferred the house of Virgin Mary's birth in Nazareth of the Holy Land,

Fig. 5.15 The Philosophical Institute in Brno in the Minorite Monastery at the corner of Janská and Minoritská streets. The building in the background is the Church of the St. John the Baptist

when it was in danger of destruction by the Turks. Imitations of the house were then built in different parts of Europe, one of them being Brno. The Church of the Saint Johns (*Janů* in Czech) gave the name to the Janská Street. Adjacent to this church, at the corner of Minoritská and Orlí Streets, is the Minorite Monastery, in which the Philosophical Institute was housed from 1808 to 1850 (Fig. 5.15), when it merged with the *Gymnasium*. The original German *Gymnasium* of Brno was founded in 1578 by the Jesuits and located in the Jesuit College near the Church of the Assumption of the Virgin Mary at the Jesuitská Street (see Figs. 5.2b, area 5b, and 5.2c, panel 5). When in 1773 the Emperor Joseph II suppressed the Jesuit order, the state took over the *Gymnasium*. Before the merger with the Philosophical Institute in 1850, the *Gymnasium* had six grades; by the merger it became an eight-grade school. In the Philosophical Institute Augustinian friars from Staré Brno taught philosophy and mathematics. F. Süsser taught philosophy from 1820 to 1835, M. Klácel from 1835 to 1844, and T. Bratránek from 1844 to 1849. A. Thaler taught mathematics from 1823 to 1833, A. Alt from 1833 to 1838, and F. Gabriel from 1838 to 1848. Agricultural and natural sciences were taught by J. A. Zeman (1780–1825) until 1825, by A. Thaler until 1832, and in Mendel's time by F. Diebl.

Of the professors teaching agricultural and natural sciences in Olomouc and Brno, four were influential in developing plant and animal breeding in Moravia: Baumgartner, Helcelet, Nestler, and Diebl. All four had approbation for lecturing at university-level institutions, though not all for the courses they were teaching. Andreas von Baumgartner (1793–1865), a native of Friedberg, Bohemia, had a degree in physics and applied mathematics from the University of Vienna, but had actually very little experience in agriculture. After his short stint at Olomouc, his career took a different turn and led him from the directorships of various industrial enterprises and presidentships of telegraph and railroad organizations to the ministry of commerce in Vienna. We include him among the four influential men only because of the part he was to play later in Mendel's life (see Chap. 6). Jan Helcelet (1812–1876) was born in Dolní Kounice south of Brno and attended universities in Padua and in Vienna, where he received a medical degree. He then practiced medicine at a hospital in Brno before accepting the teaching positions in Olomouc and Brno. What he lacked in practical experience in agriculture, he compensated for with his infectious enthusiasm, which stimulated interest in the subject in his students. In Olomouc, he taught the Czech language and literature with equal zeal, but after the events of 1848, he devoted all of his energy to politics and journalism and eventually became an important figure in the Czech nationalistic movement. Johann Karl Nestler (1783–1841) was born in Vrbno, Silesia, and studied agricultural sciences at the University of Vienna. As a lecturer in Olomouc, he could not fire up the enthusiasm of the audience the way Helcelet had. Instead, he influenced the development in his field through other means—his writings. In 1829, he published his lectures in the proceedings of the Agricultural Society[62] and so stimulated a protracted debate about the methods of animal and plant breeding, as well as about theories of reproduction and heredity. Later, at a meeting of the Society in 1836, Nestler was asked to take a stand on the issues brought up in the debate and to tell the audience what, in his view, was the most pressing problem in developing rational methods of breeding.[63] His answer was: To understand the nature of heredity. Napp, who was present at the meeting, seconded this opinion, adding that "what is and how it is inherited" was the question of the day. Mendel, who was then only 14 years old, obviously could not have been present at the meeting, but he may have read about these pronouncements later in the society's proceedings, which published the meeting's minutes. Moreover, Napp undoubtedly expressed similar views on other occasions in Mendel's presence.

The last of the four in the category of professors interested in plant and animal breeding was František (Franz) Diebl (1770–1859). He was probably the best qualified and the most influential of the four. He was born in Vřesice north of Brno and educated at the University of Vienna, where he studied natural and agricultural sciences. Behind his qualification, however, was not so much academic learning as extensive practical experience on a large estate he rented near his native village. The estate encompassed five hamlets: Křetín, Vranová, Veselka, Bohuňov, and Študlov. On it he experimented with new methods of cultivation and breeding for a number of years. Diebl had been the obvious candidate to teach agriculture and natural sciences at Brno's Philosophical Institute after Zeman died in 1825. The bishop decided, however, that he did not want these two subjects taught at the institute anymore and petitioned the authorities in Vienna for permission to abolish them.[13] But when Napp learned of this move, he organized a political counteroffensive, in which he pointed out to the authorities the importance of agriculture for the society, the importance of acquainting future priests with this subject, as well as the importance of natural sciences for the development of agriculture. The strife dragged on for seven years, but in the end Napp's party prevailed and the government allowed the two courses to continue. It also accepted Napp's proposal to appoint Diebl (who in the meantime became a curator of Brno's museum) to the professorship. Diebl authored a five-volume textbook that contained a detailed description of plant breeding methods, including methods such as artificial pollination and hybridization, which Mendel later used in his experiments.[64] Diebl also coauthored with Nestler a textbook on natural history, which Mendel must have studied carefully since he took Diebl's courses and passed, with distinction, examinations that were based on it. One can assume, therefore, that Mendel owed to Diebl much of his theoretical and practical plant breeding knowledge. But we are running ahead of our narrative, so let us now return to Mendel, where we left him last: at the end of his novitiate.

An Overgrown Path

In the fall of 1844, Mendel had completed his one-year novitiate and had become a student once again, this time of theology, which was a four-year course of specialized religious training in an institution popularly referred to as *seminary* or *alumnate*, especially if it provided students not only with education but also with board. The school's official name was *Caesario Regio ac Episcopali instituto theologico Brunensis* (Imperial-Royal and Episcopal Theological Institute of Brno), which we will abbreviate here to Theological Institute. It trained candidates for priesthood and so was under the jurisdiction of the bishop. Brno's bishopric was then still young: Pope Pius VI founded it in 1777, at the time when he upgraded the older Olomouc bishopric to archbishopric. The new bishopric was to serve the region (diocese) of southern Moravia. Its seat became a complex of buildings near the cathedral of St. Peter and St. Paul on the Petrov Hill (see Figs. 5.2b, area 6b, and 2c, panel 6).[65] For a while, priests of the Brno diocese continued to be trained in Olomouc, at the Theological Institute affiliated with the University of Olomouc. In 1805, however, Emperor Franz I granted the wish of Vinzenz Joseph Franz Sales Graf von Schrattenbach, the diocese's third bishop (in office from 1800 until 1816), to establish a Theological Institute in Brno. The institute opened in 1807 in the former monastery of the Dominican order, abolished earlier by Joseph II. The monastery now faces the Dominikánská Street on the one side and the square of the same name on the other, not too far from the bishop's residence (Fig. 5.16).

Mendel did not board at the Theological Institute since it was only some 30 min walking distance from the abbey. And so, for four years, every weekday, he trotted along the same path, in the morning in one direction and in the afternoon back in the other. Exiting from the abbey, he crossed a square then called Zámecké náměstí (which today bears his name) and entered a passageway through a building to emerge on the other side of Křížová Street, which after a short distance became Pekařská Street (see Figs. 5.2b, area 6a, and 2c, panel 6). The latter took him to the Brněnská Gate through which he entered the inner city behind the bulwarks. The road then split into triplex, the Dolní (Zadní) Brněnská, Horní (Přední) Brněnská, and Petrovská streets (today's Dominikánská, Starobrněnská, and Biskupská streets, respectively). He took the leftmost street of the three (the present-day Dominikánská), and through a portal (now walled up) on the left side of the street reached the Theological Institute (Fig. 5.16). A reverent pilgrim to Staré Brno may

Fig. 5.16 Brno's Theological Institute in the former Dominican monastery (the low building with the grated windows), where Mendel trained for priesthood from 1844 to 1848. The two towers are those of the St. Michael Church

relive the walk, albeit not without recourse to a great deal of imagination and historical reconstruction, for back then, in 1844, the area looked quite different than it does today. The Zámecké náměstí (Castle Square), now a major traffic knot, was then a place to which families from the inner city went for a Sunday afternoon stroll or picnic. It had a rustic character, with a water channel crossed by small bridges, a water mill, trees, an old-fashioned town hall, a tavern "At the Blue Lion," and of course the church and the abbey with its small botanical garden, orchards, and vineyards (see Fig. 5.3). The water channel may have originally been one of several temporary or permanent arms into which the Svratka River branched out between the Žlutý and Červený hills, before it joined the Svitava a short-distance downstream. Its existence is documented since the thirteenth century, when it had been deepened to serve as a race for the mills built on it[3,4]. In Mendel's time, the race made a loop along the square (Fig. 5.1), but later its course had been straightened out, and ultimately, in the twentieth century, it was filled and obliterated. At the

Fig. 5.17 Young Mendel—how he might have looked like at the time when he became a priest

bend of the channel, on its left bank, the Knights of the Cross owned a property, a "prebend," which gave the Křížová or Křížovnická Street its name ("cross" being *kříž* in Czech). The name of the Pekařská Street seems to be derived from a mistranslation of the original Czech name Bekyňská ulice, which referred to the presence of the Cistercian nuns, the original inhabitants of the Staré Brno convent (*bekyně* was, back then, a Czech term for a nun). From a corruption of the Czech name arose the German name Bäckerstrasse and from it, by translation, the Czech Pekařská ulice (*Bäcker* = *pekař* = baker[4]). As he walked along the Pekařská Street toward the inner city, Mendel passed a group of buildings on his right-hand side that comprised the St. Anne's Hospital complex, which a few years later would play a decisive part in steering his career. Behind the hospital was a large parklike estate, the Královská zahrada (The King's Garden). The street was steeper then than it is today; over the years, though, numerous successive reconstructions at its lower level gradually raised it, which ameliorated its steepness. Nevertheless, in his younger years (Fig. 5.17). Mendel probably negotiated the slope with only a quickening of his breath. Later, however, when he became corpulent, we can assume that he preferred to take the horse-drawn streetcar introduced in 1869.

The streetcars still run up and down Pekařská Street, only nowadays they are electric. The Brněnská gate is, of course, long gone, as are the rest of the bulwarks, having been replaced by a ring of broad avenues encircling the city's center. Where the gate once stood, there is now a square—Šilingrovo náměstí (Šilinger being one of the city's politicians). The Theological Institute can now be reached only from the Dominikánské náměstí, which in the past had, for a long time, been called

Rybný trh (the Fish Market), because when Brnoians wanted to have fish for dinner, it was there that they went to get it. The square, too, looks different now than it did in Mendel's time. The fish and meat stands are now gone, as are, regretfully, some of the historical buildings that surrounded the open space. Across from the St. Michael Church, at the corner of the Veselá Street, stood the Margrave's House with a tower and a Chapel of the Virgin Mary. In the house, Jindřich of Lípé set up a residence that rivaled a royal court. Next to it was another house that John of Luxembourg gave to Eliška Rejčka which he used to visit her after she moved to Brno. In 1908, a group of barbarian bureaucrats had the entire corner demolished. Part of the Dominican monastery situated along the bulwarks in the upper part of the square, which in Mendel's time served as a meeting place of the Moravian Estates, is now the New Town Hall.

The Road to Priesthood

Mendel may already have been enrolled in the Theological Institute in 1843, but he did not begin his studies until September 1844. His status at the abbey had thus changed from novitiate to *juniorate*; he became a candidate for priesthood. The simple vows he took at the end of the novitiate bound him to the monastery, but the bond was not quite unbreakable yet and transgressions against the rule were pardonable. After all, was it not St. Augustine, who pleaded with God: *Give me chastity and continence, but do not give it yet?* The lecturers at the Theological Institute, however, did their best to keep the thoughts of the young men in their care away from worldly matters. The four-year curriculum was heavy from the start, and the lecturers did not waste time on reviewing knowledge that the students were expected to have acquired at the *Gymnasium* and the Philosophical Institute. They plunged directly into the specific topics of theology. In the first year the students heard lectures on church history, Biblical archeology, Hebrew language, as well as an introduction to and exegesis of the Old Testament (*exegesis* being critical interpretation of the Biblical text). In the second year the load was hardly any lighter: Canon (i.e., ecclesiastical) law, Biblical hermeneutics (i.e., methodological principles of interpreting the Bible), Greek language, introduction to and exegesis of the New Testament, and pedagogy. In this year Mendel also took the non-compulsory two-semester course on agricultural sciences that Diebl taught at the Philosophical Institute. In the remaining two years, the load was somewhat lighter. The third year was devoted to dogmatic theology (as if there were any other!) and moral theology. And the fourth year wrapped up the studies with three practical subjects: pastoral theology, methodology of elementary school education, and *catechetics* (instruction on how to teach moral and religious principles; from Greek *katechein*, to teach). In the final report card, under the heading *Ex studiis ordinariis* (i.e., from compulsory subjects), a handwritten note states that Mendel took the prescribed two semesters in pedagogy already in 1843 at the Philosophical Institute in Olomouc.[66] Similarly, under *Ex studiis extraordinariis* (i.e., from non-compulsory subjects), a note credits Mendel's grades in Chaldaic, Syrian, and

Arabic languages from Olomouc and in agricultural sciences from Brno's Philosophical Institute. The report card, which the Theological Institute issued on June 30, 1848, indicates that Mendel passed all courses with the best marks possible (*prim. eminens*, highest distinction) and followed all instructions with greatest diligence (*diligentissime*) and that his overall conduct was excellent (*ad prime conformes*).[66] Once again, and for the last time, Mendel passed all examinations with honors.

Ordinarily, a candidate could become a priest only after the completion of theological studies. But the circumstances at the abbey during the time of Mendel's juniorate were anything but ordinary. Within a short time span, the institution had lost several of its members. Thaler died in 1843; Rossner left the abbey before completing his novitiate; then, Süsser died in 1847; Keller and Lang were incapacitated; and several other members were on temporary or permanent assignments elsewhere in the province. Consequently, when some of the remaining friars reported sick, there were not enough priests left to meet the demand of the pastoral duties. To steer the abbey out of this bottleneck, Napp wanted at least one of the juniorate students to assume pastoral duties before the completion of his studies. For that to happen, though, the student would have to become a priest first. Mendel, with his impressive performance at the Theological Institute, seemed like an obvious choice, for he presented little risk of having the burden of pastoral work endanger the completion of his studies. However, to shorten Mendel's road to priesthood, Napp needed the consent of the *Landespräsidium* (the provincial government) and of the bishop. Furthermore, before he could even be presented to these authorities, Mendel would have to take the solemn vows as a token of his permanent bond to the abbey. Working toward this goal, on December 17, 1846, Napp assembled the abbey's chapter and asked the body to decide whether it considered Mendel worthy of becoming a permanent member of the convent. As he expected, the chapter voted unanimously that indeed it did and that he should therefore be asked to take the solemn vows. Napp then instructed Mendel to prepare himself for the ceremony, which he set to take place the day after Christmas, on December 26. Mendel's preparation consisted of a series of spiritual exercises, the so-called *devotions* (mostly ceremonial prayers) performed on three consecutive days (December 21, 22, and 23), followed by confession (December 24) and Holy Communion on Christmas Day. Then, on Saturday, December 26, 1846, the ceremony of taking the solemn vows took place in the parish Church of Virgin Mary in the presence of the assembled members of the abbey.[67] The high point of the ceremony was Mendel's recital of the prescribed Latin text of the vow. In an English translation the vow reads thus:[68]

> *In the name of Our Blessed Lord Jesus Christ, Amen.*
>
> *In the year thousand eight-hundred and fourty-sixth of His nativity, December twenty-sixth. I, frater Gregor Mendel, son of Anton Mendel, Silesian from Hynčice, profess and promise obedience to Almighty God, to Saint Mary ever Virgin, to our Holy Father Augustine, and to You Very Reverend Father in Christ Cyril Napp, Prior forever of Saint Thomas Monastery in Brno, Abbot and Prelate of Moravia, and at the same time to Your*

successors who have been elected according to the rules of canon law; I promise to live in poverty and chastity until death according to the Rule of Our Holy Father Augustine.

Mendel then signed a document with the same text and added the date. Essentially, the solemn vow was similar to the simple vow Mendel took at the conclusion of his novitiate. The difference between the two vows was not so much in their content as in their binding power. While, after the simple vow, Mendel could still quit the abbey without serious legal consequences, quitting it after the solemn vow would not be all that simple. After the solemn vows, St. Augustine's *but do not give it yet* no longer applied.

Besides his incomplete studies, there was another snag to Mendel's early priesthood—his age. The rules stipulated that a priest must be at least 25 years old, an age which Mendel would only reach on July 22, 1847. Napp knew that he could not get this requirement waved, but in the summer of 1847, he took steps to shortcut the first obligation. On July 15, 1847, he wrote a letter to the bishop, in which he described the situation in the abbey, explained the reasons behind it, and asked to advance Mendel's ordainment one year ahead of schedule.[20] He pointed to Mendel's outstanding performance in his theological studies and vouched for his character and devotion. Five days later, not having heard from the bishop, Napp sent a similar letter to the *Landespräsidium* in Brno. It is a testimony to Napp's good standing with the civil authorities that they approved his request that very same day. Napp, in turn, wasted no time and had the permission delivered immediately to the bishop's office. The next day, July 21, the bishop, too, approved the request and set the dates for the ordainment of Mendel along with the other candidates who had completed their theological studies that year.

In the rigidly hierarchical organization of the Roman Catholic Church, the clerics pass through (or become arrested at) different tiers of the ranking order, with each tier carrying different privileges and responsibilities. The lowest tier in the hierarchy is the subdeacon, followed by the deacon, priest, bishop, archbishop, cardinal, and the pope.[20] Hence, to become a priest, the candidate must first become a subdeacon and then a deacon, both being assistants to a priest, especially a parish priest. To each of these tiers, the bishop must ordain the candidate ceremonially. A candidate may spend several years being a subdeacon and then a deacon before he is admitted to priesthood. In urgent cases, however, the passage through these two tiers can be accelerated. In Mendel's case the acceleration was taken to an extreme: The bishop ordained him subdeacon on July 22, deacon on August 4, and priest on Sunday, August 6, 1847. Mendel then celebrated his first mass (the "Primiz") on Tuesday, August 15. All of these events took place in the Dominican Church of St. Michael (Fig. 5.17), near the Theological Institute, and each was a part of a ceremony. The most solemn of the three ordainments was the last one, the ordainment of a priest, which was a part of the Holy Mass. The essential part of the ceremony, the laying of the hands on the candidate's head, derives from an ancient Jewish rite of installing rabbis. In the ordainment of a priest, the bishop, by the act of laying hands, invests on the candidate the right to administer sacraments, preach, and exercise pastoral work.

The formal distinctions of the newly ordained priest are a new title, tonsure, and clerical clothing.[20] As a priest, Mendel acquired the prerogative of being addressed as Pater (Father), abbreviated in written form as "P." before his name. The first *tonsure* (from Latin *tonsus*, past participle of *tondure*, to shear) is performed by the ordaining bishop, who at the onset of the ceremony cuts off a few hairs on each side and at the top of the candidate's head as a sign that from now on he is to eschew worldly pleasures. Later, the ordained priest wears a tonsure in the form of a shaved-off, coin-sized spot on the top of his head. A priest also acquires the prerogative to wear a distinctive type of clothing. Roman Catholic priests wear two types of garb, depending on the occasion: the non-ceremonial clerical clothing (clericals) and the ceremonial vestments. The two principal types of clerical clothing are the cassock and the clergy shirt, both worn with a distinctive clerical collar. The *cassock* (soutane) is an ankle-long, hoodless, long-sleeved garment made of a black fabric. It resembles the tunics worn underneath the toga by the citizens of ancient Rome, except that it closes up in front by a row of 33 buttons (a feature that also distinguishes it from the otherwise similar monasterial habit) and is tightened around the waist with a cincture. The *clergy shirt* resembles a secular long-sleeved black shirt, except that it is either collarless or has a band-like collar sewn on to it. Normally, a priest combines it with a conservatively cut dress. The *clerical collar* is a band made of white, tough, but flexible material; it closes around the neck of the wearer. It can either be fastened to the collarless shirt or inserted into the band of the neckband shirt. In the latter case only a small squared white area of the collar remains visible in front, an area not covered by the neckband. It is the collar that identifies a person as a priest in nonliturgical emergency situations, in which his assistance might be desired.

When preparing for liturgical services, a priest puts on *vestes sacrae* (holy garments) or vestments, the specific articles of which vary according to the type of service he is about to perform. For a mass, a priest slips on a *tunica alba* (white tunic) or *alb*, a simple, ankle-long, long-sleeved garment made of light-weighted white fabric, and gathers it around his waist with a sarn of cloth. Next, he puts around his neck a *stole*, a long, narrow silk band, choosing its color according to the season and letting its ends hang in front of his legs. And finally he puts on a *chasuble*, an ornate, sleeveless, circular garment, passing his head through the hole in its center and letting it rest on his shoulders and arms. Its color must match that of the stole. Thus, attired, he then approaches the altar, accompanied by the altar boys.

Mendel became a priest while he was still a student, and as a priest, he thought that he had acquired the privilege to put away the habit with its annoying cowl and wear clerical clothing instead. As we already know, Napp quickly reminded him that in this case his student status took priority and, therefore, he had to continue wearing the habit. One year later, however, on June 30, 1848, Mendel graduated from the Theological Institute and was finally free to leave the habit in his room and put on a clerical dress whenever he had tidings outside of the abbey.

The Troubled Shepherd of Souls

Catholic writers claiming that Mendel became a priest because he "felt he had a vocation to the priesthood"[69] are hard pressed to explain what happened next. After his ordainment, while still a student, he assumed some pastoral duties, but these were mostly of the ceremonial type, largely restricted to the parish church, and the workload was light because he still had to attend lectures and prepare for examinations. All that changed, however, in the summer of 1848 when Napp assigned him full time to pastoral work. As a *Kooperator* (= *Kaplan*, vicar), he became the right hand of the parish priest, a factotum, and a do-everything person, both in the church and in the parish. In the church he celebrated masses, delivered sermons in German and in Czech, took confessions, baptized newborns, wed couples, and buried the dead. In the parish, he visited the sick, infirm, and dying in their homes or in the St. Anne's Hospital on the Pekařská street. Founded in 1320 at a site near St. Anne's Chapel, the hospital owed its existence to the contest between the two queens—the two Alžbětas, Přemyslovna and Rejčka. Fulfilling his wife's wish, John of Luxembourg donated part of the Royal Garden to the order of Dominican nuns, to be used for the construction of a convent and a hospital right across the site where Eliška Rejčka planned the convent of the Cistercian nuns. In Mendel's time, the hospital complex included a maternity ward, a lunatic asylum, a poorhouse, and a foundling house, in addition to its various clinical wards. The hospital itself had 178 beds and admitted, on average, 1,700 patients yearly, of whom some 150 patients died there of all sorts of sicknesses.[70] The maternity ward handled 737 pregnancies per year; the asylum admitted annually 91 mentally ill persons; and the foundling house had on average 443 children of whom 63 died yearly. As the complex served an area inhabited mostly by poor people, all sorts of human misery could be encountered there. Mendel visited these wretched people to give them spiritual consolation, take their confessions, provide Holy Communion, and administer the last sacraments to the dying.

How did he respond to this entirely new situation into which life placed him? If it were true that it had been "his dream to become a priest,"[69] he could have been expected to welcome the opportunity "to serve God in a more perfect way"[69] and see it as a challenge matching his ambitions. In reality, however, his reaction was exactly opposite: He fell to pieces. For a while he did his best to continue performing his duties while struggling internally to overcome the mounting anxiety and distress, but ultimately, by the end of 1848, he collapsed both mentally and physically and took to bed gravely ill. We, of course, have no way of knowing what went on in Mendel's mind in the months leading to his breakdown. If he ever confided his anxieties to anyone, no word of them has reached posterity. So, all we can do is to speculate about the causes of the collapse.

One relatively benign cause could have been the shuttering of his aspirations to become a researcher. Although he might not have had at that time a clear idea of what kind of research he would want to get involved in, his interest in natural sciences was distinctly delineated by then. He might have dreamed of becoming a teacher, ideally a professor at a secondary-level school, where he would have time to devote himself to natural history. If so, the assignment to full-time pastoral work must have been a great disappointment to him as he saw his dream fading away. After a few weeks of parish duty, he must have realized that should this become his lot, he might as well say good-by to any serious scientific aspirations, and he might have found it difficult to reconcile himself with this thought.

Another factor could have been an inner tension building up in him, a tension stemming from the feeling that he was unsuited for the responsibilities with which he had been entrusted. No matter how hard and how conscientiously he tried to carry out his new duties, he had to admit to himself that he was not accomplishing what was expected of him.[70] His heart was not in what he was doing, and he could sense that the people whom he tried to console realized it. He just could not find the right words, proper tone of voice, and suitable gestures to gain their trust. And the feeling that he was failing these people at the time of their greatest need pressed heavily on his conscience.

The third and perhaps the most significant factor that might have driven Mendel's mental system into a crisis was his direct encounter with human suffering.[71] Witnessing so much distress, destitution, agony, and wretchedness as he had in the hospital and its affiliated institutions was more than he could bear. A part of this anguish might have been an instinctive fear for his life. Later, when he became an abbot and got himself embroiled in a controversy with the government over taxes, he pointed out, in one of his arguments, how dangerous the pastoral work in the hospital was.[72] He cited two cases, when his youngest friars actually died of a disease contracted there.

Whatever the reasons might have been, less than six months after assuming pastoral duties in the hospital, Mendel lost control of his emotional distress and collapsed. Records of the abbey's expenditures show that he remained bedridden for at least 34 days (the entire month of January 1849) and then convalesced for several additional months (at least through April).[70] During this entire period he remained confined to his room in the abbey; contrary to some claims, he was not hospitalized. The records indicate, however, that Napp judged the illness serious enough to hire a nurse, who during January kept a 24-hour watch over the patient.

The Patient Mendel

Mendel's illness in 1849 was the third of altogether five episodes in his young years. The first occurred in the late spring of 1838 in Opava, when an unspecified illness forced him to interrupt his studies at the *Gymnasium* and return to Hynčice. The second took place in 1841 in Olomouc, where it disrupted his studies at the Philosophical Institute. After the 1849 episode, two more episodes would follow in 1856: one in January and the other in May. The five episodes have certain characteristics in common: First, all seem to have had been precipitated by stressful situations, the first two by existential worries; the third by Mendel's inability to come to grips with the emotional side of the pastoral work; and the last two by the stress of examinations. Second, although we do not have a medical diagnosis of any of the episodes, from the fragmentary information that has reached us, it appears that they can all be classified as psychosomatic or somatoform disorders. In all five Mendel apparently experienced physical symptoms that did not seem to have any identifiable causes. What the symptoms were, we unfortunately do not know since no medical records of the episodes, if they ever existed, have been preserved. In only one case, the 1849 episode, we know for certain that physicians examined Mendel. The Abbey's expense record shows that in April 1849 the abbey paid 200 florins to the *Hausphysikus* (an equivalent of a family doctor) Dr. Dudei and another 100 florins to the *Hauschyrurg* Mandler for examinations and treatment of Mendel.[70] The former was presumably Dr. Joseph Dudecy, the resident physician in the convent and hospital of the Brothers of Mercy in Staré Brno, under the Červený Hill on the other side of the Svratka River. (The Augustinians in Staré Brno did not have a resident physician.) Nothing specific could be found about the *Hauschyrurg*. What conclusion the two medical practitioners reached has not been recorded, but even if it were, it would not be of much help since it is not a great exaggeration to say that medicine in the middle of the nineteenth century was only slightly better than in the time of Hippocrates. The era of great discoveries in microbiology, immunology, physiology, pathology, neurology, and all the other disciplines that would become the scientific foundation of modern medicine would begin only in the second half of the nineteenth century. In all fairness, however, if Mendel's disorder was really psychosomatic, even today's physicians would be at a loss what to do about it. As stated earlier, it presumably had physical symptoms; otherwise Napp would not have been so concerned about Mendel's state. That he was concerned is evident from the fact that he spared no expense to get him the best care available and that during the 1856 episode he informed Mendel's father, who then came together with his brother (Mendel's uncle) to visit the "dying patient."[71] The common physical signs of a psychosomatic illness are shortness of breath, chest pain, rapid heartbeat, cold and hot spells, and muscle weakness bordering on paralysis. Information that Mendel showed at least the last of these symptoms comes from Klácel's letter to Bratránek dated May 8, 1856.[73] Referring to Mendel's examination in Vienna, Klácel writes: "*...although he had the good luck of drawing easy questions, he fell so ill after the first one that he could not write...Being afraid that more such attacks would follow, he returned home.*" Apparently Mendel was temporarily afflicted by what physicians call *agraphia*, the pathological loss of the ability to write caused by muscle paralysis resulting from an impairment of the controlling nerves. To this news Klácel adds information that makes one wonder about its source: "*He seems to suffer from bad nerves in general having had several such attacks of the treacherous illness, and in his youth, they say, having been predisposed to epilepsy.*" Where indeed did this information come from? Since no other source corroborates it, it probably was nothing more than a rumor that arose in the abbey itself. Having witnessed Mendel's bouts of illness, the friars, not above the human frailty of gossiping, may have just speculated about the nature and origin of the fickle ailment. If the information had a solid basis, it would have been most unlikely that the friars would have later elected Mendel their abbot.[69] It is also improbable that Schindler would not have known about it or would not have mentioned it to Iltis if he had.

Psychosomatic illnesses are today almost as enigmatic as they were in Mendel's time. The fact that other friars looking after the patients at the St. Anne's Hospital were exposed to the same stresses, yet they did not suffer a nervous breakdown, may be indicative of a predisposition of some sort on Mendel's part. A medically trained biographer describes Mendel as a "vegetatively stigmatized" and "constitutively conditioned neurasthenic,"[70] which is one of those diagnoses physicians come up with when they have no idea what's wrong with a patient. "Vegetatively stigmatized" implies an involvement of the vegetative nervous system controlling involuntary body functions, while "constitutively conditioned" intimates a condition inherent in the constitution of body and mind (in other words, genetic). And "neurasthenia" is an umbrella word so broad that half of all mental disorders could fit under it. It refers to conditions characterized by fatigue, anxiety, headaches, impotence, depression, and neuralgia (sudden fits of sharp pain along the course of a particular nerve). Whether Mendel actually had any of these symptoms, we don't know.

In short, we have no clue about the nature of Mendel's illness, except that it was somehow tied to the condition of his nervous system. Before leaving this topic, however, we must ask the seemingly unthinkable question: Was Mendel really sick, or did he feign his sickness as a means of resolving situations that he was unable to cope with? Many people the world over feign sickness for a variety of reasons, most commonly because they want to stretch their weekend into a Monday, and doctors the world over fall for this ruse because they are either unable or unwilling to distinguish simulated from real illness. Students sometimes resort to this deception when they need extra days to prepare themselves for an examination. (On the other hand, however, the stress of an examination can be so great as to cause a nervous breakdown. Francis Galton, on whom more later, suffered one, as did several other students preparing themselves for particularly difficult examinations at Cambridge University.)[74] Could it be, then, that Mendel's sickness was in reality an act with the objective to gain more time before an examination (the 1838 and 1841 episodes), to attain a relieve from parish work (the 1849 episode), or to bail out from an examination he realized he could not pass (the 1856 episode)? This possibility seems to us truly unthinkable for three reasons: First, it is unlikely that Mendel could fool so many people (including his parents, siblings, and colleagues at the abbey) so many times. Second, Napp would not have approved large expenditures on a patient who did not show signs of grave illness. He was not the kind of a person easily fooled by a playact. And third, feigning a disease would be totally out of Mendel's character. From all we know about him, he was certainly not an actor; he was too shy for that. So, Mendel joins a club of distinguished patients suffering from mysterious nervous disorders of unknown origin. He comes into a good company, which includes his contemporary Charles Darwin. Zealous Catholic writers have a ready explanation for Darwin's malady: God's punishment! But Mendel? What would God be punishing him for?

Recently, it has been suggested that Mendel, like several other geniuses, including Isaac Newton, Charles Darwin, and Albert Einstein, had the Asperger's syndrome, which some psychologists classify as a variant of autism.[75] The latter is a poorly understood spectrum of mental disorders affecting language abilities and social relations. Asperger's syndrome is characterized by poor social skills, stilted pedantic speech, and preoccupation with certain subjects such as mathematics or physics. However, the attribution of the Asperger's syndrome to Mendel is based on a highly superficial familiarity with his life and behavior, based on popular sources. In reality, Mendel displayed neither an impaired social behavior nor an idiosyncratic mode of speaking, and his intense preoccupation with scientific work effectively ceased with his assumption of the prelatic office. The other cited characteristics, supposedly diagnostic of the syndrome, Mendel shared with legions of other scientists.

Inevitability and Serendipity: Part 3

Once more we might be tempted to engage in the game of *what if.* Of all the serendipities that have steered the course of Mendel's life, the most influential one might have been his admission to the St. Thomas Abbey at Staré Brno. The serendipities included Mendel's being the student of Professor Friedrich Franz in Olomouc, Mendel's graduation from the Philosophical Institute coinciding with the abbey's search for new novices, Napp being at the helm of the abbey, and the strong interest of the abbey in natural sciences and agriculture. It is very likely that Mendel would have become a priest even if the circumstances had not conspired to bring him in the St. Thomas Abbey, simply because this was the number one option of a poor student after the completion of their secondary education. He might have even been lucky enough to be accepted as a novice in another monastery run by another order. But the chances that anywhere else than the St. Thomas Abbey he would have had the opportunity to carry out the experiments that Napp eventually permitted him to work on seem to be negligible. Even at the St. Thomas Abbey at one point, the threat loomed over Mendel's career that he would become a priest occupied fully with pastoral duties. Inevitably or serendipitously his mysterious illness had removed this threat.

Of course, becoming a Catholic monk and priest was a Faustian deal concluded with God rather than the devil. Its essence was comfortable life in exchange for sexual abstinence combined with the termination of one's line of descent. The strictness of the adherence to this vow varied in different times and places. If you were to visit a Czech pub in which the *Stammgäste* had already lost track of the rounds of beer they consumed, you might hear them singing to the accompaniment of an accordion a popular song about a young monk who like a caged bird grieves for the girl he lost by becoming a priest. In the second strophe of the song, the chorus consoles the monk and advises him how he can gain her back (Fig. 5.18).[76]

Fig. 5.18 Like a bird in a cage

"Na Strahově pod Petřínem klášter stojí
a tam jako v kleci ptáček pláče malý strahováček,
pro dívenku svoji.
Neplač, neplač strahováčku, zanech nářků,
dobrá fara všechno spraví, vezmeš si dívenku svoji
sobě za kuchařku..."

It is only fair to state that, contrary to what some writers hint at, there is no evidence of Mendel ever breaking the vow of chastity.

References and Notes

Abbreviation: FM *Folia Mendeliana Supplementum ad Acta musei Moraviae*. Moravské zemské muzeum Brno; published since 1966

[1]Lawless G (1987) Augustine of Hippo and His Monastic Rule. p. 80. Clarendon Press, Oxford, UK. Loose translation of the Latin text: *Before all else, live together in the house, being one soul and one heart seeking God.* (The Rule of St. Augustine)

[2]Merriam–Webster's Collegiate Dictionary (1996) 10th edn. Merriam—Webster, Springfield, MA

[3]Podborský V et al (1993) Pravěké dějiny Moravy. Vlastivěda moravská. Země a lid. Muzejní společnost v Brně, Brno

[4](a) Kuča K. Brno (2000) Vývoj města, předměstí a připojených vesnic. Baset, Praha-Brno; (b) Dřímal J, Peša V (eds.) Dějiny města Brna. Vols. 1 and 2. Blok, Brno 1969–1973. (c) d'Elvert

Chr (1828) Versuch einer Geschichte Brünn. Brünn; (d) Bretholz B (1911) Geschichte der Stadt Brünn. Brünn

[5]The convent of the Augustinian nuns was then known as *Herburgian*, after its first mother superior, the Abbess Herburga. When it later began to decline, the Jesuits took it over and rebuilt it, as well as the church

[6]*Běhounská* (Runner) is a strange name for a street. And sure enough, it has no relation to anybody running up and down it, but rather it owes its designation to a linguistic corruption and mistranslation. The Germanic people settled along this street actually called it *Rheingasse* (Rhine Street), but with time it became corrupted to *Rhennergasse* and since in German *rennen* (to run) translates into Czech *běhat*, the Czech name became *Běhounská*

[7]Spelled in this form, the name is first recorded in 1277. It is generally believed that it represents a corruption of the German *Spielberg*, meaning "Game Hill."

[8]The title of Silvio Pellico's (1789–1854) book is *Le mie prigioni* (My Prisons). When it came out in 1832, it became a bestseller translated into many languages. It was said that the book was more damaging to the Austrians than a lost battle

[9](a) Saak EL (2002) High way to heaven: The Augustinian platform between reform and reformation, 1292–1524. Brill, Boston, MA; (b) Zumkeller A (1986) Augustine's ideal of the religious life. Transl. by E. Colledge. Fordham University Press, New York. (c) O'Donnell JJ (2006) Augustine. A new biography. Harper Perennial, New York, NY

[10]New Catholic Encyclopedia (2003) 15 volumes, 2nd edn. Thomson–Dale, Detroit

[11]Wolný G (1856) Kirchliche Topographie von Mähren meist nach Urkunden und Handschriften. II. Abteilung Brünner Diöcese 1. Band. Nitsch und Grosse, Brünn, pp 118–136

[12]There is a curious modern sequel to this old story. When Jošt died in 1411, he was buried in the St. Thomas Church. In 1999 archeologists and anthropologists opened his grave and microbiologists isolated bacteria from the femur bone of the deceased, which they identified, by means of genetic cloning and DNA sequencing, as a new species of *Rhodococcus*, a microbe involved in wine brewing. They named the species *R. jostii* in honor of the ancient wine drinker. It might be the only bacterium named after a king (see Takeuchi M, Hatano K, Sedláček I, Pacová Z (2002) *Rhodococcus jostii sp.nov.*, isolated from a medieval grave. Int J Syst Evol Microbiol 52: 409–413)

[13](a) Květ R, Samek R (eds.) (1984) Starobrněnské sídlo ústavů Československé Akademie Věd. Geografický ústav ČSAV, Brno; (b) Samek B (1993) Klášter augustiniánů v Brně. Augustiniánský klášter, opatství na Satrém Brně. Památkový ústav, Brno

[14](a) There is no agreement among historians concerning Rejčka's birth year. Consequently different authors claim that she was 12 or 15 years old at the time of her first marriage. (b) Květ J (1931) Iluminovane rukopisy královny Rejčky. Příspěvek k dějinám české knižní malby ve století XIV. Česká akademie věd a umění, Praha

[15]Cistercians are a monastic order founded by St. Robert, abbot of the Benedictine abbey of Molesme in France. Dissatisfied with the life there, he and a handful of monks left the abbey for a secluded, swampy place called Cîteaux, Cistercium in Latin, not far from Dijon in Burgundy. They built a monastery there in which they lived life with emphasis on austerity, simplicity, manual labor, especially fieldwork, and literal observance of the Rule of St. Benedict

[16]Rejčka's burial place lies where the axes of the central nave and the first transept of the church intersect, in front of the Altar of the Holy Cross. The site was originally unmarked, but since 1900, the tile above it is decorated with the letter "E" topped with a crown

[17]Dokoupil V (1972) Dějiny moravských knihoven ve správě Universitní knihovny v Brně. Musejní spolek v Brně, Brno

[18]A new brewery, no longer owned by the abbey, has replaced the original one. We can attest to the excellent quality of the *Starobrno pivo*, the beer brewed there

[19]Ulrich J[78] (p.8) claims that essential for Mendel's admission was a test sermon he had to deliver in front of all the assembled friars. There is, however, no evidence that Mendel visited Brno before receiving the admission letter and indeed such a visit is highly unlikely. He was

apparently accepted solely on the basis of the recommendation letter from Professor Friedrich Franz

[20]Iltis H (1924) Gregor Johann Mendel. Leben, Werk und Wirkung. Julius Springer, Berlin, An English translation by E. and C. Paul was published under the title *Life of Mendel* by George Allen & Unwin, London 1932

[21]The facsimile of the document published by Iltis[20] reveals the difference between the light and heavy handwritings of the schooled son and his hardworking parents

[22]Iltis H[20] (p.19) states that Dr. Schwarz was a municipal physician of the city of Brno, whereas J. Křížený[66] (p. 180) places the examining doctor at Odry. Since in September 7, 1843, when the examination took place, Mendel was in Hynčice, the latter placement seems to be correct

[23]Neumann AA (1930) Acta et epistolae eruditorum monasterii ord. S. Augustini Vetero-Brunae. Vol.1 (A) 1819 –1850. Sumptibus monasterii Vet. Brunensis, Brno

[24]In a letter to Hugo Iltis, Mendel's nephew Alois Schindler estimated that Abbot Mendel, while avoiding cigarettes and pipes, smoked about 20 cigars daily (letter dated January 9, 1923; see[31])

[25]This was, for example, how Professor F. Franz assessed Mendel's Czech-speaking ability in the letter, in which he recommended the graduate of the Philosophical Institute for admission to the St. Thomas Abbey

[26](a) Zlámal B. Cyril František Napp (1792–1867). Augustiner Abt in Alt Brünn. Biographische Skizze. FM 26: 67–101, 1991/1992. German translation of an article published originally in Czech in 1938. Translation by H. Kostková, commentaries and corrections by P. Sládek. (b) Zlámal B. Cyril František Napp—moravský kulturní pracovník. Vlastivědný sborník moravský 18: 460–60, 1966; (c) Weiling F. F.C. Napp und J.G. Mendel (1968) Ein Beitrag zur Vorgeschichte der Mendelschen Versuche. Theor Appl Genet 38: 1444–1448; (d) Orel V (1975) Das Interess F.C. Napps (1792–1867) für den Unterricht der Landwirtschaftslehre und die Forschung der Hybridisation. FM 10: 225–239; (e) Weiling F (1971) Zur Herkunft von Prälat Franz Cyrill Napp, des geistlichen Vorgesetzten J.G. Mendels. Südhoffs Archiv Zeitschrift für Wissenschaftsgeschichte 55; 80–85; (f) Skoupá T (2010) Opat Cyril František Napp a jeho působení v klášteře na Starém Brně. Bakalářská práce. Masaryková univerzita, Pedagogická fakulta, Katedra historie, Brno

[27](a) *Catalogus Abbatiae Eremitarum S. P. Augustini ad Thomam Apostolum Prope Brunam.* Fundationis Marchionum Moraviae Ioannis & Iodoci, anno MCCCLIII. (b) *Catalogus Conventus Emeritarum S. P. Augustini ad S. Thomam Apostolum Veterobrunae.* Ineunte anno MCMII. Fundationis Marchionum Moraviae Ioannis & Iodoci anno MCCCLIII

[28]Šebela J, Obermajer J (1991/92) The portrait of Abbot Gregor Mendel by Alois Zenker. FM 26/27:9–15

[29](a) Dvořáková Z (1976) František Matouš Klácel. Melantrich, Praha; (b) Kabelík J (1908) František Matouš Klácel. Praha; (c) Vávra, M. F. M. Klácel (1808–1882) as inspirer of students in botany. FM 20: 29–32, 1985; (d) Heřmánek Peaslee M, Orel VM (2001) (Ladimír) Klácel: Teacher of Gregor Mendel. Kosmas: Czechoslovak and Central European J 15: 31–54

[30]In a letter addressed to Tomáš Bratránek and dated July 12, 1853, Klácel writes: "I would like to ask the bishop for help in establishing a mission in Texas but I don't know how to approach him at the visitation. Our people who emigrate there in large numbers need a priest, for as I have learned from letters that they convert there to the Moravian Brethren. The Catholics cannot be indifferent to this and I would dare to return them to the right path again." (A. Neumann: Klácelovy dopisy. *Časopis vlasteneckého spolku musejního v Olomouci* 1937 (our translation from Czech). The actual date and circumstances of Klácel's emigration are uncertain. Iltis[20] (p. 26) states that Klácel left the country in 1868, shortly before Mendel's election to abbot and that the liberal interior minister of Cisleithania (roughly the non-Hungarian part of the Austro-Hungarian Empire), Carl Giskra (1820–1879), then in office, secured him a passport. But other sources indicate that Klácel participated in the election (see Vol. 1 Chap. 4) and left 1869, shortly after the election

[31]Simunek M, Hossfeld U, Thümler F, Sekerák J (eds.) (2011) The letters on Mendel GJ. Correspondence of William Bateson, Hugo Iltis, and Erich von Tschermak-Seysenegg with Alois and Ferdinand Schindler. Studies in the History of Sciences and Humanities 28. Prague

[32]The handwritten notes may represent a fragment of a draft of a lecture Klácel had intended to give in the United States. Some of the sentences are truncated and unpolished. Our translation is, we believe, closest to the meaning of the original. The fragment was found among the papers Klácel left behind in Belle Plaine. In 1906, these papers were sent to Prague, where they are now deposited in the Náprstek Museum

[33]The statue is the work of Milada Othová

[34]Loužil J (1972) Franz Thomas Bratránek. Leben und Philosophie. Bohemia, Jahrbuch des Collegium Carolinum 13: 182–210

[35]Bratránek's autobiography comprises 306 densely handwritten pages in the large octavo format. It covers the first two decades of his life, starting with his childhood in Jedovnice, where he was born in 1815, and in Lysice, where his parents later moved. It then takes the reader through his studies at the *Gymnasium* (1826/1827–1833/1834), the Philosophical Institute in Brno, and the University of Vienna, where he earned his doctor of philosophy degree in 1839. In the meantime he entered the St. Thomas Abbey in 1834, where he returned after the completion of his university studies. The coverage ends with the start, in 1841, of his professional career—his assistantship at the Philosophical Faculty of the University of Lviv (Lvov), Ukraine. He started to write the autobiography in 1858 at Kraków. After his death the manuscript passed into the hands of his sister Berta, who was taking care of him in his last years and who then donated it to the *Národní muzeum* (National Museum), Praha. It has never been published; see Jan Krejčí F. Th. Bratraneks Selbstbiographie. Germanoslavica. Vierteljahrschrift für Erfosrschung der germanisch-slavischen Kulturbeziehungen 2: 385–404, 1932/1933

[36]Vogel J (1997) Leoš Janáček. Academia, Praha, pp 40–46

[37](a) Eichler K. (1904) P. Křížkovský. Životopisný nástin. Brno; (b) Racek J (1946) Pavel Křížkovský. Pameny, literatura a ikonografie. Olomouc, Velehrad; (c) Štědron B (1946) Křížkovský na Starém Brně. Slezský sborník 44: 1–2; (d) Racek J (1955) Pavel Křížkovský, tvůrce českého sborového zpěvu. Jubilejní úvaha. Slezský studijní ústav, Opava (e) Elchler, K (1904) Životopis a skladby P Křížkovského, Knihotiskárna benediktinů rajhradskch, Rajhrad

[38]Vetterl K (1929/1930) Bohumír Rieger a jeho doba. Časopis Matice moravské, Brno

[39]Vychodil P (1898) František Sušil. Životopisný nástin. Benediktini Rajhradští, Brno

[40]"Beseda" was an informal, friendly gathering with singing and drinking, either in a pub or in someone's home. It was popular with the Czech nationalist circles in the nineteenth century and gradually developed into a more formal society staging performances of classical music, particularly choral

[41]Vogel J (1997) Leoš Janáček. Academia, Praha

[42]Winmann K (1906) History of Church Music, from its origins to the Moto Proprio of Pope Pius X, 1903. Caughlin & Reilly, Boston

[43]In a letter to Carl Nägeli dated November 6, 1867, Mendel writes: "I am no longer fit for botanical excursions, for the heavens have blessed me with overweight, which, because of gravitation, makes itself felt during longer walks but especially during climbing."[79]

[44]Ohéral J (1840) Rudolf Rohrer. Biographie. Moravia (Brünn) 3: 301–302, 305–306

[45]Rohrer R, Mayer A (1835) Vorbereitungen zu einer Flora des Mährischen Gouvernements. Brünn

[46]See[20], pp. 24–25 and Hrabětová-Uhrová A. Joseph Veselý (Wesely),1813–1892. FM 9:275-277

[47]Mendel's "Autobiographie" (1850) In[79], p. 74–77. For an English translation, see A. Iltis: Gregor Mendel's autobiography. J Heredity 45:231—234, 1954. Purkyně University at Brno published a bibliophilic edition of the autobiography under the title *Gregorii Mendel Autobiographia Iuvenilis. Ad centesimum Quinquagesium J. G. Mendel natalen.* Universitas Purkyniana Brunensis 1972

[48]Bratranek FT (1853) Beiträge zu einer Aesthetik der Pflanzenwelt. Brockhaus, Leipzig

[49]Bureš, P. Botanika na Moravě před založením Masarykovy university. http://www.sci.muni.cz/botany/historie/hist_1.htm

[50](a) D'Elvert, Ch. (1868) Zur Geschichte der Pflege der Naturwissenschaften in Mähren und Schlesien. Rohrer, Brünn; (b) Pluskal FS (1856) Zur Geschichte der Pflanzenkunde in Mähren. Verhandlungen der zoologisch- botanischen Gesellschaft in Wien. 6: 367–370

[51]Klášterský I, Hrabětová-Uhrová A, Duda J (1982) Dějiny floristického výzkumu v Čechách, na Moravě a ve Slezsku. Severočeskou přírodou, Litoměřice Supplement 1982/1:1–132, 2:133–242

[52]Oborny A. Flora von Mähren und östereichishen Schlesien, enthaltend die wildwachsenden, verwilderten und häufig angebauten Gefässpflanze. Naturforschender Verein in Brünn, Vol. 21–24, published in 2 parts, Brünn 1883–1886

[53]Vávra M (1984) Mendel's cooperation with the *Fuchsia* breeder J.N. Tvrdý. FM 19: 251–256

[54]Wood RJ, Orel V (2001) Genetic prehistory in selective breeding: A prelude to Mendel. Oxford University Press, Oxford, UK

[55]Letter to his parents dated December 28, 1851; see[31]

[56]Letters to his mother from June 7, 1857, and June 25, 1859; see[31]

[57]Weiling F (1991) Historical study: Johann Gregor Mendel 1822–1884. Am J Med Genet 40:1–25

[58]Keller A (1828) Cantalupa Melonen. Mittheilungen der k.k. Mährisch-Schlesischen Gesellschaft zur Beförderung des Ackerbaus, der Natur- und Landeskunde in Brünn 16: 127-?, 1828

[59]Franke H, Orel V (1983) Christian Carl André (1763–1831) as a mineralogist and an organizer of scientific sheep breeding in Moravia. In: Orel V, Matalová A (eds) Gregor Mendel and the foundation of genetics. Moravian Museum, Brno (Czech Republic), pp 47–56

[60]Orel V (1992) Jan Sedláček z Harkenfeldu (1760–1827) v kulturní historii Brna. Forum Brunense, Brno, pp. 81–87

[61]Nešpor V (1947) Dějiny university olomoucké. Nakladatelství Národního výboru Olomouce, Olomouc

[62]Nestler JK (1829) Ueber den Einfluss der Zeugung auf die Eigenschaften der Nachkommen. Mitheilungen der k. k. Mährisch-Schlesischen Gesellschaft zur Beförderung des Ackerbaues, der Natur- und Landeskunde in Brünn. 47: 369–372, 48: 377–380, 50: 394–398, 51: 401–404

[63]Nestler JK (1837) Ueber Vererbung in der Schafzucht. Mitheilungen der k. k. Mährisch-Schlesischen Gesellschaft zur Beförderung des Ackerbaues, der Natur- und Landeskunde in Brünn 34: 365–269; 35: 273–279; 36: 281–286; 37: 289–293; 40: 318–320

[64]Diebl F (1835) Abhandlungen aus der Landwirtschaftskunde für Landwirthe, besonders aber für diejenigen, welche sich der Erlernung dieser Wissenschaft widmen. II. Von dem Pflanzenbau. Rohrer, Brünn

[65]The bishop's residence is in the present-day Biskupská Street. Several other buildings in Brno bear the epithet "bishop's," but these belonged to the bishop of Olomouc who had palaces in both cities

[66]Křííženecký J (1965) Gregor Johann Mendel 1822–1884. Texte und Quellen zu seinem Wirken und Leben. J. Ambrosius Barth, Leipzig

[67]Sajner J (1976) Johann Gregor Mendel, Leben und Werk, Ein Bildbuch, 2nd edn. Augustinus, Würzburg

[68]The text of the solemn vow—taken from Ryan[69] with minor changes

[69]Ryan PE (1938–1941) Gregory Mendel. Abbot and discoverer of the laws of heredity I–XXXI. *The Messenger*, March 1938–November 1941. This series of articles was published anonymously in the journal of the Roman Catholic Diocese of Convigton, Kentucky, on the request of Bishop Howard. The bishop sent de Waegenaere RL to Brno to collect material for the article and then instructed Father Ryan, his secretary, to write the articles. (This information comes from a letter Professor Edward O. Dodson, University of Ottawa, Canada, wrote to Dr. Vítězslav Orel, Mendelianum, Brno. The letter is deposited at the Mendelianum.)

[70](a) Sajner J (1968) Gregor Johann Mendels Erkrankung im Jahre 1849: Eine pathographische Studie zu Mendels Persönlichkeit. Clio Medica 3:59–63; (b) Sajner J (1963) Gregor Mendels Krankheit und Tod. Südhoffs Archiv für Geschichte der Medizin und der Wissenschaften 47:377–382

[71]Napp's letter dated October 4, 1849, and addressed to Brno's Bishop Schaffgotsch. See[80]
[72]The letter is cited on pages 111 and 113 in[81]
[73]Klácel's letter to Bratránek dated May 8, 1856 (See[70b])
[74]Forrest DW (1974) Francis Galton: The Life and Work of a Victorian Genius. Tuplinger, New York, NY
[75]Fitzgerald M, O'Brien B (2007) Genius genes. How Asperger talents changed the World. Autism Asperger Publishing Company, Shawnee Mission, KS
[76]The first two strophes of the song composed by Karel Hašler are in Czech:

Na Strahově pod Petřínem klášter stojí
a tam jako v kleci ptáček pláče malý strahováček,
pro dívenku svoií.

Neplač, neplač strahováčku, zanech nářků,
dobrá fara všechno spraví, vezmeš si dívenku svoji
sobě za kuchařku.

In translation:

At Strahov, under the Petřín Hill stands a monastery
And there, like a caged bird cries small strahováček
For his girlfriend.

Don't cry, don't cry strahováček, drop your lamentation,
A good parish will fix things; you will take your sweetheart as a cook.

Explanations: *Petřín Hill* is in Prague and Strahov is a monastery there

[77]Orel V (1996) Gregor Mendel. The first geneticist. Oxford University Press, Oxford, Translated by Stephen Finn
[78]Ulrich J (1907) Gregor Joh Mendel Biografische Skizze. Rainer Hosch, Neutitschein, Reprint from the Illustrierter Neutitscheiner Volkskalender für das Jahr 1908
[79]Gregor Mendel's Briefe an Carl Nägeli. FM 40/41:55–87, 2005/06
[80]Czihak G. (1984) Johann Gregor Mendel (1822–1884). Dokumentierte Biographie und Katalog zur Gedechtnissausstellung anlässlich des hundertjtährigen Todestages mit fachsimile seines Hauptwerkes "Versuche über Pflanzenhybriden." Universität Salzburg
[81]Richter O (1943) Johann Gregor Mendel wie er wirklich war. Verhandlungen des Naturforschendes Vereines, Brünn 74(2):1–262

Into the Fourth Decade: The Failed Professor

6

Ja, derfn s'denn das?
Ferdinand I, Emperor of the Austrian Empire[1]

Eighteen forty eight was a year of crisis not only in the private life of Gregor Mendel but also in the lives of whole nations in western and central Europe.[2] In the Habsburg Empire, the preceding period from 1815 to 1848 was marked by calm and stability. The dominion was at peace, the economy prospered moderately, and the police became quite efficient in dealing quietly with any attempts to change the existing state of affairs. Klemens von Metternich, originally acting on behalf of Franz I, and since 1835 on behalf of the feebleminded Ferdinand I, saw to this. But late into the Metternich period, the stillness became oppressive, turning into the proverbial calmness before the storm. The empire began to change through developments over which Metternich had no control. One of them was industrialization, accompanied by the expansion of trade and transportation. Industrialization, however, went hand in hand with the emergence of a new social stratum—the working class—and so also with the rise of a whole new set of social problems. Unrestrained by law, the industrialists exploited their employees ruthlessly, forcing them to work long hours for low wages, often under hazardous or inhumane conditions, without any social benefits or job security. Generally there were more workers than jobs and in times of recession the numbers of unemployed swelled into discontented and restless masses. In the 1840s several developments—widespread trade recessions leading to massive unemployment, bad harvests resulting in food shortages, and increases in poverty in both urban and rural areas—came to a head at the same time all over Europe. In the Habsburg Empire, the repressive government managed to keep the rising unrest under control for a while, but the situation remained explosive. All that was needed to detonate the explosion was a spark.

J. Klein and N. Klein, *Solitude of a Humble Genius - Gregor Johann Mendel: Volume 1*,
DOI 10.1007/978-3-642-35254-6_6, © Springer-Verlag Berlin Heidelberg 2013

The Year of Revolutions

At the beginning of 1848 in Europe, sparks began to fly in two places: on the Italian peninsula and in France, the motherland of revolutions. The Congress of Vienna left the peninsula fragmented into a number of political units, each governed by a different absolutist monarch, whom the Italian people perceived as a foreigner, generally an Austrian.[3] Indeed, Austrian influence was apparent in nearly all of the states. The Habsburgs made two of these states, Lombardy and Venetia, part of their empire, the former through the Utrecht treaty of 1714 and the latter through annexation in 1815. In three other, purportedly independent states (Tuscany, Modena, and Parma), they installed rulers from their own House. In at least two additional states (Lucca and the Kingdom of the two Sicilies, the second Sicily being the city of Naples) the sovereigns owed their thrones to the Austrians. The only two states *relatively* free of Austrian influence (not counting two tiny ones, the Republic of San Marino and the Principality of Monaco), were the Papal States and the Kingdom of Sardinia, encompassing also Piedmont in the northernmost part of the peninsula. The pope (in 1848 Pius IX) ruled in the Papal States, which took up the central part of the peninsula, and in Sardinia, for nine centuries, ruled the House of Savoy—headed in 1848 by Charles Albert, the only sovereign whom the Italians might not have held for a foreigner. Naturally, the people were not happy to be governed by monarchs who usurped all rights for themselves and gave none to their subjects; rulers who, on top of everything, were foreigners. And so they rebelled, repeatedly over the decades, without achieving much. In 1848, however, it appeared as if the entire peninsula rose as a single body, and the monarchs became scared. One after another, they began making concessions to the insurgents. In principle, there are two kinds of political concessions an absolutist monarch can offer to revolutionaries: to step down (alternatively, be sent packing) or share his power with the people. In the former case the monarchy might convert into a republic; in the latter case, the sharing must be guaranteed by a written document, generally a *constitution*. In both cases, a constitution is the holy cow of a revolution. The 1848 revolutions in the Italian peninsula forced Ferdinand II, the Spanish Bourbon ruling the Kingdom of the two Sicilies, to promise a constitution. Pius IX implemented a constitution in the Papal States even before the eruption of the revolutions. Similarly, Charles Albert of the House of Savoy in the Kingdom of Sardinia promulgated voluntarily a constitution so liberal that it would later be taken over by the Italian state. Soon afterward, the smaller states followed suit. The primary aim of the uprising in the states ruled directly or indirectly by Austrians was to overthrow the foreign power. Charles Albert became the leader of this anti-Austrian movement. With his military assistance, the insurgents succeeded in forcing Leopold II to flee from Tuscany and the Austrian army to retreat from Lombardy and Venetia. All three states then proclaimed themselves republics.

The success of these uprisings was short lived, however. The Austrians quickly strengthened their military presence in the southern part of their dominion and then, under the command of field marshal Joseph Wenzel Radetzky (1766–1858), reconquered the lost territories, reinstated the deposed sovereigns, and through them rolled back nearly all the gains of the revolutionary movement. Only the Kingdom of Sardinia retained its liberal constitution and led by Charles Albert's son and successor, King Victor Emmanuel II, continued to be the driving force behind the *risorgimento*, the effort to rid the peninsula of foreign presence and to unify the political units into a single state, the Kingdom of Italy. It took another wave of revolutions and another generation of revolutionaries to achieve this goal, but by 1870 the Kingdom of Italy encompassed nearly the entire peninsula, as well as Sicily and Sardinia.

France, in the meantime, became a republic in 1792 after the revolution of 1789. This First Republic lasted, however, only until 1804, when Napoleon Bonaparte proclaimed himself emperor.[4] Following Napoleon's demise in 1814, the country returned to the monarchy it had been before the revolution, with the difference that the power of the kings was now restricted by a constitution. In 1848, the king, third after the restoration, was Louis Philippe, who came to power after the 1830 revolution on the premise of being a liberal. When the people realized that he was anything but liberal, it became only a question of time until the next revolution would sweep him away. The waiting time turned out to be longer than many might have hoped—a whole 18 years—but when it finally came on February 22, 1848, all the mightier was the explosion of the students' and workers' wrath. Two days later the king abdicated and the masses proclaimed the Second Republic. The republicans, however, had to go on the barricades several more times in order to save the republic, only to lose it ultimately. In 1852 France became an empire again, and the people had to wait another 18 years for the Third Republic to be instituted.

Time Jerked the Curtain: And the World Had Changed![5]

The news of the February 1848 revolution in Paris was the starting signal for liberals in chief European cities to go on the barricades. Barricades went up in Vienna, Prague, and Buda, and demonstrations, riots, and uprisings took place in Germany, Britain, Denmark, on the Balkan, and in Romania. The governments everywhere were in retreat and for a while it seemed that the era of absolutism was coming to an end in many parts of Europe. In the Austrian Empire, the cities Buda and Pest, as well as the capital city, were among the first to revolt.[6] The Hungarian plain, like the Italian peninsula, was another restless part of the monarchy. The diet of the province had been debating the need for political and social changes for some time prior to 1848, but could not agree on how far the demands to be submitted to Vienna should go. The news from France and the consequent spontaneous demonstration of young radical intellectuals in the streets of Pest on March 15 put an end to the bickering. The nobles quickly enacted a set of the *March Laws* and sent them to their king for signature. Ferdinand I, his hands tied with the rebellion under his windows and with the bulk of his army being in Italy, had no other choice than to sign the document. The Magyars thus got, within the span of one month, what they had been demanding for years: substantial autonomy within the empire. In addition, the laws permitted the Magyars to establish National Guards and a freely elected parliament and gave them a series of rights, including freedom of press and religion.

Unfortunately, the Magyars tainted their laws with a nocuous flaw by excluding other nationalities living in Hungary from benefiting from these laws. The Magyars made themselves the masters of the region, as if they were living there alone. Naturally, the Croats, Serbs, Slovaks, and Romanians were in no mood to leave this provocation unchallenged and started their own insurrections—against the Magyars. The Croats went farthest in their expression of displeasure in that they sought out a military conflict with the Magyars.

Words of the events in Paris filtered through to the Austrian capital at about the same time as they did to Buda, but in Vienna it was the height of *Fasching*, the Carnival season, and so a revolution was postponed for a few days. Only when *Fasching* ended on March 8 did the sobering Viennese begin to take in the news. Meanwhile reports had also arrived of the stirrings in Hungary, including the daring speech the lawyer Lájos Kossuth (1802–1894) delivered at the plenary session of the Hungarian Diet in Pozsony (now Bratislava in Slovakia). In this address, subsequently translated into German and circulated in the coffee houses of Vienna, Kossuth openly demanded all the political reforms, which the Austrian radicals had been, until then, only whispering about. Debating groups began to form in the cafés and the university corridors, and soon the first petitions to the imperial government and to the emperor himself were fired off. A flood of petitions from all corners of the empire would then follow throughout 1848 and swamp the Hofburg, the residence of the imperial family. In them, students, writers, artists, middle-class activists, peasants, and representatives of the monarchy's different nationalities demanded, requested, or humbly supplicated the emperor to institute a series of political, social, and economic reforms. The centerpiece of all these envisioned reforms was to be the creation of a democratically elected *Reichstag*,[7] a parliament, which would equitably represent the different social strata and nationalities of the empire and which would also be empowered to prepare a constitution reflecting the new order of the society. Essentially, the people were asking the emperor to give up some of his powers and to voluntarily convert the current absolutistic monarchy into a constitutional monarchy. Under the existing conditions, the only institutions that could have, theoretically, limited the emperor's powers were the *Landtagen*, the regional diets[7] of the individual kingdoms and provinces. The days when the diets could do this were long gone, however, their powers having been eroded gradually by the past emperors, in some provinces more so than in others, but in all of them to the degree that, in principle, what the emperor decided became law. Furthermore, the provincial diets were assemblies of estates, and hence, they did not represent all the people in the domain; in some domains they represented the nobles exclusively.

The call for a parliament alone was revolutionary; combined with the package of basic human rights for all the people and all the ethnic groups, it amounted to bringing down the *ancien régime*. The emperor may not have grasped this reality, but the rest of the imperial family had. Nonetheless, they decided to ignore this, in their eyes, impudent proposition. But the crowd, which had in the meantime surrounded not only the government buildings but also the imperial residence, the *Hofburg*, was not in a mood to be put down in such a haughty manner. On the next day, Monday, March 13, an even larger and rowdier crowd gathered in the inner city, and the demonstrations turned violent. Vienna, its inner city then still encircled by bulwark walls, had three armed organizations responsible for keeping order. One was the regular police, which had stations in the different quarters. Another was the *Bürgerwehr*, the Citizens' Guard of some 14,000 men recruited from the middle class; it was poorly trained and visible mainly at religious parades or at celebrations of the emperor's birthday. The third organization was an army garrison of some 15,000 men stationed in barracks outside of the city walls. On March 13, 1848, it was the army, specifically the cavalry commanded by Archduke Albert, that clashed with the demonstrators. As the unit advanced from the gate at the Scottish Monastery toward the city's center, the demonstrators pelted it with bricks and stones. The unit responded by opening fire on the crowd, and the first victims of the revolution fell to the ground, dead or wounded. The news of the bloody encounter spread quickly through the city, and the demonstrators responded by blocking the streets with barricades. Outside of the city wall, where the factories were located and the workers lived, crowds began attacking the plants, demolishing their equipment, vandalizing, and looting.

Now that the situation had spiraled out of control, the imperial court was ready to make concessions. The least painful one was letting Metternich go. The chancellor stuck to his hard-liner attitude to the end, urging the imperial family to implement harsh measures against the "rabble," as he called the demonstrators, and turn the whole army loose on them. His words fell on deaf ears though, for the family was scared that what happened in Paris might also happen in Vienna. That very same evening, Metternich turned in his resignation, and early the next morning he and his family snuck out of Vienna. They then faced the problem of where to turn. After a few days in Olomouc, they realized that nobody in the empire could vouch for their safety, and so they fled incognito to London, to join the growing crowd of deposed rulers and their officials from other parts of Europe who now made their home there. The Viennese greeted the news of Metternich's resignation and departure with jubilations, but they expected more from their emperor than a simple reshuffling of his cabinet. They pressured him to promise reforms, and when the fulfillment of those promises was not immediately forthcoming, new rounds of demonstrations followed over the next several days and months. Particularly nasty revolts erupted on April 25, May 15, October 6, and October 29–30, each provoked either by the emperor's inactivity or his actions. The May 15 revolt so scared the imperial court that on May 17 it fled to Innsbruck and did not return until August 6. When, however, the October 6 demonstrators murdered the war minister and hung his body on a lamppost, the court fled again, this time to Olomouc. Gradually, it seemed, the people were getting what they wanted. Metternich was gone, and when his successors did not turn out to be much better, they, too, were dismissed. Realizing that his delay tactics would not work, the emperor gave in on the parliament issue and allowed the elections of the delegates to be held in June. Hungary did not participate because by that time it had already proclaimed its autonomy. The rest of the empire was divided into ten districts, which together elected 383 delegates. The first plenary session of the parliament, the *Reichstag*, took place on July 22 in the winter riding school of the Hofburg. After the preliminaries, the parliamentarians began to work on the constitution, but after 52 sessions, on October 31, the emperor ordered them to relocate to Kroměříž in Moravia, ostensibly for their own safety. The lasting achievement of this first period in the *Reichstag*'s existence was the Act of Emancipation of September 7, 1948, which abolished once and for all the remnants of serfdom. It freed the peasants of all forms of the *robota* and made those who bought the land they worked on its true owners. The landlords were to be compensated for the loss of land, and the burden of the compensation was to be borne equally by the peasants and the state. The *Reichstag* then continued to work from November 23, 1848, to March 7, 1849, in Kroměříž, and during this period managed to come up with a reasonably democratic constitution.

The emperor also caved in to the demand for the creation of the National Guards, a novelty imported from France, where it was established as a paramilitary organization, a militia, for safeguarding the achievements of the 1789 revolution. The Viennese National Guards were, however, predominantly volunteers from the society's middle-class layer, while the university students formed their own Academic Legion. Step by step, the emperor also fulfilled all of his other promises. He lifted censorship and instituted freedom of press and of religion, proclaimed general political amnesty, and reduced the sales tax on food. On one demand, however, he remained evasive—granting autonomy to the individual nationalities. He got away with this tactic because most of the provinces were multinational and he could therefore play the nationalities against one another, and ultimately reverse the tide of the revolution.

Calm Spring and Hot Summer in Prague[8]

The starting point of the reversal was the crushing of the revolutionary movement in Prague on June 17, 1848. The Czechs in Prague did not go on the barricades in the spring of 1848, but rather hoped to obtain concessions from the emperor through negotiations. It was not out of cowardice that they wanted to avoid violent confrontations nor because, by nature, Slavs were "doves" (in contrast to Germanic "hawks"), as the popular stereotype would have it, but because they feared that an armed conflict might bring the empire down. And that was *not* what many of them wanted. This claim may appear absurd in view of the fact that for centuries the Czechs struggled to regain the rights they once had. Behind the reticence, however, was a pragmatic logic dictated by the political developments outside of the Austrian Empire.

In 1848, the part of Europe that would soon become Germany was fragmented into 38 states, which the Congress of Vienna had sanctioned in 1815. Although the states were formally bound together into the Germanic Confederation (which, incidentally, included the Austrian Empire), in reality they were more or less independent.[9] In the past, the sovereigns of these units guarded their autonomy vigilantly, but the 1848 revolution destabilized their grip on their lands. As a result, calls for the unification of all these states into a single, powerful Germany began to grow louder, and specific plans were made for this to happen. The first step, according to these plans, was to convoke a *Nationalversammlung* (National Assembly) in Frankfurt am Main. This Frankfurt National Assembly, as it came to be known, would consist of democratically elected delegates from the individual states and would have the task of drafting the constitution of the *Grossdeutschland*, the unified Great Germany. The elections took place in April of 1848, and on May 18, the 830 elected delegates met for the first time in a plenary session at the St. Paul Church in Frankfurt. They were mostly intellectuals—professors, lawyers, judges, and physicians—although some businessmen and administrators were also included in the corps. The assembly remained in session for the rest of the year, but most of the time it was bogged down in a theoretical discussion of principles and so failed to produce practical proposals on how to proceed with the unification. One heatedly debated question was whether Austria proper should be included in the united Germany. When the vote decided that it should not, the Austrian government recalled its candidates. Soon afterward the assembly dispersed without producing any tangible results. The birth of a united Germany had to be postponed for some 20 years. When it happened, assisting at the delivery was not a corps of 830 delegates, but a single man, Prince Otto von Bismarck (1815–1898). The birth took place in two stages. First, in 1867, the states north of the Main River united into the North German Confederation. And then, in 1871, the southern states joined the confederation to form the German Empire with Bismarck as its first Chancellor.

In the Czech lands both the Germanic and the Czech people followed the preparations for the Frankfurt Assembly, the former with enthusiasm and the latter with apprehension.[8] The Germanic people expected they would be incorporated into the united Germany under one of three scenarios: Either the entire Austrian Empire, the Czech lands, or the regions of the Czech lands inhabited predominantly by Germanic people would become part of the German state. To the Czechs familiar with the history of German expansionism, none of these three possibilities was acceptable, since in each of them a loss of the Czech national identity would be preprogrammed. To the Czech historian and politician František Palacký, representing the Austro-Slavic program in Czech politics, the best and perhaps the only viable option for the Czechs was to remain in the Austrian Empire, in which the Slavs had a clear majority and could therefore resist any attempts at their annihilation. Of course, Palacký's condition was that the Czechs would be granted considerable autonomy within the empire. So, when the organizers of the Frankfurt Assembly invited Palacký to attend it, he politely but firmly declined the invitation in a famous letter in which he pointed out that Czech inclusion in the unified Germany would amount to a political suicide. The followers of Palacký's Austro-Slavism then began a campaign of urging the Czechs to boycott the elections into the Frankfurt Assembly. Needless to say, the Germanic population did not like either the Austro-Slavic program or the boycott of the elections. The possibility that the Czechs might gain a large degree of autonomy scared most of the Germanic people, who could not imagine a reversal in the distribution of political power, which would certainly follow if the Czechs would once again become masters in their own land. The Germanic people, therefore, not only did not support the Czech revolutionary movement but did all they could to hamper it. The Czechs on their part tried to avoid any confrontations that would give the emperor a cause for rejecting their demands, hence their docility.

They petitioned the emperor as early as March 19, 1848, demanding unification of the Czech crown lands; recognition of Czech as an official language equal to German in all spheres of public life; freedom of press, assembly, and religious confession; and the abolishment of the *robota*. At first the imperial government flatly rejected the demands, but a day later, undoubtedly under the influence of the events under its windows, it softened the rejection and sent the delegation back to Prague with vague, nonbinding promises. On March 23, however, the delegation was back in Vienna with a more strongly worded petition. This time the government was more forthcoming. On April 18, it granted all the demands except the unification of the Czech lands, which it referred to the *Reichstag*. Encouraged by this response, the Czechs formed a *Národní výbor* (National Committee), which decided to hold a *Slovanský sjezd* (Slavic Congress) in Prague. The aim of the Congress was to bring together representatives of all Slavic nations of the Austrian Empire to discuss the formation of a union of Austrian Slavs, the relationship of Slavs to non-Slavic nations in the empire, and the relationship of Austrian Slavs to nations outside of the empire. The Congress was obviously intended to become a counterbalance of the Frankfurt Assembly. To the great displeasure of the Germanic population of Prague and to the rancor of Prince Alfred Windischgrätz, the commander of the armed forces in the Bohemian kingdom, the Congress began its deliberations on June 2. Windischgrätz was an old Slavophobe, archconservative, and a hawk to boot, who went voluntarily into semiretirement in disgust over the way the emperor, in his view, botched up the handling of the uprising in Vienna. Later, however, he became so upset with further developments that he resumed his function in order to wait for an opportunity that would allow him to demonstrate how uprisings should be dealt with. When the opportunity was slow in coming, he apparently felt that it needed a bit of assistance in the form of staged provocations. The strategy worked, albeit at the cost of his wife's life and a serious wounding of his son. The Prague Congress seemed like the right occasion and he provided plenty of provocation in the form of a show of force and arrogant behavior of his soldiers. The organizers of the Congress pleaded with the participants not to take the bait, but they did so in vain. On Whitmonday, June 12, after an open-air mass at the present-day Václavské náměstí (Wenceslas Square), the demonstrators marched to the commander's residence and there, in the streets of the Old Town, they clashed with the army. Barricades went up and the fighting continued intermittently until Wednesday. During the night from June 14 to June 15, Windischgrätz withdrew his units from the Old Town and on the next day he ordered his artillery to bombard its streets from the Letná and Hradčany hills. On June 17 Prague surrendered. More than 10,000 soldiers were needed to subdue some 3,000 insurgents, mostly students, journeymen, and workers, 43 of whom died and 63 were wounded. Arrests, executions, and persecutions followed: In Prague, the 1848 revolution came to an end. The Germanic people, many of whom fled the city when the insurrection started, directed their wrath at the Czechs, and the fissure between the two ethnic groups deepened.

1848 in Brno[10]

In the Moravian capital, the revolution passed through three poorly delineated stages dominated by civic, nationalistic, and social issues, respectively. When the revolution was still young, the people's main concern was with their civil rights. The euphoria over the freedoms, which the March uprising in Vienna forced the Emperor to grant, was so high that the issues that would come to dominate the two later stages were disregarded momentarily. The people wallowed in the feeling that all of sudden they could freely assemble whenever and wherever they wanted without being harassed by the police and that they could speak their minds without fear of being denounced and arrested. They attended rallies at which they debated politics openly, listened to proclamations, sang revolutionary tunes, wrote petitions, and marched and danced. Czechs and Germans, burghers and factory workers, rich and poor were united in their jubilation for a while. The most popular places of assembly were Brno's two large squares, today's Dominikánské and Moravské náměstí. There they could be sure that their voice would be heard and their mood sensed by those representing the political power in the province—the assembly of the estates and the executive arm of the imperial government. For, at the former square stood the *Landtag* building (today's New City Hall) and at the latter the Governor's Palace, which was once part of the old St. Thomas Abbey, but was then occupied by Count Leopold Lažanský, the governor of the Moravian province. Often, they would march from one square to the other, sometimes in the evening with torches and candles for a greater effect. It was also there where they could hear the latest news about the developments in Vienna, Budapest, and Prague. The people of Brno, both Germanic and Czech, supported unreservedly (in contrast to the Czech radicals in Prague) the Viennese uprising and went in their backing as far as sending a detachment of National Guards to Vienna to aid the revolution when it seemed it was short-manned.

Once the initial euphoria had worn off, however, the nationalistic and social issues came to the fore. Like Prague, Brno had a large Germanic population, whose relationship with the Czechs had been less than harmonious. The two ethnic groups were like adjacent plates of the earth's crust, rubbing against each other and accumulating tension, which on occasions released itself in the form of a confrontation rather than an earthquake. The 1848 revolution was such an occasion. The friction centered on the question: What should become of Moravia and Silesia, should the Austrian Empire disintegrate, as it then seemed likely to happen? Most Germanic people wanted the provinces to be incorporated into the German Empire, for which the Frankfurt Assembly was supposed to prepare the blueprint. The Moravian Czechs were united in their opposition to this idea, but divided about the alternatives. Some wanted Moravia to join Bohemia in forming a separate, independent state, but others wanted Moravia and Silesia to form a state on their own. These frictions impressed strong nationalistic overtones on the second stage of the revolution. The nationalistic strife revealed itself early on when the *Landtag* decided to revamp itself to become more representative of the people in the province. The question then was: What should be the proportion of Czech and Germanic members and how should the different social classes be represented?

When the word about the deliberations inside the *Landtag* reached the crowds outside, demonstrators tried to influence the voting of specific representatives by non-peaceful means (see next section).

As for the social issues, one must realize that the revolution of 1848 in general profited the middle class primarily and to a lesser extent the peasants, whereas the working class was coming out of it empty-handed. Although the workers took part in the uprising in a not insignificant way, as the year progressed, their lot had actually worsened rather than improved. As a result of the revolutionary upheavals, trade slowed down all over Europe, goods began to accumulate in warehouses, and factories were forced to cut down production, which their owners did by reducing the workforce. Fearful of the growing masses of unemployed, the city administration tried to encourage the nonlocals to return to their homes by providing them with free one-way tickets. To increase the employment opportunities, it also initiated public works, such as straightening the course of the Svitava River through the town. Neither of these measures alleviated the worsening employment situation, however. The wrath of the unemployed, underpaid, and overworked then turned against the machinery the factories had been installing to increase productivity. On April 1, workers attacked the factory of the Popper Brothers in Staré Brno and demolished its equipment. The government sent in the National Guards to reestablish order, but when they failed, the army was called in and it suppressed the rebellion brutally. Staré Brno also witnessed a demonstration of peasants against the St. Thomas Abbey, the largest landowner in the area. When widespread hunger set in, house-to-house begging became so common that the police was ordered to stop it. This act, in turn, provoked attacks on bakeries and grocery stores on June 13 and 14, and the National Guard had to intervene again.

The most violent confrontation between the guards and the demonstrators took place in October, in response to the events in Vienna and Buda. On October 7, a large crowd gathered at the Cabbage Market to show sympathy for the Viennese revolutionaries. It elected a delegation of 22 guardsmen and dispatched them to Vienna with a mission to find out what exactly was happening there and to help the revolutionaries if necessary. When it became known that the *Reichstag* itself called for help, masses of demonstrators in Brno forced the National Guards, against the will of their leadership, to send an additional 600 men to Vienna to aid the exhausted fighters on the barricades. At the same time, however, reports were reaching Brno that at railway stations throughout Moravia government forces were disarming guardsmen heading for or returning from Vienna. This intelligence, together with the news that the armies of Windischgrätz and Jelačič[11] were converging on the Austrian capital, mobilized, on October 18, the largest mass of demonstrators Brno had seen to that date. With tensions high, an outbreak of violence was thwarted only by the means of a ruse. The army commanders let themselves be persuaded to pull all soldiers back into the barracks and have them replaced by members of the National Guard. The maneuver gave the impression that the army had retreated and the demonstrators dispersed believing they had achieved their goal.

By that time, however, many of Brno's burghers, who initially sympathized with the revolution and even participated in it actively, began to change their attitude toward it. Scared by the show of force demonstrated by the masses, fearing for their property, concerned about the adverse impact of the revolution on trade and commerce, and sensing that a reversal of the revolution's fortunes was in the air, they began switching sides. Under the pretense of mediation, they made advances toward the imperial government in Olomouc and sent delegations to Windischgrätz. The *Landtag*, too, all but lost its enthusiasm for the revolution, and the peasants, having attained what they wanted, abandoned it altogether. Only the poor, the workers, the students, and other democratically minded intellectuals kept on pressing for reforms and demonstrations. Enraged by the news that the armies of Windischgrätz and Jelačič had surrounded Vienna and were bombing it demonstrators gathered in large numbers on Brno's squares on October 29 to protest the infraction. They demanded that the National Guard be sent to Vienna and attack the rear of Windischgrätz's army in order to aid the burning city. But Brno's National Guards, now composed largely of middle-class members, were no longer prepared to fight for what they considered a lost cause. On that morning the Guard's commander still managed to evade the issue by claiming that he had to obtain first the *Landtag*'s approval. When, however, the afternoon came and there still was no action, the demonstrators decided to take the matter into their own hands. Armed with sticks and clubs, they attacked the city hall, in which the city fathers barricaded themselves. Failing to gain entry, they overran police stations and barracks and seized a small number of weapons. From then on, the situation deteriorated rapidly. Large scale looting began as the mob attacked food stores and restaurants. The rampage lasted through the night, and on the morning of October 30, the demonstrators again demanded weapons from the National Guards, but failing to get any, they raided butcher shops and armed themselves with knives and axes. By noon the whole city was in panic. The bells in all churches tolled alarm; innkeepers, landlords, and city officials barricaded themselves in their houses; and all shops, offices, schools, and plants closed down. The demonstrators, however, headed for ammunition and weapons factories, where they intended to arm themselves and then leave for Vienna. At the gate of the factory on Malá Křenová Street, they encountered units of National Guards, which ordered them to disperse. When the demonstrators ignored the order, the guardsmen opened fire, killing several persons and wounding many others. After this incident, the guardsmen, the army, and the police retained their control over the city, suppressing all additional attempts at further demonstrations. In Brno, too, the revolution was over.

Revolution Knocks at the Gate of the St. Thomas Abbey

Outside of the city walls, Staré Brno was one of the foci of revolutionary activity. Several factories resided there that were the target of workers' demonstrations, and it was also the seat of a major landowner—the St. Thomas Abbey—and so it became the place for the peasants to vent their grievances. For we must not let ourselves be fooled by the façade of holiness that the abbey with its parish church presented to the world. The truth of the matter was that the abbey was a feudal institution and Napp a feudal lord. Moreover, the Germanic population, as we will explain later, had reasons to regard the abbey as a hotbed of Czech nationalism. All of this together turned Staré Brno, in 1848, into a gathering storm, with the abbey acting as its convenient lightening rod. Ironically, however, at least some of the friars and the abbot himself were among the early supporters of the revolution. Here is how they reacted when they learned about the uprising in Vienna.

On March 12, 1848, as the friars were having their regular noon meal in the refectory, someone burst in with the news that in Vienna the students had taken up arms in order to chase out Metternich. Spontaneous jubilation broke out and the exhilarated assembly called on Klácel to write a celebratory poem. When he obliged, Křížkovský promptly set it to music[12] and the friars, led by Napp, sang *On March 12* repeatedly as the party continued until it was time for the afternoon snack.[13] A few days later, two friars from the abbey, Phillip Gabriel and Tomáš Bratránek, joined a group of Brno's burghers on a fact-finding mission to Vienna. Later, when the Viennese reciprocated by sending a party of students and guardsmen to Brno, the abbey played host to them. Napp made himself popular with the visitors by sporting, at their first lunch in the refectory, a Calabrian hat and a student pipe.[14] Napp also demonstrated his sympathy with the Viennese revolution by celebrating on March 18 in the Minorite church a memorial mass for its victims, at which Křížkovský conducted Cherubini's Requiem. At the mass, Klácel delivered a brief sermon in which he expressed hope that the revolution would bring rights to the Czechs, improved conditions to the working class, and democratic constitution to the country. The sermon was said to have made a deep impression on the assembled crowd.

Both Napp's and Klácel's enthusiasm for and their later involvement with the revolutionary movement got the abbey into trouble, however. As mentioned earlier, one of the outcomes of the movement in Brno was the decision to reform the *Landtag*. After the old *Landtag* commissioned a 24-member committee to prepare a proposal for the composition of the new diet, the numerical representation of the various ethnic groups and social classes became a hotly debated issue. The Germanic middle class was particularly vocal in the debate, demanding a high representation, ostensibly as a protection against the growing influence of the Czechs. On March 30, Napp stood up in the *Landtag* against these demands, arguing that the composition of the new diet should be proportional to the ethnic composition of the population in the province on the principle of equal rights of Czech and Germanic peoples. His stand irked the Germanic burghers to the extent that they instigated a stone-throwing mob to attack the abbey.[15] Fortunately, the walls encircling the monasterial ground prevented the demonstrators from penetrating the abbey itself, so the incident ended with only a few broken windowpanes. The final composition of the new diet, when it first met on May 31, consisted of 67 major landowners, 77 middle-class burghers, 103 small holders ("peasants"), and five representatives of the University of Olomouc. Because of the high number of smallholders represented, the assembly was nicknamed "the peasant diet." Indeed, one of its first legislations was the abolition of the *robota*. During the October uprising, when the abbey came under attack again, Napp, fearing for his life, fled to Jevíčko and returned only in early November, after the army had restored order in the streets.[16]

For his pro-Czech stance, Napp may have become the temporary target of enmity, but the Germanic burghers still regarded him as one of their own, if only tainted by ideas of the freethinkers surrounding him. The real thorn in their side was, however, Matouš Klácel. Of his position in the Czech-Germanic conflict, there could be no doubt. After all, Czech nationalism was one of the reasons for his dismissal from his teaching post in 1844. Although after his return to Brno at the end of May 1845, he lay low in the lee of the abbey, licking his wounds and only publishing innocuous literary works and even these under his friend's name,[17] as his nationalistic past had not been forgotten. As might have been expected, the March events of 1848 revived Klácel's spirit, and his activities made him one of the most visible figures in the Czech nationalist movement. Since Czech nationalists were coming and going in Klácel's cell, German nationalists were pointing fingers in a very unfriendly manner not just at him personally but at the whole abbey. Understandably, some of the friars did not welcome the sudden notoriety the abbey had achieved through Klácel's activities. As brothers in Christ and disciples of St. Augustine, the friars should have been above any ethnical strife, but in reality they were as divided along nationalistic lines as the world outside of the abbey's walls. Inside the walls, Napp managed to prevent any open hostilities between the two camps but outside, arbitration between the nationalities had become difficult. By April the situation in the abbey and outside of it reached a boiling point at which time Napp felt it best to temporarily relocate Klácel to some other place. Klácel agreed and chose Prague, where the Czech nationalist movement was much stronger than in Brno.

In the Bohemian capital, Klácel found accommodation in Veith's house and immediately plunged into politics.[18] He managed to get elected into the National Committee and in this function he took part in the preparation of the Slavic Congress. He shed the habit, hoping that this change would become permanent. He began making plans for the completion of his doctorate, applying for a teaching position at the university, and cutting his ties to the abbey once and for all. In this state of mind he wrote to Napp on April 16 asking him for permission to stay in Prague indefinitely. Napp, however, urged him to return to Brno, where his contribution to the Czech national movement would be needed. Klácel stalled, however, and even the prospect of becoming the editor of Brno's first Czech language daily, could not sway him. But then came the June uprising with all its consequences. Klácel participated in the Slavic Congress, but there is no record of him being involved in the rebellion. Since he was against violence of any kind, such an involvement is most unlikely. Nonetheless, because of his political engagements, his friends urged him to leave Prague, which he did on June 16, joining Veith in Liběchov. What followed then we already know from the preceding chapter. When he ultimately returned to Brno, he apparently did not give up hope that some of the revolution's achievements could still be salvaged. In this spirit, he continued to be politically active. One product of this activity was a document that apparently did not lead to any action, but that nevertheless is important for the understanding of the situation in the abbey of that period.

The Revolt of the Friars

The document[19] has the form of a petition written in German and addressed to the *Reichstag* in Vienna. Since no English translation of the petition has not been published, as far as we know, we provide it in full below.

Supreme Imperial Assembly!

At the time when our fatherland is undergoing political restructuring, the undersigned, in the name of humanity, consider it their duty to draw the attention of the Supreme Austrian Diet to the fact that an entire, not insignificant, class of people in the Austrian states has been excluded from the March and May achievements, so that it can only with sadness partake in the jubilation of the nations blessed with liberty.

According to the civil law, a regular monastic is one step below a common criminal in that the latter's civil rights have merely been suspended, whereas the former has no civil rights at all. A regular is deprived of civil rights; in the language of the law he is proclaimed for civically dead. Every civil contract he closes is declared invalid, just like that closed by a lunatic, a child, or a person under guardianship. He is not allowed to be a guarantor or witness before the court of justice, a witness at a wedding, and a godfather at baptism or confirmation. He can neither inherit a property nor bequest his trifling, hard-earned possessions to his commonly needy relatives. From him and in his name inherits solely and legally only the religious order.

The current statuses of the absolute monarchy favor the clerical organization to the clear detriment of the individual conventuals. Unfortunately, the latter come to realize the weighty practical implications of these statuses only after years of unhappy experience. These statuses remain in effect even after March 13; they have not been suspended, nor has, to this day, the constitutional right of the regular monastic been acknowledged. In France the monastery is regarded as a free association, in which the citizen must not vanish in the monk. The monk's civil right remains inviolable at all times; it is recognized as being above and nobler than the private contract with the religious order, and as such it is protected by the state.

When one examines without prejudice the ecclesiastical and social role of the nineteenth century Austrian monasteries, one comes to the conclusion that these houses and institutions of Christian love are nothing more than seminaries for forced morals. They are mere almshouses for the poor and deluded young men. One is further compelled to conclude that the enforced isolation from the common people, the exclusion from the family circle, the absorption into self, the inadequate and one-sided education, the conspicuous distinctiveness in dress, attitude, and behavior, and the insistence on absolute obedience, slay the citizen in the monk and lead to a state of deepest humiliation.

Since priests of a religious order are robbed of the rights a free citizen has, since they are looked at as mere instruments of the ecclesiastical hierarchy, and since the state endorses all this, to the intelligent part of the nation they must appear as a dead, unimportant limb of the society. Monks are viewed as naught, without rights, without own will, excluded from the great honor book of free, dignified citizenship. They are debarred from all elections into provincial and imperial diets, as well as the German Parliament. They are denied of active and passive voting rights into all committees, rights granted even to a most destitute worker.

Should in the great Constitutional State of Austria, in the State of free citizens, emerging before the eyes of the European nations, the priests remain subjugated, contempt-exposed slaves, at whom the free and cultured nations of the world will be pointing with ridicule and disdain? It would be a shame for Austria if the colossal architectonic construction of the constitution would sanction these dungeons of citizenship, these graves of constitutional freedom. Surely, before long the mightily resounding trumpets of the first Austrian parliamentary constitution will call forth to resurrection and civic activity the monks buried alive in their cells.

Consequently, the undersigned professors and pastoral workers of the order of Saint Augustine in Old Brno take the liberty of appealing to the imperial parliament to grant them constitutional civil rights, and request to be allowed to devote their entire efforts, according to their abilities and their past services, to public teaching institutions and to free, united, and indivisible citizenship. The undersigned make it respectfully their mission to promote science and humanity in accordance with the spirit of constitutional progress.

Brno, August 8, 1848

Fr. Mattheus Klácel, former professor of philosophy

Dr. Philipp Gabriel, professor of mathematics in Brno; Head of the Countess Thurn Institution

Josef Lindenthal, Kooperator at the parish church in Old Brno

Benedict Fogler, professor of French language & literature & accredited teacher of Italian language

Gregor Mendel, Kooperator & teaching candidate

Chrysostomus Cygánek, teaching candidate

Two controversial questions have arisen regarding this document: Who wrote it? And what exactly did its signatories request the Austrian Diet to do? In regard to the first question, we must distinguish between two senses of "writing"—the sense of formulating or composing the text of the petition, and the sense of physically penning down the final document. Here we shall use the word "compose" for the former sense and "write" for the latter. Two names have been proposed as the writers of the document, in either one or both senses of writing: Klácel and Mendel. The three suggested possibilities are these: First, Klácel both composed and wrote the petition and was the first to sign it.[20] Second, Klácel composed a draft of the petition and gave it to Mendel to render it in his calligraphic handwriting into an official document, which he then signed and circulated among other friars with the option of adding their signatures.[21] And third, Mendel both composed and wrote the petition in its final version.[22] The decisive argument in choosing among the three possibilities, as far as the *writing* of the document is concerned, must be the results of a handwriting analysis. Here the answer is unequivocally "Mendel."[21,23] He, however, had not one but two handwritings,[21,23] one in making notes and drafts for his personal use and another in official letters and documents. The former handwriting was a horrible scribble, which even experts have a hard time deciphering. The latter was a calligraphic handwriting in the so-called *gothic* style, in which the scribe was relatively free to embellish individual letters with minute decorative elements. These embellishments distinguished the scribes and so facilitated their identification. In either case, the handwritings of Klácel and Mendel were not too difficult to differentiate.[21] Similarly, in deciding who composed the petition, the style of the composition must be examined. It so happens that the writing styles of the two candidates are so different that the choice between them is unequivocal as well. In the petition there are certain words and phrases that Mendel would never use but that point at Klácel's authorship. The style in which the petition is written is that of late Romanticism (see Vol. 1 Chap. 1): pompous (*in the name of humanity*), bombastic (*buried alive*, *mightily resounding trumpets of resurrection*), and replete with strong words and phrases (*prisoners, slaves, dungeons, criminals, lunatics*). It is passionate, but wordy, repetitive, and vague. It is vintage Klácel and so unlike Mendel!

As for the second question, the prevalent opinion is that the petition was "a plea for freedom of teaching,"[19]) but this seems to be a too restrictive interpretation of the document's aim. Although, the freedom of teaching must have laid heavily on Klácel's mind, the other signatories had no such ax to grind. And anyway, this subject is not the centerpiece of the petition. The centerpiece concerns the friars' civic (constitutional) rights.[22] Since the emergence of the city-states in ancient Greece, citizenship has meant a full membership of an individual in a community such as a city, a nation, or a particular governmental system. In its full form, citizenship entails both rights and duties. It was the civic or constitutional rights that the friars demanded from the Constitutional Assembly. They wanted the Assembly to grant them the same rights that all the other citizens of the monarchy enjoyed. In return, they offered to serve the monarchy as teachers for the rest of their lives.

Is it important to know, who composed and who wrote the petition? We think it is because it provides clues to the personalities of both Klácel and Mendel, but especially of the latter. Was Mendel a rebel or even a revolutionary, as some historians suggest?[22] If he not only wrote but also composed the petition, one might perhaps be justified to think of him in those terms, even though nothing else we know about his character and behavior supports such a contention. On the other hand, if he merely rendered in a calligraphic handwriting what someone else (Klácel) composed and then signed it, he might, perhaps, be taken for a rebel but not a revolutionary. He might be thought of as a person participating in a street demonstration carrying a placard that somebody placed in his hands (Fig. 6.1), but not as a person on a barricade with flag in one hand and gun in the other. The fact that he signed the petition reveals at least two things about Mendel: that he had courage and that he was unhappy about ending up in an abbey.

In its content, if not its form, the petition is a remarkable document. In essence, what the six friars declare in it is this: We became friars and priests because we wanted to be teachers and researchers, but saw no other possibility of attaining our goal than entering a monastery. We find both our position and the monastic way of life degrading, humiliating, and dehumanizing. As friars, we are deprived of rights all other ordinary citizens have, we are forced to lead a life we resent, and we are not allowed to serve the society the way we want. We demand a change in the laws of the country that would make us equal to all free citizens. We want to devote ourselves to teaching and research without being burdened by duties that have nothing to do with our chosen occupations.

The six friars were no angels, but like the angels of Christian mythology, they revolted against no lesser authority than God Himself.[24] For, although the petition was addressed to a civil authority, it concerned the *Reichstag* only to the extent that it pertained to questions of a relationship between church and state. Its real addressee should have been an ecclesiastical authority, ultimately the pope, as God's representative on earth. In the Austrian Empire, a Christian had to obey two different sets of law, the civil law of the secular government and the canonical law of the Roman Catholic Church. While the canonical law affected an ordinary citizen only peripherally, it dictated the entire lifestyle to a cleric. A monk's life was, in addition, regulated by the monastic rule of his particular order. All those things that the

Fig. 6.1 The revolt of the friars

petitioners complained about (confinement to a cell, dress, exclusion from family life, obedience) were prescribed by either the canonical law or the monastic rule and had nothing to do with civil law. Even things like the denial of the right to serve as a witness or the ban on inheritance were a matter of concern for the church or the order and not for the civil authorities. The church could have had all of these laws and regulations changed, if it wanted to, but it showed no such desire. Hence, the friars should have petitioned, not the *Reichstag*, but the pope, if anybody. Similarly, their argument about being misled into a life they then loathed would not have held water legally. For, not only were they adults when they joined the abbey, but they even had a yearlong trial period, after which they could still have backed out, had they realized that the monastic life was not for them. And finally, there were then and probably always will be devoted individuals who *want* to lead a life of a Roman Catholic priest or monk with all its clauses and restraints, and they should have the right to do so. The petitioners' real complaint is not, however, about the status of the clergy and monastics in general, but about the specific circumstances under which they ended up being priests and friars. What the petition did not state explicitly, but implied implicitly, was that the state should make provisions for talented, but impoverished young persons to have an access to higher education without the necessity of signing their souls to the church. This message, however, did not get through to the clerks who handled the petition in the *Reichstag*, and so they assigned it to the committee for the state-church relationship, rather than to one concerned with social affairs.

From the church's viewpoint, the petitioners had committed a crime: They called into question several dogmas, on which the church was founded, the dogma of celibacy among them. Moreover, instead of taking their grievances to the nearest ecclesiastical higher authority (the abbot or the bishop in this case), they went to the civil authority and so broke the code of monastic behavior. In the past, challengers of ecclesiastical dogmas could have been burned at the stake; hence, by signing the petition, the six friars took a great risk. They could have reckoned, perhaps not with an autodafé, but with severe repercussions nevertheless. We must presume that the signatories must have been aware of the danger, yet they chose to ignore it: Why? The six friars differed in age (the youngest being the 23-year-old Cygánek and the oldest the 43-year-old Klácel), the length of residency at the abbey (ranging from five to 21 years in the case of Cygánek and Klácel, respectively), place of birth (Bohemia, Moravia, or Silesia), ethnicity (Czech or Germanic), and their interests and experience. They did have some features and concerns in common, however. They joined the abbey not because of a religious calling but because it seemed to offer them the opportunity to pursue their secular interests. They were all interested in teaching. And they might all have been discontented with the monasterial tedium and may have been looking for ways of either escaping it altogether or at least minimizing it. In August 1848, Klácel still nurtured hopes of landing a professorial post at a university and cutting his ties to the abbey; he achieved the latter ultimately by immigrating to the United States. In 1851, Gabriel, who was about to become the director of Brno's *Gymnasium* fused with the Philosophical Institute, welcomed his transfer to Těšín to become a *Gymnasium* director there—a position that kept him away from the abbey for most of the year. Facing the prospect of having to return to the abbey after his retirement, he chose to separate from it legally and in 1868 became a secular (i.e., non-monastic) priest. As such he retired to Vienna in 1874 to take on the function of a *k. k. Schulrat* (imperial-royal school inspector). Fogler busied himself with his teaching duties at Brno's *Gymnasium.* Cygánek taught religion in Brno from 1851 on, but died ten years later at the age of 36. Lindenthal similarly sought refuge in teaching religion and helping Mendel with his experiments. And Mendel? In August 1848, he was, as we already know, entering into a crisis elicited by his assumption of pastoral duties. Taking care of souls was not why he joined the abbey, and so he was in full agreement with the content of the petition. Although he had no plans to leave the abbey, he hoped he would be allowed to devote himself to teaching and research, and he expressed this hope by adding *Lehramts-Candidat* (teaching-profession candidate) under his signature.

In 1848, there were nominally 16 friars in the Old Brno Abbey and since August was vacation time for those involved in teaching; they all should have been present and have had the opportunity to sign the petition. Ten of them, however, chose not to do so. Why? Undoubtedly, each friar had his own reasons. Napp, for example, could not have signed it, even if he had agreed with it (which he probably didn't), because his name on it would have made it an official document of the abbey. Nevertheless, and this is significant, he did not stop the petitioners, since it is quite unthinkable that they would have sent it without his knowledge.[25,26] The nonparticipation of the other abstainers was probably motivated by one of three reasons. Some may have abstained because for them service to God and everything that came with it was their calling. Others may have been sympathetic with the cause, but were put off by the petition's aggressive tone. Even in the revolutionary atmosphere of 1848, it must have been too orotund for suave stylists like Bratránek. Still, others might have been attuned to the petition, but were simply afraid to sign it.

The petition was mailed, received in Vienna on August 11, 1848, duly registered under No. 318, taken to Kroměříž when the *Reichstag* moved there, discussed in the committee for church-and-state affairs, and forgotten.[27] That the *Reichstag* took no action on it is hardly surprising. As stated earlier, it was really not in its competence to meddle in what must have been seen as an internal affair of the church. Moreover, the *Reichstag* had more pressing issues to deal with than the discontent of six monks who realized too late that they made a mistake by taking the habit and becoming priests. The committee for church-state relationships concerned itself with issues ranging from the influence of the church on the educational process to the abolition of religious orders. The views of its members on these issues varied so greatly that reaching a consensus on any of them proved to be all but impossible.[28] What *is* surprising, however, is that the church, after it regained its power following the quashing of the revolution, made no attempt to punish the rebels of St. Thomas Abbey. Either the church authorities were unaware of the petition's existence or found it so embarrassing that they did not want to draw attention to it by making a scandal out of it. The first possibility seems rather unlikely, considering that the church was (in) famous for its spy net. Indeed, it might not have been a mere coincidence that after regaining their power, the church authorities wanted to close the Old Brno Abbey, as we shall describe later.

The Bach Decade of Darkness

The 1848 revolution stunned the Habsburgs. Their first reaction to it was one of disbelief famously epitomized by the inane uttering of the feebleminded Emperor Ferdinand when they reported to him that people are erecting barricades in the streets of Vienna: "But are they allowed to do that?" The next reaction of the imperial family was to run, and they ran like a flock of scarred hens who have noticed a hawk hovering in the sky above them. Only when they calmed down somewhat did the members of the inner family circle put their heads together and begin to think and act rationally. Their first rational decision, reached in November 1848, was that Ferdinand had to go and that the person to replace him was young Franz Joseph, the son of Archduke Franz Karl of Austria and Ferdinand's nephew. Franz Joseph (1830–1916; see Fig. 3.3)[29] was 18 years old when he ascended the throne in December 1848, ambitious, energetic, and conservative. Not feeling bound by the concessions Ferdinand made to the revolutionaries, he began rescinding them one after the other. Since it was Ferdinand and not he, Franz Joseph, who swore to respect the decisions of the Constitutional Assembly, he felt free to ignore them. He was in full support of his hard-liner ministers, who began to use harsh methods of dealing with the rebels. Quickly the conditions in the empire began to return to their pre-1848 state. The military subjugation of the uprising in Prague, Buda, and Vienna and the dissolution of the *Reichstag* signaled an end of the revolution in the Austrian Empire. True, Metternich was gone, Ferdinand abdicated, peasantry was freed of serfdom, and a few inconsequential concessions the imperial court was forced to make remained in place, but other than that not much had changed. Prince Felix Schwarzenberg (1800–1852) and Baron Alexander von Bach (1813–1893) took over Metternich's place, the former as prime minister and the latter as minister of interior, and both were at least as reactionary as their predecessor. If the people called the feebleminded Ferdinand "good-natured," they were hard-pressed to find any strain of goodness in his successor. Although he was only 18 years of age, Franz Joseph was as impervious, inflexible, and uncompromising. His form of governing is sometimes called "neoabsolutism," but the only thing new about it was that he, even more than any of his predecessors, wanted to control everything that went on in his empire. To consummate this aspiration, he voided, in 1851, the octroied constitution, which he offered initially in place of the democratic constitution prepared by the *Reichstag* and which was entirely toothless in terms of limiting his authority. Then he abolished all regional diets and set up a government that he ran via Schwarzenberg and Bach, both of whom were attuned to his idea of governing the country. Ironically, Bach, a lawyer by profession, was initially a revolutionary, who participated in the March uprising in Vienna, but shortly afterward changed sides. In Franz Joseph's government he first became minister of justice and then, after Schwarzenberg's sudden death in 1852, minister of interior. In the latter function he controlled an elaborate bureaucratic apparatus of "Bach's hussars," so nicknamed in reference to the methods they used to extract taxes, recruit young men in the army, and exercise judicial and administrative duties.[30] Backing the "hussars" was the overt and the secret police and the army. So oppressive was the entire interval from 1849 to 1860 that historians call it "Bach's period of darkness." The three characteristics of the period were restrictions of civil rights, aimed at quashing people's democratic aspirations; vigorous Germanization as an antidote to all national movements; and the return to the Catholic Church all of the powers it had before the reforms of Joseph II. The *Concordate* of 1855 empowered the church to intrude into all aspects of people's lives and to fully control the educational process. It is in light of these developments that an affair termed an *Apostolic Visitation* of the St. Thomas Abbey must be viewed. Before we turn to it, however, we must make two digressions, the first explaining briefly how the empire's name came to change in 1876 and the second introducing Brno's bishopric, which played a central part in the *Visitation*.

The Birth and Death of Kakania

The Habsburgs had learned very little from the 1848 uprising. They refused to take into account that the empire was a conglomerate of nations brought together involuntarily and unwilling to lose their identities through forced Germanization.

The repressive measures that Franz Joseph gradually reinstituted had an opposite effect from what he expected: They strengthened rather than weakened the nationalistic movements in the different lands, especially in the largest of the nations—Hungary. The Magyars were dead set to get, if not their independence, at least their autonomy within the empire. Recovering from the military defeat of the revolution, they turned to methods of passive resistance in the form of noncooperation with the Austrians such as the refusal of collecting taxes for them. The methods worked in that the Magyars were achieving gradually various degrees of autonomy. Their great moment came, however, in 1866 when the Prussians defeated the Habsburgs in the so-called Seven Weeks War, and the weakened government in Vienna was in no position to reject the demands of the Magyars. A compromise was then worked out, which restored Hungary to its pre-Habsburgs borders and so included several non-Magyar peoples. Hungary became an equal state to the rest of the Austrian Empire, the two states sharing a monarch who was the emperor of Austria and the king of Hungary. The two states also had a common army and common ministries of war, foreign affairs, and finance. In all other respects, the two states were autonomous, with the Hungarians being governed by their own parliament. From then on the Dual Monarchy was to be called Austria Hungary and all issues pertaining to it to be referred to as *kaiserlich und königlich* or "Imperial and Royal," the abbreviation of the German title being *k. und k.*

Now, in German the letter "k" is pronounced "kha" as in Kalamazoo, so that by leaving out the "and," the "kk" is pronounced as "khakha." In German, as well as in several other languages, *kaka* is a children's word for excrement. Knowing all this, the Austrian novelist Robert Musil (1880–1942) set his unfinished three-volume masterpiece *Der Man ohne Eigenschaften* (The Man Without Qualities) in *Kakanien* (the Kakaland) on the eve of the First World War.[31] As the novel's main character lives in Vienna, Musil leaves little doubt that the true identity of *Kakanien* is the Austrian-Hungarian Monarchy, and of his rather low opinion regarding the latter. He was not alone in his ironic attitude toward the monarchy, but by the time he wrote his final novel, *Kakanien* was gone, swept from the stage of history by the first great, worldwide conflagration. Through it, the nations of the former empire, including the Czechs, Moravians, Silesians, and Slovaks, finally found their freedom and independence.

The Bishop's Throne in Brno

The Roman Catholic Church divided the world into ecclesiastical territories, which did not always match the land's political divisions.[32,33] It organized the territories hierarchically in correspondence to the ranking order of its clergy. The smallest unit of the division was a *parish* with its spiritual center, the *parish church*, administered by the *parish priest* (*Koordinator* or *Pfarrer*) and his assistant (*Kooperator* or *Kaplan*). A group of parishes formed a *diocese* headed by a *bishop* and two or more dioceses made a *province* supervised by an *archbishop*. The system of provinces was presided over by the *pope* and his advisers, the *cardinals*. Another word for diocese is *bishopric*, which, however, stands also for the bishop's office or his residence. The bishop had his seat, his Episcopal throne or the *cathedra* in the form of an armrest chair, in the apse, behind the high altar, of the largest and usually most ancient church in the diocese—the *cathedral church* or *cathedral* (the adjective becoming a noun). Early on in the history of Christianity, a bishop generally resided in the capital city or *metropolis* of a particular political province. Later, however, the popes also established bishoprics in nonmetropolitan cities. The *metropolitan bishop* or *metropolitan* then became an archbishop, to whom the *suffragan bishops* in his province were subordinated administratively. The archbishop had direct authority over his own diocese (i.e., *archdiocese*) and supervisory authority over his suffragan bishops. He also presided over the meeting of bishops, the *synod*, which dealt with administrative matters, discipline, and judicatory issues. In addition to parish churches and cathedrals, there were also several other kinds, of which we should mention only, the *collegiate church*, administered by a group of clerics living together under a particular ecclesiastical rule. Both the group and the building in which it lived came to be called *college*. Similarly, both the rule and the priest living under it became the *canon*. Since reading a chapter from the book of rules customarily opened the meetings of the group, the groups became known as *chapters*. And since similar groups of clergy were also associated with cathedral churches, a distinction had to be made between *collegiate chapters* and *cathedral chapters*. The chief dignitary of a collegiate or cathedral chapter was the *provost*. The canons lived on *benefices*, an income awarded for the services rendered (essentially administration of sacraments). The two main sources of income were foundations and estates donated to the church. The latter source was called *prebend* and the post supported by it *prebendary*.

The first bishopric in the Czech lands was founded in 973 in Prague. The diocese encompassed both Bohemia and Moravia and was part of an ecclesiastical province administered by the archbishop of Mainz, in the state of Rheinland-Pfalz, which, like the Czech lands, was then part of the Holy Roman Empire. Moravia became a separate diocese in 1063, when the pope founded a bishopric at Olomouc. Both the Prague and the Olomouc bishoprics remained, however, under the control of the archbishop of Mainz. Entreaties of the Czech rulers to the pope to establish an archbishopric in Prague remained unfulfilled until 1344, when Charles IV with his diplomatic skills succeeded where his predecessors failed. The Prague archbishop controlled both the Bohemian and the Moravian dioceses. It then took 433 years before another pope gave his blessing to the foundation of a bishopric in Brno and simultaneously promoted the one at Olomouc to archbishopric.[34,35] The first of eight bishops in Brno (and their periods in office) in the time interval from 1777 to Mendel's death in 1884 were Mathias Franz Reichsgraf von Chorinsky (1777–1786), Johann Baptist Lachenbauer (1787–1799), Vinzenz Joseph Franz Sales Graf von Schrattenbach (1800–1816), Wenceslas Urban von Stuffler (1817–1831), Franz Anton Gindl (1831–1841), Anton Ernst Schaffgotsch (1842–1870),[36] Karl Nöttig (1870–1882), and Cardinal Franz Salesky Bauer (1882–1904).[37] Although all of them, with the exception of Gindl, were born in either Moravia or Bohemia, they came, as their names suggest, from Germanic or

Germanized families. This fact may have been one reason why Brno's bishopric was generally on the Germanic side in nationalistic disputes. Paradoxically, the one exception, the Austrian-born Gindl, was most sympathetic to the Czech cause. He not only demanded that all clerical candidates learned Czech, but he himself spoke the language fluently. The political orientation of Brno's bishops ranged from a mild antiliberalism to vigorous ultraconservatism. This tendency may have had to do with their origin from noble or at least affluent families. In this regard, however, they did not differ from the higher clergy elsewhere in the Austrian Empire. Here again, however, Gindl was an exception in that he was of a rather tolerant disposition. As long as Gindl was in office, the relationship between the bishopric and the St. Thomas Abbey of the Napp era was quite good. It cooled off rapidly, however, when Schaffgotsch succeeded Gindl. Schaffgotsch was in many respects the opposite of his predecessor and of course also of Napp. A scion of an old aristocratic family, which branched out into different parts of the Austrian Empire, Schaffgotsch (Fig. 6.2) was arrogant, intolerant, and reactionary. Two recorded incidents illustrate his character. In November 1856, it was reported to him that a miracle occurred in Brno's Sisters of Mercy Institution. A blind woman was claimed to have regained her sight by kissing a vial containing oil purportedly used by St. Walpurga. The gullible bishop pounced on the incident and on the basis of mere hearsay produced a document certifying that a genuine miracle had taken place, which would henceforth be celebrated yearly on its anniversary. The church's officialdom kept their hands off of the affair, but to the bishop any means were acceptable as long as they served the promotion of religious beliefs. The second incident concerned a celebration in Slavíkovice, a small village in southern Moravia near the Austrian frontier. It was in this village that Joseph II demonstrated his qualification for the title of "the people's emperor" by taking over a plough from a peasant and ploughing a few furrows. The village honored the event by erecting a statue depicting the event. In 1869 the village organized a celebration to which it invited Archduke Karl Ludwig, as a representative of the imperial court, and Bishop Schaffgotsch, as a representative of the church. The archduke showed up, but the bishop, to whom Joseph II was a heretic, did not. The court interpreted the incident as an affront to the imperial family and the bishop had to apologize for his behavior. The bishop was a man for whom promoting religious fraternities and societies was more important than setting up schools and distributing books for public education.[37] He was also a man who would attack monks for earning a few *Kreuzer* of pocket money (see below), while he himself accumulated a considerable wealth, in part through gambling (lottery), and then bequeathed much of it to his relatives rather than to needy humanitarian organizations. We must also keep in mind that it was Schaffgotsch who dismissed Klácel from his teaching post and banned Křížkovský's public performances. The bishop may have had a sheep in his coat of arms (German *Schaf* = sheep), but he certainly was no lamb.

Even for the smooth and diplomatically skilled Napp, it had been difficult to get along with Schaffgotsch. The bishop had absolutely no empathy for Napp's efforts to make the abbey a center of enlightenment within the framework of a religious creed. To him, the sole mission of a monk and a priest was a spiritual communion with God and leading others to such communion. He beheld therefore

Fig. 6.2 Anton Ernst Schaffgotsch (1804–870), bishop of the Brno diocese (1841–1870)

the St. Thomas Abbey for a breeding ground of liberalism and a playfield of misguided attempts to secularize a religious institution. It may seem that Napp should not have been concerned about what the bishop thought about the abbey because they both held equivalent positions in the ecclesiastical hierarchy. In reality, however, Napp depended on Schaffgotsch in matters such as the ordaining of his friars. The bishop also liked to regard the abbey as being under his jurisdiction since it was located in his diocese. Although formally, the Abbot's only higher ecclesiastical authorities were the order's general and the pope and not the bishop, a decree issued by Franz I in 1802 could be interpreted as giving the bishop some supervisory power over the abbey from the standpoint of the civil law. The question then was whether the emperor's decrees could override the pope's edicts; the church, until then, had always insisted that they could not. Before 1848, the bishop's hostility toward the St. Thomas Abbey restricted itself to friars like Klácel and Křížkovský, who fell directly or indirectly in his sphere of influence. But after the revolution, once the imperial government got the situation under control, the bishop began to mount a full-scale attack on the abbey, in congruity with the general trend toward the restoration of the power the church once had. The goal of the trend was to bring back the good old times before Joseph II wreaked havoc onto the Church Empire. Its two specific aims were, first, to make provisions against similar intrusions of secular rulers into the internal affairs of the church and, second, to repair the damage the meddling of Joseph II and Franz I had caused. A formal expression of the attainment of the first aim was the 1855 *Concordat* between the pope and the emperor, by which the latter essentially promised not to interfere with the ecclesiastical activities of the church in his empire. One part of attaining the second aim was to reform the monasteries.

Recall that Joseph II abolished all religious orders committed solely to a contemplative lifestyle and allowed only those orders that were involved in activities benefiting the general public to remain. Later Franz I decreed that certain orders, the Augustinians among them, must take upon themselves the teaching responsibilities in the lower and middle levels of the empire's school system. The orders complied and in the process "secularized" themselves. The St. Thomas Abbey provided a good example of what secularization meant. Although its friars were Augustinian Hermits, in the nineteenth century nobody in his right mind expected them to lead a life of hermits. The church, however, did expect them to remain confined to their cells except when performing pastoral duties in the parish. Obviously, this expectation clashed with their teaching responsibilities, which required some of them to live outside of the abbey for most of the year and others to spend most of the day in the secular environment of the school. Napp did his best to fulfill the emperor's order and could be (and was) proud of his accomplishments. But by doing so, he brought upon himself the displeasure of the church authorities, which viewed his accomplishments as a violation of the monastic rule. Similar conditions, if perhaps not as extreme as in the St. Thomas Abbey, also existed in other Austrian monasteries, and after the revolution, the authorities began taking steps to rectify the situation.

An Apostolic Visitation

A cardinal, Prince Friedrich Johann Jacob Celestin von Schwarzenberg (1809–1885), the younger brother of the Austrian prime minister Felix Schwarzenberg (1800–1852) and a relative of many other Schwarzenbergs in high positions (as well as Alfred Windischgrätz, his brother-in-law), became the central figure in the anti-secularization movement.[27] Friedrich Schwarzenberg chose a clerical career out of a religious devotion and under the influence of his mentors. His pedigree, which extended back to the eleventh century; his political connections; and his religious zeal all conspired to assure him a steep ascent in the church hierarchy. Ordained a priest at 24, he was appointed bishop of Salzburg, Austria at 26, elevated to cardinal at 33, and installed as an archbishop in Prague at 41. His religious fervor, combined with the power of his office, made him the driving force of the postrevolutionary church revival crusade. He revitalized bishopric conferences (synods) and transformed them into a vehicle of the crusade. It was at these conferences that the bishops formulated the crusade's program and agreed on the steps to be taken for their realization. The first three conferences met in Vienna in short succession, the first from August 31 to September 12, 1848, the second from April 29 to June 20, 1849, and the third from April 6 to June 17, 1856. In attendance were 29 bishops and four proxies at the first two conferences (which bishops from Hungary and the Italian provinces did not attend because the political situation in their lands was too unstable) and 66 prelates at the third meeting, which included not only bishops from all the lands of the empire but also representatives of the Greek and Armenian Churches. One issue brought up in the program of the first conference was what the bishops regarded as a general decline of the monastic spirit. What the bishops meant by this was that the monks lived less and less according to the rules they had vowed to follow, devoting themselves more and more to the execution of civil duties, to the detriment of their pastoral activities. In the bishops' opinion, the monasteries were gradually turning into institutions of the Austrian government instead of being establishments of the Roman Catholic Church. To underpin this claim, for which each of them had plenty of anecdotal evidence, but no solid support, the bishops decided to petition the pope, asking him to order a general apostolic visitation of the monasteries in the empire. The pope, Pius IX, took his time to respond to the request, but after he assured himself that the emperor, Franz Joseph I, not only did not object to such an act but welcomed it, he ordered the visitation of all 380 monasteries in Austria proper, Bohemia, and Moravia. And he empowered Cardinal Schwarzenberg to execute the order.[38]

Naturally, the cardinal was not going to waste his precious time on inspecting 380 monasteries himself, so, as an archbishop, he delegated this assignment to his bishops. For the bishop of Brno, he reserved the task of inspecting monasteries of the Augustinian Hermits in the Prague and Brno provinces and the Krakow diocese of present-day Poland. The list of the monasteries to be inspected included the St. Thomas Abbey of Staré Brno. Whether the inclusion was legitimate is contentious. On the one hand, the bishop, at best coequal to the abbot in the pecking order of the hierarchy, had normally no authority to investigate the latter; on the other hand, the pope could presumably empower him with such authority. The problem was, however, that the pope empowered the cardinal, who then authorized the bishop: Was this latter step legally sound? Napp thought not and as soon as the bishop served him the notice of the visitation in February 1853, he protested.[39] On April 9 he sent a letter to the cardinal, in which he politely, but firmly, objected to a visitation by a bishop. He pointed out all the special privileges that past popes granted the St. Thomas Abbey and how that placed the abbey in a unique position among all the monasteries of the empire, including the Augustinian Hermit monasteries in Rajhrad and Nová Říše in Moravia. Specifically, he called the cardinal's attention to the abbey's exemptive status, which excluded it from visitations by any provincial clerical authorities. He also reminded the cardinal that the abbey became heavily involved in teaching and research on direct orders of the emperor, and stressed that these activities are in full compliance with the mission of the Augustinian order.

The letter remained unanswered and the bishop proceeded with his plans for the visitation. Prior to actually visiting the abbey, he requested that all its members answer in writing a series of questions and confirm the truthfulness of their answers by signing the questionnaire. The questions were inquisitive, intrusive, nosy, and humiliating. How much time do you devote to secular activities, pastoral duties, and prayers? Do you own any private possessions? If you earn money, what do you do with it? Have you kept your chastity vow? If not, how and how many times did you breach it? What do you think of the way the abbot is running the monastery? On the bishop went, as if he were dealing with a group of unruly teenagers and not adults some of whom had distinguished themselves profoundly through their accomplishments. For some reason, the date of the visitation had to be postponed until the summer of the next year, but on June 7 and 8, 1854, the bishop and his entourage descended on the abbey. The word "visitation" has two meanings: it stands not only for "an official visit with the purpose of an inspection" but also for a "severe trial or affliction coming down on a person or an institution." Undoubtedly, the friars of the St. Thomas Abbey interpreted the bishop's visit in the latter sense. If the bishop was, as the Christian mythology asserts, the successor of the apostles, the friars might have wished to be visited by the devil's disciple instead. The bishop interrogated all friars individually and as a group, inspected their living quarters, and went through the accounting books demanding receipts for all expenditures, no matter how small. In general, from the way he conducted the inquiry, the friars must have realized that he was bent on dissolving the abbey and was looking for pretexts for such an act. To counteract this anticipated move, on June 8, on the second day of the visitation, Napp assembled the friars and proposed to petition the cardinal for a change in the order's denomination from Augustinian Hermits to Augustinian Canons (*Chorherren*). Since he surmised the inquisitor's main charge against the abbey would be that the friars had become too involved in worldly activities and so ceased to be hermits, the change Napp was suggesting would take the wind out of the bishop's sails. As the Canons Regular of St. Augustine, they would retain the rule of their order, but without the requirement for withdrawal from worldly affairs. The friars approved the proposal, and on the same day, Napp penned a letter to the cardinal, in which he restated the arguments he put forward in his previous letter and added the request for changing the order's status as a way of circumventing the apparent conflict that arose from the emperor's decree to take on teaching responsibilities and the order's regulation to eschew worldly contacts. All 16 friars then signed the letter and the Abbot sent it off. In the meantime, Schaffgotsch prepared his report on the situation at the St. Thomas Abbey and on September 7, 1854, sent it to the cardinal. Space limitations do not permit us to provide here the full English translation; instead, below we provide a conspectus of the report to illustrate the attitude of the high clerical authorities toward this remarkable institution at Staré Brno.

In the introduction, the bishop informs the cardinal that he carried out the visitations at the Augustinian Hermit monasteries in the Prague and Olomouc provinces, as well as in the Krakow diocese, as instructed. He can substantiate the cardinal's suspicion that in these monastic communities, the devil (*Menschenfeind*) had sowed a profusion of weed in the wheat field. There is therefore an urgent need to pluck out the evil with its roots so that the monasteries can once again become the strongholds guarding the struggling church. Regarding the St. Thomas Abbey, the bishop has this to report. Its Abbot, Cyril Napp, is certainly intellectually an outstanding man, as his many titles and public functions attest to. He has, however, become so preoccupied with his political responsibilities that he has little time left for running the monastery. Neither he nor his friars have the desire to adhere to the order's rule they vowed to follow. In the Abbot's view, the mission of the abbey is to conduct scientific research along with pastoral work. Under his leadership, the monastery has reached the apogee of secularization, so much so that it has now applied for a change of its status to Augustinian Canons. Such change, however, would be contrary to the decree of the Holy Council of Trent, which specifically prohibits a transfer of monastics from a more to a less austere order. The Abbot apparently believes that just by substituting one dress for another he and his friars would gain greater liberty. Under a superior with such views one can hardly expect to find a sound religious community. The Abbot admits that neither he nor his friars strictly adhere to the order's constitution in certain points, for example, the dress code, because they regard them as overhauled by the general progress of the society at large, and he believes they should be exempted from these regulations. The Abbot claims that he has had no cases of insubordination: naturally, when he sanctions behavior contrary to the regulations! Each of the friars claims insolently that he does not have any private possessions and is unaware of anybody in the community who does. Yet, it is generally known that at least some of the friars in public functions are on fixed salaries while others get paid for their irregular public activities. The friars are apparently free to spend these revenues any way they please without any accounting obligations. They thus violate Chaps. 9 and 10 of the 1589 Council of Cambrais, which prescribe a vow of poverty to monastics as the source of indescribable sweet peace of the soul.

Unable to extract from the friars any admission of infractions against the chastity vow, the bishop focused on the observance of the canonical hours. As described in the preceding chapter, Napp had instituted a rather lax attitude toward this prescribed set of prayers for each ecclesiastical division of the day. Some prayers were included in the morning mass, while others were left at the discretion of the individual friars. This relaxation was necessary if he expected the friars to do something more than spend most of the day in prayers. He could hardly expect a school to adjust its schedule according to the canonical hours of a monk, for example. Yet, in his report, the bishop harangues about the friars' violation of the hours as if the monks were committing a mortal sin. Similarly, he devotes a whole page to complaints about the friars' limited knowledge of the order's constitution and virtual ignorance of the various decrees issued by the popes regarding monastic life. All they know, he says, is the so-called praxis, which is a mere synopsis of the major rules, omitting many detailed regulations. As a consequence, the friars run around in civil clothing most of the time, fasten rarely, ignore the rule of monasterial silence altogether, and violate many other regulations prescribed for Augustinian monks.

The bishop then goes through the list of the 14 members of the community (not counting Napp), beginning with the four full-time teachers stationed outside of Brno (Wieser, Alt, Gabriel, and Bratránek). Of these four, the bishop has this to say: "The four…would undoubtedly be quite unhappy if they would be recalled back into the monastery and were forced to live life according to the rules and constitutions of the Augustinian Hermits. The last three named would surely quit the monastery and leave as if leaving the squalor of a prison." And as far as Wieser is concerned, "he never learned the responsibilities of his monasterial profession and cares little about them." In this assessment, the bishop was probably right. Turning to the six priests consigned full-time to pastoral care (presumably Vorthey, Šembera, Winkelmayer, Cygánek, Lindenthal, and Rambousek), the bishop notes that about 12,000 Roman Catholics live in the parish in which they serve and that there are three schools and one general hospital. Consequently, the six are taxed heavily by all the difficult tasks they face. The bishop then adds: "They believe to be doing their best by taking care of the parish, entrusted to them, in the style of secular clergy and by staying out of trouble." He apparently insinuates that this is not enough. Because they are so overburdened with pastoral work, while the other friars study, teach, give concerts, or spend time on other worldly activities, the six have no time left for fulfillment of their monasterial obligations. Consequently, they violate the rules of monasterial life just like the other friars. Of the remaining four "other" friars, he mentions two specifically—Fogler and Mendel—the former because of his full-time teaching obligations at Brno's secular institutions, and Mendel because "he studies profane sciences at a worldly establishment in Vienna, at the expenses of the monastery, to become a professor of such sciences at a state institution." The bishop saves his most damaging ammunition for last, however. He explodes it, by listing the transgressions of past and present friars of the St. Thomas Abbey, transgressions for which they had to be censored formally by the bishop's office. Rambousek had to be reprimanded for "bathing in a public place almost naked before the eyes of strollers." Křížkovský, about whom the abbot rants that he "conducts works to a great applause of the public" and had to be reminded to restrict his activities to church music. Klácel "was forced to quit the (Philosophical Institute) because he laced his lectures with pantheistic fantasies. In 1848, (he) participated in the Slavic Congress in Prague during the Whitsuntide days, and afterward edited *Moravské noviny* (Moravian News), which had repeatedly provoked bishopric condemnation." Philipp Neděle "...who died in the meantime, had to be relieved of his professorship in Biblical studies at Brno's Philosophical Institute for interpreting Holy Scriptures according to erroneous ideas of the rationalists." And Aurelius Thaler, now also deceased, "had to quit his professorship in mathematics at Brno's Philosophical Institute because he uttered blasphemic words in an intoxicated state in front of his students."

The bishop concludes his report with a harsh condemnation: "In a word, in the house tending the Rule of St. Augustine reigns a secular spirit which the few lappets of the Augustinian habit fail to cover up. Any hopes that the spirit could be exorcized and the order returned to a conscientious observance of its rules and constitutions must be given up. With its current individuals, it would be impossible to turn the community around to form a real monastery. Among these individuals, not one would be found who could be entrusted with training novices and so educate a new generation of Augustinians" He recommends, therefore, to dissolve the monastery and use its yearly revenue of some 30,000 florins to support another religious order. The current abbot should be retired on a decent pension, admission of new novices stopped immediately, and the fate of the remaining friars decided later.

The report as a whole was a testimonial to a clash between two personalities empowered with equal authority and rubbing shoulders territorially: the bishop asserting jurisdiction over his diocese and the abbot defending his privileges. It was, however, also more than that; it was a confrontation of two opposite outlooks, one held by an inflexible, dyed-in-the-wool conservative and the other by an open-minded, progressive scholar. A point on which the two differed was the function of a monastery. While both protagonists agreed that in the case of the parish-entrusted St. Thomas Abbey, the monastery must provide adequate pastoral services to the parishioners, they disagreed on the issue of its other functions. The bishop stood fast to the traditional, medieval view that the purpose of a monastery was worship of God in the form of leading a life according to the rules of the particular religious order. From this standpoint, his critique of the abbey was not unwarranted; all the transgressions listed in his report were indeed in violation of the Augustinian Rule. The enlightened abbot saw things differently, however. In his view, the function of the abbey was a service to the society in the form of teaching, research, and participation in public life. Realizing that such a service was not possible without bending the rules and cutting corners in the prescribed form of life, he relaxed the abbey's discipline. He may have believed, although he would have never admitted it openly, that contributing to human knowledge was a more proper way of worshipping God than reciting mind-numbing incantations exacted by the canonical hours. Oddly enough, both Schaffgotsch and Napp could ground their views on church authorities: the bishop on the decrees of the Holy Council and popes' edicts and the abbot on the teachings of St. Augustine. Unfortunately for Napp, the church authorities of his time were on Schaffgotsch's side.

The cardinal studied the report (which took him almost one year) and endorsed it. In July 1855, he sent letters to the monasteries informing them about their shortcomings established by the visitations and asking them to take a stand to the criticisms. The letter to Napp recited all the deficiencies listed by Schaffgotsch, but said nothing about a closure of the abbey. Napp and ten of his friars responded to the letter only after a reminder from the cardinal. The tone of their response was very different from their previous letters. The latter were polite, but argumentative, defiant, and almost recalcitrant, haggling on the basis of the special privileges granted to the abbey. In the new letter, though, there was no trace of such a spirit. The letter did not even attempt to rebut the charges. Instead, the undersigned accepted humbly the criticism and then in 12 points described how they were going to remedy their failings. Apparently, Napp must have gotten wind that this time the situation was quite serious. He immediately instituted an army-like discipline in the abbey, and life became much tougher for all the friars in residence. The cardinal, however, disregarded the show of goodwill from Staré Brno and on December 27, 1855, sent the report along with the rest of the material concerning the visitation to the Holy Council in Rome. In an accompanying letter, he seconded all of the bishop's recommendations, including the proposal to abolish the abbey. To these he added his own proposal to disperse the friars (with the exception of Napp, who should be retired) to different monasteries with a tougher discipline. Had these recommendations been consummated, the discovery of the principles of heredity would have, undoubtedly, been delayed by 35 years or probably more, since Mendel would not have had any opportunity to carry out his experiments. It is good to keep this in mind when reading or hearing church authorities taking credit for Mendel's accomplishments.

The Vatican's reaction to the visitation report is not known, since no records of it have been found in the archives. Overtly, the pope and the Holy Sea did not react at all. No dissolution order, not even a reprimanding letter, had ever reached the St. Thomas Abbey. Why the Vatican did not act on Cardinal Schwarzenberg's recommendations can only be speculated on. Perhaps the Holy Sea came to the conclusion that the legality of the visitation was not quite in order. Perhaps the pope realized that the secularization of the St. Thomas Abbey was the consequence of decrees issued by Emperor Franz I and that, therefore, its closing might send a wrong signal to Emperor Franz Joseph who was in the middle of important negotiations between the Vatican and the empire. Perhaps Napp was a too well-known personality with influential friends who might have interceded on his behalf. Or, perhaps, the church did not want to risk a scandal so soon after the revolution. Whatever the case may have been, in the abbey, the friars braced themselves for the worst following the visitation, but life gradually began to return to normal when nothing happened. We have, however, run ahead of our story; to resume it, we must now return to Mendel himself.

Napp's Dilemma

To reiterate, on June 30, 1848, Johann von Zelinka, the headmaster of the Theological Institute, sent Mendel the certificate of graduation and Mendel became a full-time *Kooperator* at the parish church. The pastoral duties, however, taxed Mendel's labile nervous system so heavily that he fell ill and remained bedridden throughout most of the winter of 1849. This development placed Napp in a difficult position: What should he do with a priest who was obviously not suited for his profession? Mendel knew what he wanted to be—a teacher and a researcher—but for Napp the situation was not that simple. He was aware that the Bishop Schaffgotsch was watching the developments in the abbey with a censorial eye, waiting for an opportunity to intervene. Turning another friar into a teacher, only one year after petitioning the bishop to accelerate Mendel's ordainment on the grounds that additional priests were urgently needed to service the parish, might give the diocesan the pretext he needed to launch an attack on the abbey. Napp therefore searched for a way of resolving the situation in a manner that would give the bishop no other choice than to approve of the move. Like in a bad play, at this critical moment in Mendel's life, *deus ex machina* interceded on his behalf once again. Recall that one of Napp's many functions was overseeing all of the *Gymnasien* in Moravia and Silesia. The headmasters therefore often turned to him when they needed new or substitute teachers. It so happened that in the summer of 1849 the city fathers of Znojmo in southwestern Moravia decided to upgrade their *Gymnasium* by adding to the existing six grades a seventh in the next school year and an eighth in the following year.[40] But the expansion required an extra teacher and so they turned, undoubtedly on the headmaster's recommendation, to Napp and at the same time to the *Landespräsidium*, the provincial government at Brno, for help. Napp realized immediately that via this request Fortune had offered him a way out of his dilemma: if the *Landespräsidium* were to officially appoint Mendel as a teacher in Znojmo, there would be little the bishop could do about it. Napp had high regard for Mendel and continued to be favorably inclined toward him despite the debacle concerning his pastoral duties and his mysterious incapacitation.[41] He knew, of course, that Mendel had no teaching experience and no formal qualifications for the post, but he had no doubt that Mendel could handle the job. And so he recommended him for the position. The administrations in both Znojmo and Brno accepted Napp's nomination and moved quickly through the necessary formalities, as the new school year was to start in a few days. As a result, already on September 28, Mendel received a message, written personally by the governor, Count Lažansky, ordering him to report immediately to the teaching staff of the Znojmo *Gymnasium* as a substitute teacher[40] of Latin and Greek languages as well as mathematics. The school would reimburse his travel expenses and pay him a salary in the amount of 60 % of that of a humanities professor.

Mendel must have been overjoyed, as he states in his curriculum vitae, where he refers to himself as "the respectfully undersigned": *After completing the theological studies in 1848, the respectfully undersigned received permission from his prelate to prepare himself for the philosophical Rigorosum. In the following year at the time when he was about to undergo the examination, he was asked to accept the position of a substitute teacher at the Imperial Royal Gymnasium in Znaim, and he followed this call with pleasure.*[42] On October 4, while Mendel was packing, borrowing money, and making other arrangements for the trip, Napp wrote a letter to Bishop Schaffgotsch, in which he informed the diocesan about the appointment and explained the circumstances that led to it. He pointed out that Mendel *leads a very retired, modest, virtuous, religious life fully befitting his standing. He is very diligent in the study of sciences, but much less fitted for work as a parish priest, because he is seized by an unconquerable timidity, when he has to visit a sickbed or to see anyone ill or in pain. His timidity made him so dangerously ill that I found it necessary to relieve him from pastoral service.*[40a]

It may appear odd that in an empire, in which life was codified by governmental rules and regulations, Mendel could be assigned to a teaching job in a public school without having the required certification and qualification. Three circumstance help to understand why this was possible. First, Mendel's appointment was that of a *substitute* teacher, which meant it was temporary and arising from an emergency situation. For such appointments, it was tacitly tolerated to bend or even disregard the rules. Second, the school assigned him to teach subjects for which he might have been deemed sufficiently qualified on account of his education. And third, at that time, elementary- and secondary-level schools were run, directly or indirectly, by the church and the employment of unqualified priests as teachers was a common practice, especially at schools run by religious orders.[43]

Znojmo: One Year in Haven

Mendel arrived at Znojmo on Sunday, October 7, 1849. By what means he traveled is not known, but it could not have been by rail, because the railroad connecting Znojmo with Brno (and Vienna, as well as Prague) was not built until some 20 years later. So, either a *Kutscher* (coachman) from the abbey drove him there in a horse-drawn *Pritsche* (light carrice) or, more likely, he took the stagecoach. Znojmo lies in a region imprinted by the Dyje River, whose one arm, the Moravská Dyje, originates in the Czech-Moravian Highlands and the other, the Thaya, flows from the forest of Lower Austria. From the point of confluence of the two arms, the Dyje meanders along the Moravian-Austrian border like a suture on skull bones. Near Znojmo, it carves a picturesque winding valley in the earth's crust as it negotiates its course around the rock formations in its way. The region's southern location, the tilting of the plate on which it sits, and the relief of its surface conspire to create a warm microclimate conducive to grape, vegetable, and fruit growing. The wines of southern Moravia are among the best the Czech lands produce, *Znojemské okurky* (Znojmo cucumbers) have been famous for centuries, and the region's peaches are unsurpassed in their sweetness and aroma. The warm climate of the Znojmo region inspired some etymologists to derive the city's name from the Old Slavic *znoj*, meaning heat, passion, or sultry weather. Others argue that this derivation is inconsistent with the presence of the "m" in the name, and so, like many other old geographical names of central Europe, Znojmo's (Znaim in German) remains in the unresolved category.[44]

The city itself is among the oldest in Moravia, having existed long before it was first mentioned by name in a document from 1048.[45,46] Walking its streets is like taking an art history course. A Romanesque chapel in the castle, Gothic houses along the streets, a Renaissance City Hall with a 76-meter tall tower, and a Baroque monastery are all on display within a short walking distance of one another. The chapel is renowned for its wall paintings of the "Přemyslid pedigree"—the portraits of eight Přemyslid and 19 Moravian rulers. As incredible as it may sound, this art treasure was almost lost because of neglect and ignorance, when the rotunda lost its function, and was used subsequently as a storage place, pub, workshop, and at one time even as a pig pen. Restoration efforts lasting over several decades were then needed to salvage the artwork. Not far from Znojmo is the village Přímětice, where the Premonstrantesian parish priest Prokop Diviš (1698–1765) invented the lightening rod in 1754, two years after, but independently of Benjamin Franklin. Superstitious villagers destroyed the contraption, which Diviš erected in the parish garden, blaming it for the 1760 draught and famine. Historians like to point out certain parallels between Diviš and Mendel: Both became priests to be able to devote themselves to the sciences, both made great contributions to their fields of interest, and in both cases, their contemporaries showed no interest in their contributions. Up to the first quarter of the thirteenth century, Znojmo's inhabitants were nearly all Czechs. In 1226, however, after turning it into a royal city, Přemysl Otakar I opened its gates to foreign, mostly Germanic immigrants. They came in large numbers and gradually changed the ratio of Czech to Germanic inhabitants in favor of the latter. In 1849, the ratio was approximately 1:7.

Znojmo's *Gymnasium* (Fig. 6.3) was founded in 1624 by the Jesuits and moved from its original site to the former convent of the Sisters of St. Clare (the Clarisses) in 1824.[45] It was there where Mendel reported for his first teaching duty on Monday, October 8. Although Znojmo had several monasteries and Napp demanded that friars

Fig. 6.3 The entrance gate to the *Gymnasium* in the former convent of the Clarisses at Znojmo, where Mendel taught Greek and mathematics as a substituting teacher in the school year 1849/1850

living outside of the abbey find a monasterial accommodation if at all possible, Mendel rented a room in a private house at Böhmgasse 50. Whether he did so because there was no occupancy available at a monastery or because he wanted one year of freedom from the monasterial tedium and risked Napp's ire to get it, we don't know. At any rate, the location was convenient because the house was within a short walking distance from the school. Settling down into the new circumstances was, however, not predicament-free. The problem was—like during his student years—money. Although he borrowed small sums from Klácel and Cygánek before he left Brno and had Rambousek send him 15 Fl to Znojmo during his first month there, it was not enough to pay the rent and all of his other initial expenses. Fortunately, Václav Šembera[47] happened to be in Znojmo shortly after Mendel's arrival there, and when Mendel confided his predicament to him, the subprior lent him 50 Fl. This was enough to sustain Mendel until he received his first salary of 30 Fl one month later. Immediately, however, he began to worry about paying back his debts, especially the larger sum he owed to the subprior. He designed a scheme and executed it with the help of Rambousek. In an October 31 letter to him,[48] he asked Rambousek to approach Napp on Mendel's behalf for an advance payment of his "pocket money" (a monthly remuneration of eight florins to each friar for small expenses) from April to September 1850 as well as his *Kleidergeld* (an annual payment of 24 Fl for the purchase of clothing). Mendel calculated out that the whole advance should come to 66 Fl, of which Rambousek forwarded 15 Fl to him already. Of the remaining 51 Fl, he asked

Rambousek to give 50 Fl to Šembera and 1 Fl to Cygánek. With Klácel, Mendel had already prearranged that he would be repaid from Mendel's pocket money for the months of January, February, and March of 1850. The letter reveals also that to save money, Mendel was sending his dirty linen for washing to the abbey, presumably via the stagecoach. It delivered the package to a specified place in Brno, and there a servant from the monastery picked it up; the washed linen was then sent back to Mendel the same way. The letter affords us three insights into Mendel's personality. First, Mendel obviously disliked having debts and if forced by circumstances to borrow money, he went out of his way to repay the loan as quickly as he could. His aversion toward indebtedness might have been an instinct, a part of the package of inherited properties passed on from generation to generation along his peasant lineage. Second, Mendel was highly parsimonious with money, giving it out only when absolutely necessary. This trait must have been imprinted on him in his young years on the farm, where there never had been an extra *Groschen* to spare, and then during his student years, when his pockets were chronically empty. And third, he was not above condoning innocent small lies; he was but a human being. An example illustrating this side of his personality is provided by the letter, in which he implores Rambousek not to tell Napp that he (Mendel) borrowed money from the subprior, but rather claim that the advance will be sent to Znojmo.

Upon his arrival in Znojmo, Mendel assumed his teaching duties almost immediately as the school year had already started.[40a] On Monday, October 8, he attended a staff conference at which the ten professors divided among themselves their teaching responsibilities. In its original letter to Napp, the city council specified that the open position was for a seventh-grade teacher. Count Lažansky, however, informed Mendel that he would be teaching Latin, Greek, and German literature in the fifth grade and mathematics in the sixth. At the conference, the professorial staff decided that in view of Mendel's pedagogical inexperience, it was more prudent to assign him to teaching the lowest grades, specifically arithmetic in the first four grades and Greek in the fourth and fifth grades.[40a] Altogether, he had to teach 20 hours per week, which was a rather heavy load for a beginner. Of the ten members of the staff (including the principal and Mendel), only one professor had a heavier (21 hours) schedule and another had an equally heavy load; all other members taught fewer than 20 hours per week. The staff consisted of five permanent and five substitute teachers. The first-grade curriculum consisted of arithmetic; Latin, German, and Czech languages; natural history; geography; religion; and nonobligatory drawing. To these came the Greek language from the second grade on, and history replaced natural history in the third and fourth grades. The professors graded the students' performance on a scale from 1 to 5, with 1 being the best and 5 the failing mark. The *Gymnasium*'s reputation was that of a very demanding school. In the 1849/1850 school year, for example, of the 25 students enrolled in the first grade (two of whom were repeating the grade), five had to repeat the final examinations, another five had to repeat the grade, and two had to leave the school because of inadequate performances. Similarly, of the 36 students in the second grade, nine failed and several others left the school because of poor performances. In the higher grades the number of failings was lower, presumably because the lower grades acted as a sieve that eliminated the weaker students.

By all accounts, Mendel fit into the school's environment quite well.[49] His adjustment period was quite short, undoubtedly because he was not quite the greenhorn for whom the staff had initially held him. True, he never stood in front of 20–30 teenage boys whose attention he had to retain for a whole hour, knowing full well that their minds had a strong tendency to drift toward more interesting subjects than the cold logic of algebraic deductions or the rules of Greek grammar. He had, however, faced on several occasions even larger and more distracted congregations, which followed with critically appraising eyes and ears his every gesture and word during his sermons from the pulpit of the parish church. And as for explaining things, he had some experience from his student years, when he tutored in Opava and Olomouc. Also from his own experience as a student, he knew that the best way to hold a pupil's attention was for the teacher to make himself understood. He actually stated this first rule of a good pedagogue, when he said that a teacher had to make *all efforts to present his assigned subjects to the students in an easily comprehensible manner*.[42] Also, let us not forget that he took the course for school candidates and private teachers at the district teachers' seminary at Opava and was highly recommended in the qualification report.[42] The rest of his pedagogical success in Znojmo must be attributed to his natural talent. (Recall that one of Mendel's ancestors taught not only himself but also the children of Hynčice the elementary school material.) His scientific activities and literary verbalizations attest to Mendel' being a superb logician. Although logical thinking does not always go together with clarity of oral expression, in Mendel's case it did. And lucidity of expression is the *sine qua non* of a good communicator.

Biographers associate the Znojmo chapter of Mendel's life with a story of his alleged encounter with Bishop Schaffgotsch. It is claimed that during the school year 1849/1850 the bishop made a visitation to the Znojmo *Gymnasium* during which he impressed the staff with his corpulence and lean intelligence, impelling Mendel to utter: "That one hauls more fat than brain on himself." Naturally, since denouncements were in full swing in those days, someone reported this remark to the bishop, and since that time, Mendel was in bad graces up on the cathedral hill. As far as we can determine, the man who sent this story into circulation was Mendel's first major biographer Hugo Iltis.[43] He does not name his source and we could not find it in any independent source. Moreover, the story is contradicted by everything that is known about the relationship between Mendel and Schaffgotsch. As another Mendel biographer has pointed out, there is no evidence for the existence of sour grapes between them.[50] On the contrary, when the St. Thomas chapter elected Mendel abbot, the bishop, apparently pleased with the choice, wrote a laudatory letter about him, praising even his scientific accomplishments.[51] Furthermore, in the report on the 1854 visitation of the St. Thomas Abbey, Schaffgotsch did not attack Mendel personally the way he did the other friars. And finally, in all other matters concerning Mendel, the bishop approved all of Napp's supplications without a word of criticism. Considering the bishop's personality, all this would not have happened had the story been true. We suggest therefore that the story might be apocryphal.

Preparing for an Examination

In 1849 it was still possible for secondary schools to hire unqualified instructors like Mendel on a temporary basis as substitute teachers. During that year, however, the government, as part of an overall tightening of its grip on the society, began to clamp down on the laxity in the educational sphere. On November 15 the *Minister des Cultus und Unterrichts* (minister of culture and education) decreed that beginning with the next school year, *Gymnasien* would no longer be allowed to employ substitute teachers who could not present a certificate that they were qualified to teach their particular subject.[52] To obtain the certificate, candidates had to pass a qualifying examination administered by the *k.k. wissenschaftliche Gymnasial-Prüfungskomission an der Universität Wien* (Imperial-Royal Scientific Gymnasial Examination Committee at the University of Vienna), established by the ministry. The ministry urged all unqualified substitute teachers to apply for admission to the examination immediately. At the Znojmo *Gymnasium*, where substitute teachers comprised half of the teaching staff, the decree was bad news, since three of the five substitute teachers were unqualified. Its director Ambrosius Augustin Spallek knew he could not conjure up three new qualified professors with the snap of a finger, so he pressured his substitute teachers to apply for admission to the examination. Spallek was particularly keen on keeping Mendel, for not only was he developing into a very competent educator, but he was also well liked by both students and colleagues. Spallek was well aware that pedagogues with these three qualities were by no means plentiful. Mendel, however, vacillated. On the one hand, becoming a permanent professor at the *Gymnasium* was one part of his dream. He missed the intellectual atmosphere of Brno, but if this was the price to be paid for his freedom from the monasterial routine, he was willing to pay it. On the other hand, however, the prospect of an examination frightened him. Although, in the past, he had managed numerous examinations quite well, he realized that the one in Vienna would be a different affair altogether. In the past examinations, he knew the examiners, he knew the subject matter of the examination, and he knew also how to prepare for it. He had, however, no such knowledge about the teachers qualifying examinations; in fact, he knew very little about examinations at the university level. If he had wanted to continue teaching mathematics and classical languages, he would not have hesitated to sit the examination, because he had sufficient training in both to face an examining board. These, however, were not the subjects that he wanted to teach for the rest of his life. His interests lay elsewhere—in natural history and physics—in which he hoped also to carry out research, besides teaching. And if he were to be honest with himself, he would have to admit that in these two subjects, his knowledge was quite inadequate, primarily because he had no formal training in them. In his time, natural history (*Naturgeschichte*) dealt with three realms of nature—animals, plants, and rocks and minerals—each of which was the subject of a separate subdiscipline: zoology, botany, and geology, respectively. Natural history, physics, and chemistry comprised the natural sciences (*Naturwissenschaften*). The term "biology" (*Biologie*), although introduced early in the nineteenth century,[53] did not take root until the 1870s; it was then that it acquired the meaning of a branch of knowledge concerned with all living organisms and all vital processes and hence encompassing zoology, botany, and much more. Mendel knew that each of the three disciplines hidden under the name natural history covered a wide world of knowledge of which he had received only scraps during his studies at the *Gymnasium* and the Philosophical Institute.

Gymnasien were traditionally humanity-oriented institutions. Although at the time of Mendel's studies natural history and physics were already part of the *Gymnasium* curriculum throughout the Austrian Monarchy, they acquired a Cinderella status among the subjects taught. Only two hours per week were devoted to them, and if for some reason lessons had to be cut or if there was no instructor to teach them, they were eliminated altogether. Since there was no standard textbook, the choice of the material to be covered was left up to the teacher, often with disastrous consequences. The teachers of the two subjects were frequently not qualified and so the level of instruction was generally low. Consequently, natural history acquired a reputation of being an easy subject in which good grades could be earned without much effort. The situation began to improve only in the second half of the nineteenth century, mainly because of the efforts of one teacher, Dr. Alois Pokorný, an outstanding pedagogue and director of the *Leopoldstädter Gymnasium* in Vienna.[54] Through his relentless agitation, the ministry of education doubled the number of hours reserved for natural history in the curriculum and introduced a standard textbook for all the *Gymnasien* in the monarchy. It was authored by Pokorný, covered all three realms of nature, was of high quality, and was translated into all of the languages spoken in the monarchy. The decree requiring the teachers of natural history to pass a qualifying examination also contributed to the improvement of the quality of teaching in the subject.

All of these changes, however, came too late for Mendel to profit from them as a student. As we described in Chap. 4, at the Opava *Gymnasium* Mendel had not received any education in natural history, except that which he might have learned at the museum and the library. At the Olomouc Philosophical Institute, he had one aborted semester of natural history and one year of physics. The rest of the natural history course fell out because of the professor's illness. At Brno's Philosophical Institute, he took Diebl's course on agriculture and natural history, which was heavily oriented toward practical aspects of the subject. Mendel tried to fill these gaps in his education by studying on his own from books, from plant and mineral collections at the abbey, and from Klácel and Bratránek. The outcome of his efforts, however, remained patchy and fragmentary at best. It was certainly not enough to qualify him to be a teacher of natural history and physics. Nevertheless, against his better judgment, he let himself be persuaded to go ahead with the application and try his luck. He cast his die, bravely submitted the application, and plunged into his studies in order to prepare for the examination. Here, however, he encountered another problem—access to the necessary literature. The library at the Znojmo *Gymnasium* contained 867 volumes, of which 26 dealt with physics and mathematics, but only four with natural history, and even those were hopelessly out of date.[54] The few stuffed animals, a drawer with shells, and another with minerals in the natural history cabinet were not of much help to him either. Under these circumstances, his chances of qualifying were not overwhelming. He must have surmised as much, and it is therefore quite doubtful that he would have applied had the director not pressured him to do so. Spallek might have argued that Mendel had nothing to lose: If he passed, all would be well; if he failed, well, he could try again later. In reality, however, Mendel had quite a lot to lose, as further developments will show. To ease the burden on the Examination Committee, the ministry reserved the right to turn down candidates at any stage of the multistage examination process. In the first stage, the candidate had to supply evidence that he was qualified for *admission* to the examination. This was accomplished first by having the school apply on the candidate's behalf and, second, by appending to the application a variety of documents attesting to the applicant's qualification and morality.

Jacta Alea Est!

In a letter he wrote to the Examination Committee on Mendel's behalf,[55] Spallek first described the circumstances under which the Znojmo *Gymnasium* hired Mendel and then had this to say about him: "Since he assumed his teaching post, he has continually developed the most meritorious qualities of an exemplary and thorough teacher of youth. Through his clear and vivid presentations and indefatigable devotion. . .he has demonstrated daily that not only is he fully trained in the subject matter he teaches, but also strives to impress virtuous morality and religiosity on his pupils. The results of his pedagogical efforts are outstanding in regard to both students' progress and their moral conduct... (He) also deserves to be praised for having high moral and religious principles and for his patriotism." In addition to Spallek's letter, the staff, too, supplied a character reference, signed by seven of the ten professors including the director.[56] Since it bears the date of May 25, 1850, this document could not have been part of the package that Mendel sent to the committee on April 17. The document again first recounts the course of events that led to Mendel's appointment and then continues: "...all his behavior has always been virtuous and irreproachable, as expected of a priest. In his conversations, he has never uttered a word that could have been considered unbecoming of a religious or objectionable in respect to religious principles or political ordinances. On the contrary, he has always taken great pains to remain calm, modest, and reserved. He does not seek any other company than that of his colleagues. In his free time he visits only the local book club [library], with the exception of six visits to the theater, always in the company of his colleagues. On our request, the local civil and religious authorities have confirmed that all of the above is true." Such were the times: Hardly a year had elapsed since the revolution and the spying and informing on people, the snooping into their private lives was back in force. His colleagues, "the civil and religious authorities," and who knows who else watched over Mendel's every step: recording how many times and with whom he went to places, what he said to whom, and what his moods were in different situations. And all of these efforts, mind you, were spent on a person of whom there was no reason to suspect any illegal doings. How much more effort must have been spent on someone who *had* been suspected!

Mendel's own application to the committee consisted of his own letter and 13 attachments (Mendel lists only 12, because the thirteenth was the envelope Spallek attached to the package before it was mailed). Mendel's letter,[57] dated April 17, 1850, states that he wished to be certified for teaching natural history in all grades and physics in lower grades of the *Gymnasium.* He gave three grounds, which he believed qualified him for the teaching position. First, he attached enclosures, which he hoped would attest to his theoretical education, and promised to double his effort to make up for any shortcomings in it. Second, he pointed out his experience in both private tutoring and teaching at his present post at the Znojmo *Gymnasium.* And third, he stated that he prepared himself for teaching the specified subjects by studying privately in his spare time. The enclosed attachments included his baptismal certificate; certificate of domicile; certificates of his final examinations from the Opava *Gymnasium*, Olomouc Philosophical Institute, the agricultural course at Brno's Philosophical Institute, and the Theological Institute; as well as his curriculum vitae, now commonly referred to as his "autobiography."[42] The examining committee returned most of the documents to Mendel after the completion of the examination; what happened to the rest of them has never been clarified. If Mendel was familiar with Shakespeare, he may have thought—as he posted the application on April 17, 1850—of these lines in Act 5, Scene 4 of Richard III:

I have set my life upon a hazardous cast,
And I will stand the hazard of the die.

If not, he certainly must have known what Caesar said after crossing the Rubicon.

The Examiners

The application arrived in Vienna on April 22, 1850, and was duly registered with all of its 13 attachments in the *Einreichungs-Protokoll*, a register kept at the committee's office.[58] The Examination Committee decided that Mendel was indeed eligible for admission to the first part of the examination, the *Hausarbeit* (homework), and on May 2 and May 5 sent him the themes on which he was to write essays in natural history and physics. Mendel confirmed receiving both themes on May 12. From the themes he also learned who his main examiners in these two subjects would be: Andreas von Baumgartner (1793–1865) in physics and Rudolf Kner (1810–1869) in natural history (Fig. 6.4). In Baumgartner, Mendel was to find a backer on the Examination Committee—and it soon became apparent that he needed one. Perhaps it was Mendel's autobiography that struck a chord with Baumgartner, reminding him of his own boyhood in the village Frymburk at the foothills of the Šumava Mountains in Southern Bohemia. Even though with a baker for a father, he did not have to endure the hardships that marked Mendel's boyhood; he nevertheless was in a position to imagine what life was like for this farmer's son. His own fond memories were of taking a cow to the pasture, where he would meet other boys from the village for a day of exploits, play, divertissements, and adventure. He could also

Fig. 6.4 Mendel's examiners at his first attempt to acquire accreditation for teaching natural history and physics at *Gymnasien*: (**a**) Andreas von Baumgartner (physics). (**b**) Rudolf Kner (natural history)

well recall, however, that on most of the days, the farmer's sons could not join him because they had to toil on their fathers' fields. Baumgartner, better than any of the other members of the Examination Committee, could therefore appreciate the struggle the candidate described in his CV, his determination, and the seriousness of his striving. To Mendel, on the other hand, Baumgartner must have appeared an awesome figure, for not only was he an esteemed university professor but also a well-known public figure[37,59] with interests and accomplishments to his credit in several different domains. While attending elementary school, under the influence of the principal, the composer of church music, organist, and *Regenschori*, Johann Nepomuk Maxandt (1750–1838), Baumgartner had a go at a musical profession. But at the *Gymnasium* in Linz, he came under the spell of physics and chose that discipline for his career. After a brilliant performance at the University of Vienna as a student and assistant, he landed, at the age of 24, a professorial position at the *Lyceum* in Olomouc and then, in 1823, at his alma mater. It was at the latter position that he became a chairman of Mendel's Examination Committee. In physics, Baumgartner was not a great theoretician nor an accomplished experimenter, his main contributions being in the application of the discipline to the solution of practical problems. He was a gifted lecturer, successful author of textbooks, devoted editor of a physics journal, and well-liked popularizer of science. When a throat affliction forced him to give up teaching, he expanded his interests into new domains. First, he had himself appointed (in 1833) as director of the Imperial-

Royal China Factories. Then (in 1847), he took over the management of a state-owned tobacco factory, the producer of the popular Virginia cigar.[60] At the same time (1846), he supervised the construction of the telegraph line running along the Vienna-Brno railway. Additionally, he also oversaw the construction of the railway itself in certain parts of the monarchy (in 1847). To top all of these activities off, he also ventured into politics and became a statesmen, first (in 1848) in the function of the Minister of Public Works and later (1851) as the Minister of Finances, as well as Minister of Commerce, Industry, and Public Construction. Also, in 1851, the Imperial-Royal Academy of Sciences in Vienna (of which more will be said later) elected him as their president, which he remained until his death in 1865. He took all of these commitments seriously (so seriously that, as the head of the tobacco factories, he took up, at the age of 54, smoking cigars) and was apparently quite successful in all of them. He was an extremely busy man and yet, in 1850, he found the time to examine Mendel and so unwittingly joined up the cortege of key players in the friar's life drama.

Kner, on the other hand, was the opposite of Baumgartner in many ways, of which we will mention three. As a son of a well-placed government official in the Austrian city of Linz[37, 61] he didn't have to graze cows on a pasture nor was there any question about what he was going to be. Complying with his father's wish, he studied medicine at the University of Vienna and became a practicing physician and surgeon. Simultaneously, however, he followed his heart's desire and spent more and more time at the *k.k. Naturalien-Kabinett* with Professor Johann Jacob Heckel (1790–1857). The *Kabinett* was a collection of specimens brought from the various expeditions sponsored by the imperial government and Heckel was the man in charge of its zoological section, primarily fishes. Kner was helping Heckel to catalogue, characterize, and describe the fish specimens, initially on an unpaid part-time basis, but ultimately as a full-time assistant on a salary of which he said, "it was too little to live on, but too much to die on." He left Vienna in 1839 when the University of Lvov offered him a position as a professor of natural history. Ten years later, however, he was back in Vienna to occupy the newly created chair of zoology at the university. All in all, Kner's career was predictable and its path fairly smooth. Therefore, he would not have had the compassion of a Baumgartner, who started small and made it big, a compassion for someone who was trying to reach higher than predetermined by the circumstances of his birth. Another difference between Kner and Baumgartner was that the two had different expectations of what a teaching candidate should know. As stated earlier, Baumgartner was a man of many interests and of broad vision. In his view, concepts rather than details and logical thinking rather than memorized knowledge were important. Kner, on the other hand, was a taxonomist for whom the cataloguing of nature was the backbone of natural history and he, therefore, expected a teacher of natural sciences to have memorized the catalogue. The different expectations of the physicist and the taxonomist became apparent in the questions they asked Mendel during the examination and from the way they evaluated his answers. Baumgartner's questions probed the candidate's ability to generalize from facts, and when encountering evidence of this ability, the examiner was satisfied even if it became apparent that Mendel had gaps in his knowledge. Kner, by contrast, demanded the reproduction of a section of a catalogue and was displeased when Mendel tried instead to put the examiners questions into a broader context. The third difference between the two examiners was in their professional focus. Baumgartner found stimulation and challenge in a great variety of activities, excelled in all of them, but left a lasting impression on none. Kner's interest was from the beginning in the taxonomy of fishes, extant and extinct, and he developed into one of the foremost ichthyologists and fish paleontologists of his time. Of the two, Baumgartner was more famous in his lifetime in the Austrian Monarchy, whereas a more modest but longer lasting reputation was reserved for Kner. Baumgartner, however, achieved immortality of a different kind. He lives on as one of the key figures in Adalbert Stifter's novel *Der Nachsommer* (Indian Summer),[62] for which the author used Baumgartner as a model (see Vol. 1 Chap. 1).

The Essays

The homework part was the first of the three stages of the actual examination. The candidate's task was to write an essay on each of the two questions he received. Mendel had six to eight weeks for the writing and was free to consult any books or any other sources of information available to him. The purpose of the homework was not so much to test the candidate's knowledge, as it was to determine his ability of working with sources and of expressing himself in writing. The essays were, therefore, evaluated not only for their content by experts in the specific subjects (Baumgartner and Kner in Mendel's case) but also for their form by a linguist (in this case Theodor Georg von Karajan,[63] professor of German language and literature at the University of Vienna). Mendel wrote one essay in physics and another in natural history. Both have been preserved and represent a watershed of information in the otherwise arid archival documentation of Mendel's intellectual development.[64] The essays have been analyzed by Mendel's biographers, the one on physics briefly[65] and the one on natural history in detail.[61,65,66,67,68] At the end of each of the two essays, Mendel lists the sources from which he drew information; in the physics essay he only lists the authors' names,[69] whereas in the natural history essay, he also specifies the books.[70] Both essays are divided into hierarchically organized sections designated by capital Latin letters followed by Roman numerals, lower case Latin letters, and Greek letters. The theme of the physics essay was this: "Describe the mechanical and chemical properties of the air and from the former explain the origin of wind." Mendel divided his essay into three sections that corresponded to the three parts of the assigned theme. He relied heavily on the sources he listed. Oddly enough, one book on his list, the textbook authored by Baumgartner and Ettingshausen,[69a] covered very little of the essay's topic. Was this an attempt on Mendel's part to flatter the examiner by citing his book? Every smart university student would resort to this stratagem, but would Mendel? No matter, one can surmise that Baumgartner was not displeased by the acknowledgment. The first two sections were fairly descriptive but the third part of the theme gave Mendel an opportunity to demonstrate a bit of original thinking. This section concerns the origin of wind and thus meteorology. Some authors[65] have argued that it was this part of the essay that stimulated Mendel's interest in meteorology, which he retained for the rest of his life. Others,[71] however, have pointed out that Mendel was probably exposed to meteorology already in Opava, where one of his professors carried out weather observations regularly, and in Olomouc, where he took Baumgartner's physics course. Be this as it may, while working on the essay, Mendel must have given deeper thought to this subject as evidenced by his suggestion to improve weather forecasting by establishing a network of meteorological stations over Europe and linking them up via telegraph. He did not get this idea from the books on his list but he may have read about it in the newspapers since it was a possibility that was then gaining currency. His mentioning it may have again pleased Baumgartner who was very much taken by the telegraph at that time. Overall, the style of the essay is light, the writing smooth and fluent with one sentence connecting logically with the next, sprinkled with a metaphor here and there. It is a text that could have been written by a popularizer of science—or a good teacher. Mendel's essay did indeed please Baumgartner. Here is what the examiner wrote about it:[72] "In view of the fact that the candidate seeks a teaching permit for the lower *Gymnasien* grades only, the submitted essay deserves a high commendation. The conditions of the atmospheric air, together with the relevant experiments demonstrating them, are clearly discussed and neatly analyzed and applied toward the explanation of the wind. The whole essay is written in a plain, unadorned, and clear language; it is lucid and well organized; and its style and presentation enables one to grasp it at a glance. If the outcome of further examination will be of similar quality, the candidate will deserve to be granted a highly favorable certificate."

Unfortunately, Kner did not share Baumgartner's enthusiasm. He assigned Mendel the following theme: "Describe the essential differences between rocks formed by water and those formed by fire; then list and briefly characterize the main formations of the neptunic strata according to their age; and finally review in a like manner the igneous formations, both plutonic and volcanic." Thus, the theme consisted again of three parts and Mendel accordingly divided the essay into three sections. In the first section, he approached the question concerning the difference between sedimentary and igneous rocks from a somewhat unconventional, but for several reasons, very interesting angle. What Kner presumably expected from him was a simple list of differences between the two types of rock, but what he got was something quite different. To Mendel it may have seemed too trivial to copy a catalogue from a textbook: surely the examiner must have had something different in mind, something more profound! Anyway, to him the question made sense only if posed in the context of the rock's origin. And this was, therefore, how he answered the question. He began his discourse with the description of the earth's origin from the ball of gases as the nineteenth century imagined it. He then explained which rocks formed first and which later, why this was so, and how this circumstance led to their distinctive properties. His explanations were simplistic, but rational and logical. When reading this part of the essay, one begins to realize that the director of the Znojmo *Gymnasium*, Mendel's colleagues, and his pupils did not exaggerate in calling him a good teacher. The lucidity and rationality of his thinking shines through the essay, and this is one reason why it is such an important document, because it gives us an insight into how Mendel's mind operated. Another reason is that it reveals him as a highly progressive thinker. In the Austrian Monarchy one could probably count on one's fingers the number of Roman Catholic priests who would have been prepared to declare so openly their opposition to the Biblical account of the origin of the earth. In fact, there are probably not many more even today in some parts of the world. The third reason why the essay is so interesting lies in the following two sentences in his overview of the earth's geological history: "As in the course of time the earth acquired the conditions necessary to support organic life, first plants and then animals of the lowest kind appeared." And "Plant and animal life grew richer and richer, its oldest forms partly disappearing to make room for newer, better ones." These could have been sentences from *On the Origin of Species*, if the publication of Darwin's *magnum opus* were not nine years in the future. Note Mendel's choice of words: He does not say that plants and animals were *created*, but rather that they *appeared* (*traten auf*). Similarly, he does not propose that *all* organic life disappeared (e.g., as a result of a catastrophe) to make room for a new round of creation, but it disappeared in part (*zum Theile*), that is, some forms disappeared gradually over time as new forms emerged. Luckily for Mendel, his essay was apparently not brought to the attention of Bishop Schaffgotsch or of any other clergy. One can imagine the consequences for Mendel in the post-1848 years, if it had been. Let us remember the two sentences, for there are, as we shall see later, historians of science who claim that Mendel undertook his hybridization experiments to disprove Darwin's theory of the origin of species. Why would he want to do this, when he apparently had no problem in imagining species coming and going in the history of the earth? As to where Mendel's remarkable thoughts on the development of life may have come from, one could imagine Klácel leaning over the essay writer's shoulder as he was penning these sentences. Not literally, of course, because the two friends were separated geographically by a day's carriage ride. At any rate, the two sentences are the third reason why anybody who is interested in Mendel should read his natural history essay.

The rest of the essay is an overview of the three principal types of rocks: igneous, sedimentary, and metamorphic. This section is also interesting but for different reasons than the first part. Bearing in mind that Mendel had less than eight weeks to write the essays and was writing them in his spare time, the product is quite remarkable in its breadth and depth. Even if he had read up on natural history and studied the mineral collection at the abbey before coming to Znojmo, he could not have learned all of the details described in his essay. An ordinary university student would require more than eight weeks in order to read the four books on his list alone, and an instructor would spend at least a semester on covering the topics that Mendel dealt with in the essay. The fact that Mendel not only assimilated this amount of material in such a short time but also managed to organize it into a logical system and present it in an intelligent and intelligible way indicates that Mendel was not an ordinary student. It also suggests that with a mind like this, Mendel would have had no difficulty teaching natural history at a *Gymnasium* and teaching it well.

Mendel, however, did more than synthesize information extracted from four books—he also managed to *interpret* it. At the turn of the eighteenth century a controversy raged in earth sciences between the so-called neptunists and plutonists. The former, led by the German geognosist (as geologists were then called) Abraham Gottlob Werner (1749–1819), claimed that all rocks of the earth's crust arose from material deposited in the primeval ocean (the realm of the god Neptune) which once covered the globe. By contrast, the plutonists, led by the Scottish geologist James Hutton (1726–1797), insisted that the earth's crust originated by the solidification of molten masses upwelling from the underworld (the realm of the god Pluto). By the middle of the nineteenth century, at the time of Mendel's examination, the controversy was largely over, resolved in favor of a concept postulating three different origins of rocks. One kind of rock (the sedimentary) was formed by the consolidation of sediments. The second (igneous, from Latin *ignis*, fire) kind arose from the solidification of molten magma deep in the earth's crust (plutonic rocks) or from lava issuing from a volcano (volcanic rocks). The third kind, the metamorphic rocks, originated by the transformation of preexisting rocks under the effects of high temperature, pressure, or chemical reactions, in the depth of the earth's crust. Some of the older or more conservative compendia that Mendel consulted may have given him the impression that the debate was not quite settled, and so, if he wanted to present a synoptic view of the subject, he had to take sides. He did this so well that to some historians of science, he seemed to have had as a guide the three-volume *Principles of Geology* (1830–1832) by Charles Lyell (1797–1875), the major proponent of the progressive synthetic view. Mendel, however, did not include Lyell's treatise in his list of references, and it is indeed unlikely that he had access to its German translation at Znojmo. The essay thus reveals another aspect of Mendel's mind: In addition to being logical and synthetic, it was also critically analytic.

The foregoing appraisal of Mendel's natural history essay now shared by his biographers was unfortunately not the opinion of the man who mattered most in evaluating it—Kner, the examiner. He said[73] that he expected a precise and clear overview emphasizing the characteristics of the rocks, an overview that would reveal the extent and accuracy of the candidate's knowledge. But that's not what he got. In his opinion the candidate covered many things but he did not do so concisely and clearly, he often missed salient points, his characterizations were anything but sharp, and some of his assertions were wrong. In the answers to the first part of the question, instead of focusing on the differences between igneous and sedimentary rocks, the candidate wasted space on a description of how the earth formed, which had no relevance to the question. Nonetheless, he answered this part somewhat better than the rest of the question. In the discussion of the sedimentary rocks, the candidate devoted too much space to the description of the mineralogical components, which had little relevance to their characterization. On the other hand, his description of the various layers was rather dry, unclear, blurred, and wanting in the discussion of the leading fossils. The description of igneous rocks was marred by a failure to consider their structural properties. The language of the essay was generally satisfactory but spoiled by the use of exaggerated or inappropriate expressions. The candidate used only a few high quality works in writing the essay, but the way he interpreted them revealed that he had not as yet absorbed geognosy to the extent that would make him sufficiently qualified to teach the highest grade of the *Gymnasium*. Nevertheless, he should be admitted to the next stage of the examinations in the hope that he would perform better in other areas of natural history.

Biographers from Iltis[43] onward have accused Kner of harshness, unfairness, and even prejudice toward Mendel. Are these accusations justified? Each of these three judgments is relative and hence subjective, so it is difficult to judge the judges. Of the three, harshness on the part of Kner is probably the most defensible judgment. As a university professor, Kner had the right to set his own standard of what to expect from a student. Was this standard "unduly exacting"? Not, if the examiner made it clear to the students what the standard was. But Mendel was not Kner's student, and he could not have known what exactly was expected from him. Perhaps, Kner should have taken this circumstance into account, which he obviously did not. And for this, he now stands accused of harshness by posterity. As for unfairness and prejudice, that's a different matter altogether. There is no evidence whatever that Kner's judgment was "marked by injustice, partiality or deception."[74] He treated Mendel the same way he did all of his students and so from his point of view he was fair to him. The unfairness was in Mendel's situation which did not allow him to have been Kner's student, but that was of course neither Mendel's nor Kner's fault. In regard to prejudice in the sense of "an irrational attitude of hostility directed against an individual,"[74] Iltis[43] raised the possibility of anticlericalism among intellectuals brought to the surface by the 1848 revolution. He could not, however, provide the slightest indication substantiating this possibility in Kner's case. To sum up, the only thing that Kner can be accused of is a lack of leniency toward Mendel. As for his criticisms, the points that he found disagreeable in the essay are either a matter of personal preferences or trifles attributable to Mendel's inexperience, use of out-of-date sources, and insufficient time to master the material. One can in fact, turn the tables and consider the essay a test of the examiners perspicacity. In that case one must come to the conclusion that Kner flunked the test, in that he failed to see the signs of an original mind at work.

Behind Locked Doors

For the time being, Mendel remained blissfully unaware of the dark clouds gathering over his attempt to gain an accreditation as a *Gymnasium* teacher. The Ministry of Education informed him and the other two candidates from Znojmo that the rest of the examination would take place in one-week intervals starting on July 15, so he must have assumed that the examiners adjudged his two essays as satisfactory. With this letter, however, began a comedy of errors, which did not decide but certainly influenced the outcome of the examination. Since it affords a glimpse of the bureaucratic ineffectuality in the Austrian Empire, we outline it briefly. The details of what followed the mailing of the letter are hazy,[75] which explains why the accounts of the events differ in the various books and articles on Mendel. The account that follows is essentially based on Iltis[43] but is supplemented by our own deductions. The ministerial letter irked two agencies, the directorate of Znojmo's *Gymnasium* and the Examination Committee in Vienna. At Znojmo, Spallek fired off a letter to the Ministry pointing out that the school year at the *Gymnasium* ends on July 31 and that the last month is therefore the most hectic one. Final examinations are taking place, certificates and reports are being written, grade catalogues are finalized, and many other administrative matters are taken care of in this time. The school's teaching staff could not possibly be expected to accomplish all of these tasks if three of its members were absent during this critical period. He therefore requested respectfully to either move the examination date from July to the first half of August or allow him to end the school year on July 20 (the logic of the latter request is somewhat obscure). The ministry approved the first of the two alternatives. In Vienna, the Examination Committee was upset that the Ministry was arranging examination dates without consulting the committee members first and was doing it even before the candidates were officially *admitted* to the second part of the examination.[76] Responding to these complaints, the Ministry then informed Mendel in a second letter that the date of the examination had been changed from July to August, presumably[77] to August 12. The Examination Committee didn't like this date because the academic year was scheduled to end on it, which would have meant that the examinations would have to take place in their vacation time. In August, it was generally unpleasantly warm in Vienna and the faculty, therefore, tried to leave the city for the countryside not a day later than was absolutely necessary. Baumgartner, therefore, wrote another letter to Mendel informing him that the examination had been postponed until the beginning of the next academic year. Alas, Mendel, who in the meantime had returned from Znojmo to the abbey, left Brno for Vienna either before the letter arrived or simply without having received the letter at all. And so, on Monday, August 12, he reported for the examination at Baumgartner's office. Undoubtedly, Baumgartner must have been quite surprised when the secretary ushered Mendel into his office and even more so when he learned the reason for his visit. Presumably, he tried to persuade Mendel to go back to Brno and return in a couple of months, but Mendel must have put forward some good arguments as to why the examination should not be postponed. At any rate, he got his way and the examination was hastily arranged. The examiners could not have been thrilled at the prospect of having to stay in Vienna several days longer just because of a stubborn monk, and it was certainly unwise of Mendel not to go along with the postponement. In the end, however, it probably would not have made any difference in terms of the examination's outcome.

On Tuesday, August 13 and Thursday August 15, Mendel took the written tests behind a locked door (*Klausurarbeit*) in natural history and physics, respectively.[78] The tradition at the University of Vienna had been to sequester the candidates of the written test in a room, give each of them a sealed envelope with the assigned theme of the examination, and lock them up together with an overseer and a guard at the door.[66] The overseer's function was to make sure that the candidates did not communicate with one another after they had opened their envelopes and did not use any aids such as books or notes they may have smuggled into the room. The guard's duty was to ensure that candidates who had to leave the room during the examination would not communicate with anybody on the outside. The test lasted up to three hours. We can assume that because of the improvised date of Mendel's examinations, he took both of them alone. In his natural history examination, the sealed envelope contained the instruction "list and briefly characterize mammalian orders and their main subdivisions, identify in each order animals which are useful to humans, and explain in what way they are useful in commerce and pharmaceutics." As a result of an odd slip of Kner's pen, Mendel should have been well prepared for this theme because he encountered it before when he was working on his take-home essay. Earlier, Kner wrote this question on the same sheet of paper that had Mendel's homework assignment on it, but then crossed it out. It's nearly impossible that Mendel had not noticed it since it remained clearly legible.[79] Some biographers[43] have speculated that Kner might have done this on purpose to give Mendel a clue, but Kner had no reason to do so and the harshness with which he treated Mendel otherwise indicates he had also no inclination to make the examination easier for him. The sheet with the assignment may have been a draft on which Kner scribbled possible questions for Mendel's homework; he then crossed out the question he was not going to use. Later he forgot about this and assigned the alternative question to Mendel at the written examination, without being aware that the examinee must have seen it before. One would think that a student preparing himself for an examination and noticing a crossed out alternative question would also prepare himself for the alternative as well, just in case it would come up during the examination. Oddly enough, Mendel ignored the alternative and came to the second stage of the examination totally unprepared for the assignment he got. He ignored the helping hand Lady Luck had extended to him and paid for this gaffe by losing the chance of becoming an accredited *Gymnasium* professor. It is possible, however, that Mendel actually did notice the crossed out theme and prepared himself for the possibility of being asked this question. If so, part of the blame for his, let us say, unspectacular performance on the written examination must go to the source from which he picked up the information that he used. He specified this source as "*Gistel*," presumably it was *Naturgeschichte des Tierreichs für höhere Schulen* (Natural History of the Animal Kingdom for Higher Schools), published in Stuttgart in 1848 and authored by the German naturalist Johannes von Nepomuk Franz Xaver Gistel (1809–1873). Indeed, some of the passages in Mendel's examination paper follow Gistel's textbook rather closely, which indicates that he must have not only read the book but also memorized parts of it. The problem with his source was that it was obsolete. Gistel was an entomologist and so while he was up to date on insects, he was a much less so on mammals. One year after the publication of Gistel's book, Kner, an expert on vertebrates, published his own textbook on zoology,[80] in which he presented a more modern classification of mammals. Mendel, of course, should have studied from Kner's rather than from Gistel's book, but it was not available in Znojmo; he probably did not even know of its existence.

Ever since Illtis, biographers have bemoaned or ridiculed Mendel's examination paper, but if one wants to evaluate it objectively, one must do so by taking into account the state of mammalian classification in the middle of the nineteenth century. Only then do some of the ridiculous statements in Mendel's paper become understandable. Others, however, remain quaint and must be attributed to lapses in Mendel's memory. Following Gistel, Mendel divided mammals into six orders: *Händetiere* (animals with hands), which are now the primates; *Pfotentiere* (animals with paws), today's rodents and lagomorphs, such as hares; *Flatterfüssler* (animals with flapping feet), today's bats; *Krallenfüssler* (animals with clawed feet), today's carnivores; *Huftiere* (hoofed animals),[81] a combination of three orders in modern classification schemes, perissodactyls (such as horses), artiodactyls (such as pigs and bovides), and proboscideans, the elephants; and *Ruderfüssler* (animals with webbed feet), the present-day cetaceans, the whales (Mendel erroneously included in this order walruses, sea lions, and seals, which are actually carnivores). To a present-day biologists both the terminology and the classification look weird and arcane; indeed, some of its elements go all the way back to Aristotle. Yet, it was still in use in the middle of the nineteenth century, although by then more progressive zoologists were taking steps to modernize it. Unluckily for Mendel, the only zoology textbook available to him in Znojmo (Gistel's) was written in the traditional spirit as far as the classification of mammals was concerned. It did, however, contain further subdivisions of the individual mammalian orders, which Mendel failed to memorize and so he could not answer this part of Kner's question. Mendel also only vaguely remembered the characteristics of the orders that Gistel provided and consequently did not do well on this part of the question either. One might have expected that Mendel's analytical mind would take a comparative (if not outright evolutionary) approach to the characterization of the orders and extract the essential idiosyncratic features of each order from the large amount of descriptive material, but apparently he did not have the time for that. What he presented instead was a pell-mell of fragmentary knowledge, which he dug out randomly from the depths of his memory, mixing quintessence with trivia. The same is true about the list of species and their utility to humans. These two parts are so dilettantish that today a knowledgeable person would find them entertaining to read. They are the parts that Mendel's scholars like to poke fun at. Yet, what the scholars often do not realize is that not all of the errors should be attributed to Mendel's ignorance. Thus, for example, Mendel has been ridiculed for grouping the kangaroo with rodents. The fact is, however, that in 1850, the division of mammals into three major groups—monotremes, marsupials, and placentals—was by no means paradigmatic. Indeed, in Gistel's textbook kangaroo is classified as a rodent, though of a different family. Similarly, scholars deride Mendel for attributing a tripartite stomach to the horse: A farm boy who didn't know the difference between a cow and a horse! But here again, placing the horse in the same order as the cow was then a common practice in many zoological textbooks including Gistel's. Although Mendel did list the tripartite stomach among the characteristics of the order, he never said that the horse had it. The confusion arose from the fact that perissodactyls and artiodactyls were then included in the same order, and Mendel mistakenly elevated the tripartite stomach to the status of an attribute of the whole order. This is not to say that Mendel did not make mistakes in his written examination; he did (e.g., placing the llama into Mexico rather than Peru), but they were rather unimportant ones. Considering the circumstances, this part of his essay is not as bad as it may appear to present-day biographers.

Nevertheless, the paper did not fulfill the requirements for teaching candidates, and Kner stated it clearly in his evaluation.[82] According to him, Mendel used an outmoded classification system, failed to subclassify the orders, characterized the orders inadequately, described the utility of animals in a dilettantish way, and availed himself of colloquial instead of scientific nomenclature. In Kner's opinion, Mendel was not qualified to become a teacher at the *Gymnasium*, but the examiner once again did not object to giving Mendel one last chance by admitting him to the oral examination. Here, too, Kner was correct in all the points of his critique and so he should not be accused of undue harshness and unfairness.

After a one-day respite, on Thursday, August 15, Mendel was again handed a sealed envelope, provided with blank sheets of papers, a goose feather quill,[83] and an ink jar, and locked up for three hours in a room to write his *Klausurarbeit* in physics. He was asked to describe how one magnetizes a steel bar and what laws govern the distribution of the magnetism thus induced. When he read the question, his self-confidence, fairly eroded by his struggle with mammalian classification on Tuesday, had returned to him. It may seem odd, but the discoverer of one of the most fundamental biological laws was more at home in physics than in biology. One reason for this oddity was, however, that he received some formal instruction in the former but virtually none in the latter; another was that his mind was more attuned to the abstract and logical structure of physics than to the largely descriptive and retentive nature of nineteenth century biology. The difference in his attitude to the two different written examinations shows in the organization of the thoughts he committed to the paper. The natural history paper betrays nervousness, lack of concentration, and a poor organization of thoughts.[68] The physics paper, by contrast, reveals a hand led by a calmer mind and a memory that could recall whole passages from Baumgartner's book.[69a] The well-organized presentation is once again vintage Mendel. One might suppose that this time he may have been content with what he had written, even though he did little more than merely skirt the second part of the question. If he was content, he deluded himself, as he was to learn as early as the next day. For, after reading the paper, Baumgartner had to agree with Kner that the candidate did not know enough, to be certified as a teacher.[84] He found Mendel's answer to the first part of the question essentially correct (after all it reproduced what he himself wrote in his book), but lacking in depth, and the second a total failure because Mendel did not answer it at all. What Baumgartner expected from the candidate was a mathematical treatment of the problem, but this Mendel could not provide, possibly because he studied physics from a book that did not contain it.[85] Thus, this time both examiners were dissatisfied with Mendel's performance, yet both agreed to let him take the oral examination.

Viva Voce

On the next day, Friday, August 16, 1850, the whole committee assembled to examine Mendel orally. The committee consisted of eight members. In addition to Baumgartner, the chairman and physics examiner, Kner, the natural history examiner, and Theodor Georg von Karajan, the language assessor, it included the physicist Christian Doppler (1803–1853), the philosopher Franz Karl Lott (1807–1874), the historian Wilhelm Heinrich Grauert (1804–1852), and the philologist Hermann Bonitz (1814–1888), all of whom were professors at the University of Vienna. Present at the examination was also a *Gymnasium* professor, whose name is not quite legible on the documents (probably Enk). A protocol, presumably prepared by one of the committee members,[86] is the only source of information about the examination. According to this document, the chairman began the examination by asking Mendel a series of questions from physics. Mendel's answers "fully confirmed the impression gained from the candidates previous performance, especially his written examinations, to wit, that he had acquired fragments of knowledge, which he had failed to put together so as to get a clear understanding of what he has learned." And so Baumgartner came to the conclusion that "there was no other choice than deny him, for now, the teaching certificate for the lower *Gymnasium* grades." Baumgartner then excused himself from the examination (presumably because of other engagements), and Kner took over the chair, to examine Mendel in natural history. "The outcome of this part of the oral examination was more satisfactory than might have been expected from the written examination. The candidate demonstrated that he had been diligent and thorough in his studies. He obviously lacks neither talent nor will power, but has suffered from inadequate opportunity to study and from using inferior sources of information." Essentially, what the committee was saying was that a self-taught man from the province had little chance of passing an examination requiring the level of knowledge as taught at the University of Vienna. It therefore came to the conclusion that there was hope for the candidate. If only he could be educated at a proper institution, there was every reason to believe that he would become qualified to teach natural history at least at the lower grades of the *Gymnasium*. Thus, the oral examination harmonized the initially somewhat discordant opinions of the two chief examiners. Baumgartner, who was originally impressed by Mendel's rationalizing to the extent of being willing to turn a blind eye to the lapses in factual knowledge, cooled off as a result of the candidate's performance at the oral examination. And Kner, who was highly critical of Mendel's essay, warmed up to the candidate at the *viva voce*. Baumgartner and Kner both (as well as the other members of the Examination Committee) came to the same conclusion: Mendel had the intellect, talent, and enthusiasm, but had serious gaps in his factual knowledge. They recommended, therefore, that he amends the insufficiency by taking the pertinent courses at the university and then try for the teaching certificate again. An unbiased observer must conclude that not only the first part but the *entire* examination was fair in that the examiners treated Mendel like any other candidate. A biased observer might want to argue that the unfairness of the examination was exactly in this point; that the autodidact Mendel was *not* like the other, university educated, candidates; and that the examiners should have taken this fact into account (which, in fact, they did—by admitting him to the second and third stages even though they felt that his performance in the first two stages was not satisfactory).

What happened after the examination we do not know. Presumably, either still on that Friday or more likely the next day, Mendel was told that he failed. Baumgartner called Mendel to his office, where he gave him the bad news. In all likelihood, he sweetened the bitter pill with the suggestion that Mendel should take the appropriate courses at a university and then try again. The sweetness of this proposal, though, was not only in the prospect of escaping the monasterial tedium for another couple of years. At this stage of his life, Mendel wanted not only to be a teacher but also a researcher and to become the latter he may have felt handicapped without a formal university education. He may have hoped, therefore, that Baumgartner's recommendation would be incorporated in the written report on the outcome of the examination and would prod Napp to act on it. All the same, the impact of the failure on Mendel's psyche must have been devastating. It was his first failed examination, and the fact that it occurred in his mature age and in his favorite subjects made it so much more humiliating. To what extent did he think that he deserved the flunking? He studied hard, without realizing that he studied from the wrong books. He knew little what was expected of him. And he may have thought that his performance was not as bad as to earn him a failing grade. So, he may have vacillated between blaming the examiners and blaming himself for the outcome. Heaviest on his mind may have been the thought of facing his fellow friars, especially Napp. He probably felt that he let them all down. It must have been a rather disheartened friar traveling on the train that took Mendel back to Brno. Only the flickering hope that Napp might send him to Vienna to study at the university may have protected him from another nervous breakdown.

Back in the abbey, Mendel reported to Napp what had happened in Vienna. He probably mentioned Baumgartner's suggestion, as Napp's correspondence with Znojmo seems to imply. Shortly after Mendel's return, the abbot received a letter (dated August 30, 1850) from Anton Buchberger, the mayor of Znojmo, imploring him to allow Mendel to teach at the *Gymnasium* in the coming school year or, if that was not possible, to send another qualified friar. He preferred Mendel, though, who "because of his altogether praiseworthy manners, but especially his excellence in teaching,. . .enjoys the respect of the entire local population." The mayor went even as far as indicating that in recognition of Mendel's qualities he would be prepared to increase the friar's salary from 360 to 400 florins. In his answer dated September 7, 1850, Napp expressed regret that he could not fulfill the mayor's wish because he had other plans (*bereits anders verfügt habe*) for Mendel and has nobody else whom he could send to Znojmo.[87] Presumably, the other plans were to send Mendel to Vienna. If so, it is puzzling why he then waited one whole year before actually realizing these plans. One possibility is that he waited for the official report of the Examination Committee and especially for Baumgartner's judgment. Another reason might have been that by that time it was already too late to make all the necessary arrangements before the beginning of the academic year.

The report Napp may have been waiting for had indeed been on the way, but it was on a rather oblique path. The Examination Committee issued the report on October 17, 1850, after its return from vacation. It sent the document, signed by all its members, not directly to Mendel, but to the Moravian Regional Educational Authority (*Landesschulrat*), which then forwarded it to Brno's *Gymnasien* Directorate, which then sent it, finally, to Mendel. This entire transaction took nearly one year and thus stands as a monument to imperial-royal bureaucracy. When Mendel received the document on August 9, 1851, he must have been disappointed, because it did not contain the eagerly expected recommendation for university studies, at least not in so many words. Moreover, the document contained a report on only the oral examination in physics and was based on the protocol described earlier.[88] Kner prepared a protocol on the entire examination in natural history, dated November 11, 1850, which Mendel apparently never received.[89] Puzzled by the delays in receiving the official examination report, Napp finally, in the summer of 1851, decided to inquire of Baumgartner about it. Neither Napp's letter nor Baumgartner's response has been preserved; we know about them only from a concept of a letter, in which Napp thanks Baumgartner for the information. From this document,[90] it follows that Baumgartner spoke well of Mendel and highly recommended his enrollment at the University of Vienna. Only then did Napp spring into action, and when he did, events followed one another quickly so that Mendel could belatedly be allowed to start the winter semester. Before we turn to those events, let us complete this part of the story and describe what Mendel did during that one year of waiting, from the summer of 1850 to the summer of 1851.

Substitute Teacher at Brno's Technical Institute

For the first half of that year, we have no information about Mendel's activities. Presumably he spent most of his time in studies and in occasionally helping out with the innocuous forms of pastoral work. In the second half of the year, he got a chance to have a very short stint at teaching again—ironically in the very same subject that he flunked at the Vienna examinations. Here is how it happened. The escalating industrialization of Brno in the nineteenth century brought to the fore a problem which increasingly affected the whole monarchy, but was felt more acutely in the Moravian metropolis than elsewhere. The problem was that the city's schools became hopelessly out of step with the city's social and economical development. Because of their focus on humanities and neglect of all practical subjects, they brought up graduates who were not prepared for the challenges of the industrial age. They could read Latin, but not English or French; they could quote the *Ecloges* or *De rerum natura*, but they knew next to nothing about sheep breeding nor could they tell wheat from rye; and the function of any machinery was simply an enigma to them. Increasingly aware of this disengagement between the society and its schools, the city officials stepped up their efforts to launch new types of schools in which the students would get a solid education in science, agriculture, and technology. In Moravia, the first step taken in this direction was the establishment of the

Estate Academy at Olomouc in 1725 (see Chap. 5). Although its curriculum was outmoded because it had hardly changed over a period of more than 100 years, it was still better than nothing at all, and so the officials attempted to transfer it to Brno. When this effort failed, they proposed, in 1834, to the authorities in Vienna to establish a similar school in Brno in association with Franzen's Museum, which was founded in 1817. Protracted negotiations followed, but the project faltered on lack of money and the inflexibility of the imperial bureaucracy. But after the 1848 revolution, things began to move. In 1849 Emperor Franz Joseph approved the opening of a *k.k. technische Lehranstalt* (Technical Institute) in Brno, which would offer courses in science, agriculture, and technology, taught by a staff of 12 professors and supported by the government.[91] The science courses included physics, chemistry, and natural history. The teaching was supposed to be conducted in both German and Czech, but in reality, all courses were taught in German exclusively. The school opened in 1850 and its first professor of natural history was Franz Diebl. After Diebl's retirement one year later, the post went to Jan Helcelet, a medical doctor, brought to Brno from Olomouc. You may remember that while Mendel attended the Olomouc Philosophical Institute, he missed instruction in natural history because Helcelet was sick at that time and the school failed to find a substitute teacher. Well, in April 1851, Helcelet fell ill again, but this time the new school was determined to find a substitute for the period of his sickness. As the protocol of the school's staff conference on April 3, 1851 reveals,[92] when the question of the substitute teacher came up, Professor Bedřich Kolenatý[93] brought up the name of Gregor Mendel. The professor knew that Mendel had no formal education in natural history and that, in fact, he had failed the examination in this subject quite recently, but this small handicap did not seem to bother him. The staff must have known better than the examiners in Vienna what Father Mendel's real qualities were. They approved the proposal unanimously. The school director then empowered another staff member, professor Václav Hrubý (1813–1889) to approach Mendel with the offer of about one florin of conventional currency per one hour of instruction. And so by a strange quirk of fate, Mendel was offered to teach the very same subject which Helcelet had failed to teach him. How could he have declined? The salary was decent, the challenge exciting, and the boost to his morale invaluable. The offer indicated that the failed examination did not damage Mendel's reputation in Brno and that he was beginning to get recognition as a naturalist and as a teacher. He accepted, of course, and began teaching on April 7, 1851. So, once again, Mendel trotted four or five times a week from the abbey, this time to a different part of the town, and not as a student but as a professor. The Technical Institute was located temporarily in a shabby building, No. 24 on Trnitá Street, outside of the inner city (Fig. 6.5). Nine years later it would move to a new building on today's Komenského Square and would develop into a Polytechnic Institute, and later into a university. But this part of its history has little to do with Mendel.

After two months, Helcelet recovered from his illness and resumed his teaching responsibilities at the school. On June 6, director Schindler wrote a highly laudatory letter[50] to the governor's office about Mendel's substitute teaching. On the same day

Fig. 6.5 The Technical Institute (*Technische Lehranstalt*) *Trnitá* Street at No.24 in Brno. Here Mendel taught natural history for two months in April and May 1851, substituting for Professor Jan Helcelet, who fell ill

he also sent a letter[50] on the school's behalf to Mendel thanking him for his help and expressing great satisfaction with his work. The otherwise highly critical Schindler heaped praises on Mendel for his great zeal, excellent instruction, for being most considerate to his pupils, and for his goodwill to all the members of the staff. On June 13, Mendel wrote a letter[50] to the director asking for the remuneration of 25 florins conventional currency for his two-month period of work, thereby bringing the brief episode of his second substitute teaching to a close. Shortly afterward the Natural Science Section of the Agricultural Society admitted Mendel as an extraordinary member;[50] presumably, Mendel's success in teaching played a part in his election.

Rushing to Vienna

Missing school year's beginnings seems to have been Mendel's lot. At Opava he started the *Gymnasium* two months later than all the other students. At Znojmo, the teachers of the *Gymnasium* had to reschedule their assignments because the freshly baked professor was two weeks late in reporting for duty. And now in Vienna, Mendel was poised to ask Baumgartner to pull his strings with the administration to allow him to matriculate at the university five weeks after the beginning of the winter semester. In none of these cases was Mendel's late arrival his fault. And in all of them, he managed to catch up with the proceedings already in progress. In the last case it was Napp's fault. To reiterate on August 9, 1851, Mendel finally received the report about his failed examination. In that same month, Napp wrote a letter to Baumgartner who responded by recommending Mendel for university studies.[43] Thus, reaffirmed in his assessment of Mendel's potential, Napp informed on October 2, 1851, Bishop Schaffgotsch about his plan, explaining again that Mendel proved to be unsuited for pastoral duties, but apparently good at teaching natural history, which he had been studying on his own with great zeal. Napp pointed out that "Herr Minister" (meaning Baumgartner) himself had recognized Mendel's talent and had recommended university study to complete his education. He (Napp) therefore decided to follow this recommendation. Napp reassured the bishop he would make certain that Mendel would not stray from the monastic discipline during his studies and that to this end he had already contacted the prior of the convent of the Brothers of Mercy at Vienna with a request to accommodate Brother Gregor for the two years.[43] Although the bishop replied within a few days, the bishopric curia (administrative body) needed 11 days to expedite the letter.[94] In it,[43] the bishop expressed his consent with the abbot's decision provided that, indeed, steps would be taken to assure that at Vienna Mendel would lead the life of a religious. On October 22, Napp did contact Pater Auremundus Jahn of the convent of the Brothers of Mercy at the Tabor Strasse, Vienna, asking him to provide board and lodging for the two years of Mendel's study, adding that because of the generally rising cost of living the abbey would remuneration the convent for the expenses.[43] The prior answered promptly on October 26: He regretted it very much, but he was unable to grant the request. The convent was already so overcrowded that two monks had to board per room and visitors passing through the city constantly occupied the three guest rooms.[43] Napp had no other choice, as he noted on the back of the prior's letter, than to send Mendel to Vienna with the instruction to find accommodation in some other monastery or a religious house. And so, on October 27, 1851, Mendel boarded the night train to Vienna to start a new chapter of his life.

The Imperial-Royal Capital and Residence: Vienna[95]

After Lipník, Opava, Olomouc, and Brno, the 29-year-old Mendel was on his way to a two-year residency in a city that was farthest away from his birthplace, was older than the other four cities he had lived in, was certainly largest of them all, and was both politically and culturally the most sophisticated one. Traveling to Vienna, Mendel was again following the ancient amber route,[96] this time to a place some 40 kilometers west from the point where it crossed the Danube River at Carnuntum. In Roman times, the latter was the larger of the two settlements on the Danube River, but now only a few stone walls and archeological artifacts remain of it. The prehistoric settlement that developed into the city of Vienna was on the right bank of the Danube on the second lowest of the four terraces constituting the *Wienerwald* (Vienna Woods), the foothills of the eastern Alps. Spreading from the lowest terrace before the settlement was the *Marchfeld* (Morava Field), the site of one of the decisive battles which determined the history of central Europe 573 years before Mendel's arrival in Vienna. The plain is part of the Vienna Basin reaching to Moravia in the north, Slovakia in the northeast, and the Hungarian Puszta in the southeast. The strategic location of the settlement on the intersection of the north-to-south and west-to-east trading routes made it a highly contested place and so a place that changed hands frequently. From the seventh century BC on, Scythians, Celts, and various Germanic tribes may have claimed it as theirs, until the Romans took it over at the end of the first century BCE, calling the fortified camp they erected on the site, Vindobona. The names Wien (German), Vídeň (Czech), and Vienna (English) all derive from it, but whether "Vindobona" derives from "good wine" or from a more ancient Celtic name remains contentious. Of the earlier inhabitants few traces remain, but a visitor to the city is reminded of the early Roman presence by a street sign (*Mark Aurel-Strasse*, named after the Roman emperor, who allegedly died in Vindobona), the outlay of the streets in the inner city (it reflects the orientation of the ancient Roman walls), and an assortment of excavations. After the Romans left in the fifth century CE, waves of invading Huns, Goths, Avars, and Slavs passed through the territory, until in the ninth century, Charlemagne made it the eastern margravate (*Österreich*) of the Holy Roman Empire. Subsequently and successively, three noble families claimed the city, which was granted municipal privileges in 1221, as their own: the Babenbergs (from 976 to 1246), the Přemyslids under Otakar II (from 1230 to 1278), and the Habsburgs (from 1278 to 1918, with minor interruptions). The Habsburgs, as the longest ruling family of Austria left the most enduring mark on Vienna, especially from the period of 1648 on. To wit, it was only after the 30 Years War that the Habsburg emperors of the Holy Roman Empire, the Austrian Empire, and, finally, the Austrian–Hungarian Empire used Vienna as their permanent residence and began to beautify it. They baroquized the city so extensively that now only a few buildings in the inner city (most prominently the *Stephansdom*, the Cathedral of St. Stephan) remind visitors of its original Gothic character (Fig. 6.6). The prominent Baroque buildings from the seventeenth to eighteenth centuries include parts of the *Hofburg* (the residence of the imperial family) and the *Karlskirche* (Charles' Church).

The city had undergone the grandest (and most expensive) facelift under Franz Joseph in the second half of the nineteenth century. A precondition to the remodeling was the removal of the two rings of fortifications encircling the city. The ring had enclosed the inner city since the twelfth century and when it proved its worth during the Turk siege in 1529, it was rebuilt and strengthened in 1548. It again served the city well by keeping the Turks out when they returned in 1683. Since both sieges devastated the villages outside of the ramparts, a second fortification ring, the *Linienwall*, was erected in 1704 around them. By the end of the eighteenth century, however, both rings had not only become obsolete, but also an impediment to further growth of the city. Napoleon demonstrated twice to the Viennese the ineffectiveness of the walls in a confrontation with a modern army, first when he took the city in 1805 and then again in 1809. Obviously, sooner or later, the walls had to go. In the Austrian Empire, however, when something needed to be done, later was the preferred option. And so, when Mendel arrived in Vienna in 1852, the walls were still standing. It would not be until 1857 that Franz Joseph would finally issue an order to begin with their demolition. What followed was an unprecedented burst of urban renewal, which changed the city's appearance from month to month and year to year. Although the burst occurred after Mendel's completion of his studies he nevertheless witnessed it on his later frequent trips to the city. On each visit he would encounter something new, a new street, a new

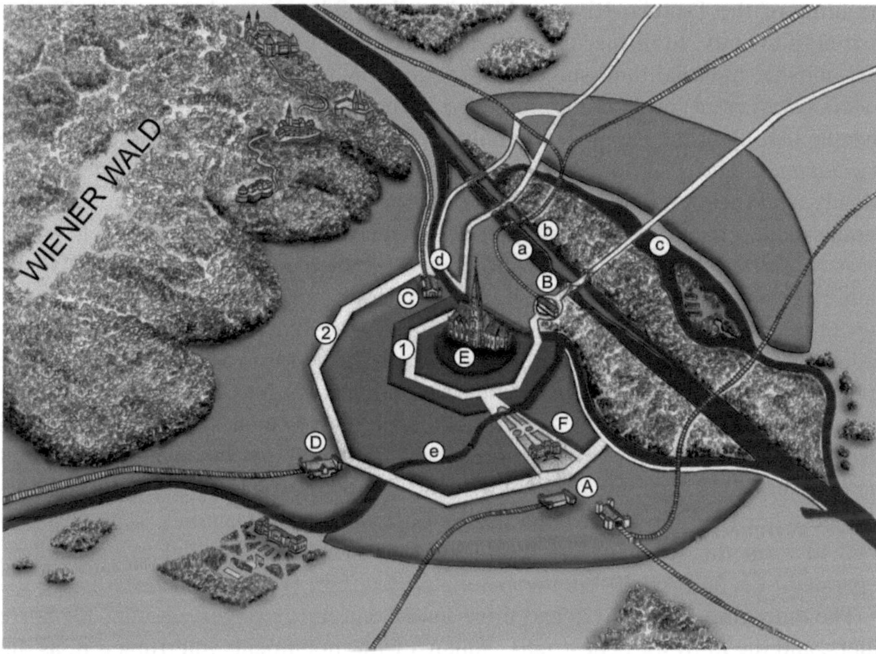

Fig. 6.6 Basic topography of GreaterVienna. *Railway stations*: (A) *Südbahnhof*, (B) *Wien* Mitte, (C) *Franz-Josefs-Bahnhof*, (D) *Westbahnhof*. *Waterways*: (a) *Donau* (Danube), (b) *Neue Donau* (New Danube), (c) *Alte Donau* (Old Danube), (d) *Donaukanal* (Danube Channel), (e) *Wienfluss* (Vienna River), (1) Ringstrasse, (2) Gürtel (Based on *Freytag-Berndt Führer durch Wien und Umgebung*. Verlag Freytag-Berndt und Artaria, Wien 1967)

building, a new park. The ring of the inner fortifications became the center of the initial activity. Their demolition from 1858 to 1860 proceeded with verve to the tune of the *Demolierpolka* (Demolition Polka) composed by the rising star on the Viennese musical haven, Johann Strauss the Younger (1823–1899). The massive walls themselves took up much space, but between them and the suburbs were also fields, the *glacis* (from French *glace*, an embankment sloping down gradually from a fortification), which were left undeveloped in order to afford a better view of the movement of potential enemies. With the fall of the walls all this space became available for urban development.

Franz Joseph entrusted the development to the most renowned architects he could get and they acquitted themselves of the task quite sensibly. Unlike Haussmann, who tore down old Paris in order to build a new one in its place, Franz Joseph's architects left the historic inner city virtually intact and integrated it harmoniously into the constructions encircling it. The backbone of the new construction was the *Ringstrasse*, a broad boulevard embracing the center. Not really a ring, but rather an irregular polygon, the *Ringstrasse* connected to the main streets of the center centripetally and the avenues of the suburbs centrifugally (Fig. 6.6). Along one side of the polygon ran the *Donaukanal* (the Danube Channel), one of the arms of the Danube River, while the other sides flanked the newly formed district of the city. In former times, the area between the inner city and the Danube was marshy, malarial, and subject to flooding. When major floods did occur the waters overflowed into the city. Included in the construction plans was, therefore, also the taming of the Danube, beginning with the arm closest to the inner city, the *Donaukanal*. The 17-kilometer-long arm was straightened and widened in 1598 already, and then regulated again in the 1860s, so that it began to look more like a human-made channel rather than a river (hence its name). The channel, which formerly skirted the inner city now flows through the modern city and returns to the main river at the city's periphery. It thus demarcates an island, on which lie two of the 23 districts of the city. Another waterway, the *Wienfluss* (Vienna River) or simply *Wien*, which originates in the *Wienerwald*, skirts the inner city on the opposite side to the *Donaukanal*, and eventually flows into it. The main marshy area of the Danube with its many streams and branches was regulated in 1870–1878. All branches except three were eliminated and the swamps filled. Of the three branches, one, the *Alte Donau* (Old Danube), was left bending into a bow and the other two, the *Donau* and the *Neue Donau* (New Danube), were straightened to form the string of the bow (Fig. 6.6). They run parallel to each other, separated by a narrow, 20-kilometer-long *Donauinsel* (Danube Island).

The emerging *Ringstrasse* had three lanes, one for carriages, another for horse riders, and the third for pedestrians. Squares with monuments interrupted it, as did parks, which contributed to Vienna's reputation as being one of the "greenest" cities in the world. They then added to the parks already in existence: Prater, Augarten, Belvedere Park, Schönbrunn Park, and others. Prater, the largest of them, was formerly a hunting ground of the imperial family until Joseph II opened it to the public; it is located on the island formed by the *Donaukanal* and the Danube. Several other parks, Belvedere, Schönbrunn, and others, derived their names from the palaces with which they were associated. Because of all of these changes, the large, hilly, and woody area of the *Wienerwald*, which was once at some distance from the city, is now at its doorstep.

In the remaining space on the *Ringstrasse* arose, in the period from 1861 to 1890, monumental buildings, each in a different style, according to its function. Actually, the construction of one of these began in 1856 already, in response to an event, about which Mendel wrote to his parents: *You must have heard about the attempted murder attack on the Emperor and the fortunate warding off of the peril. Before I left Vienna I saw the Emperor fully recovered again. The murderer named Libeny was hanged on the 26 of last month.*[97] The assassin attacked Franz Joseph as he was strolling on the city ramparts and attempted to stab him in the neck. Fortunately, the Emperor's collar button deflected the blow and he escaped with a mere scratch. Attributing his good fortune to God's holding a protective hand over a Habsburg, Franz Joseph made a vow (*votum* in Latin) to have a church built at the place near the site where the attack took place, commemorating this "miracle." Thus originated the *Votivkirche* (Votive Church), erected in a neo-Gothic style to evoke the medieval faith in God. Next to arise on the *Ringstrasse* was the *Staatsoper* (State Opera House), built from 1861 to 1869 in a style difficult to define because it incorporates elements of several past trends. Following the *Staatsoper* were the twin buildings of the *Naturhistorisches Museum* (Natural History Museum) and the *Kunsthistorisches Museum* (Museum of Art History) constructed from 1871 to 1890 in a combination of Baroque and Renaissance styles; the *Neues Rathaus* (the New City Hall), built from 1872 to 1883 in the style of Flemish late Gothic, the style in which cities first asserted their independence; the *Universität* (the University) put up between 1873 and 1883 in a neo-Renaissance style judged appropriate for a place of learning; the *Reichsrat* (the Parliament) built between 1874 and 1884 in a neoclassical style, as an allusion to the origin of democracy in classical Greece;[98] and the *Burgtheater* (Castle Theater) rebuilt between 1874 and 1888 in the Italian Renaissance style as a reminder of the revival of the arts in that country. Opinions differ on the aesthetic value of the great buildings along the *Ringstrasse*. Some pundits take the *Ringstrasse* for little more than a nineteenth century high-class version of Disneyland, only more pompous and theatrical. Others, while deploring its unabashed eclecticism, hold the entire assembly of different styles for a new *Ringstrasse* style. As for the Viennese they love what they got. Verily, the *Ringstrasse* style is well attuned to their tastes and the tastes of the period. Similar, though not quite as showy, constructions went up also at the belt bared by the removal of the *Linienwall*. There, along the *Gürtel* (Belt), sprang palaces (*palais*, in the parlance of the Viennese) of the aristocracy and of the *nouveau riche* bourgeoisie. These developments united the old city with its suburbs—the *Vorstädte* (those between the *Ringstrasse* and the *Gürtel*) and the *Vororte* (those outside the *Gürtel*) into one greater Vienna. The unification necessitated a subdivision of the city into *Bezirke* (Districts) numbered by Roman numerals and named by their traditional epithets. The old city became *Bezirk* I, the *Innere Stadt* (inner city), followed by *Bezirk* II (*Leopoldstadt*), Bezirk III (*Landstrasse*), and so on, to *Bezirk* XXIII (Liesing) of modern Vienna. Parallel to the expansion of the city grew also its population. In 1850 it reached 431,000 and in subsequent years it swelled at an ever-faster rate. Its composition diversified as well. In the cities, in which Mendel lived before coming to Vienna, there were essentially only two ethnic groups: the Germanic people and the Czechs. In Vienna, one could encounter individuals from three dozens of ethnic groups encompassed by the Austrian Empire. They differed not only in their languages but also in their behavior, customs, and culture. Each of these groups contributed in its own specific way to the cuisine, fashion, traditions, and culture of the city, and so to the city's atmosphere. Although they clashed occasionally, on the whole the groups got along remarkably well in this fabulous melting pot of nationalities.

The University of Vienna[99]

The university, at which Mendel was about to enroll, had a long and distinguished history. A Habsburg, Rudolf IV, *der Stifter* (the Founder), Duke of Austria (1358–1365), founded it in 1365, some say not because he was particularly fond of learning, but because of a rivalry with his father-in-law, Charles IV. The latter founded the first university in central Europe in Prague 17 years earlier. Rudolf founded the *Alma mater Rudolfine*, but did not provide any building for it in which to conduct the instructions, so for a while, the professors had to teach their students in private houses and churches. It was not until 1384 that another Habsburg, Duke Albrecht III (1349 or 1350–1395), acquired the first house for the university—the *Collegium Ducale* (the Duke's College), facing the Dominican Monastery on what is today the *Postgasse* (Fig. 6.7), not far from Stephan's Dome. Later, the university acquired other buildings in the same area, the *Collegium Iuristarum* (Lawyer's College) and the House of Physicians. By the fifteenth century it owned several buildings which were scattered all over the inner city and were intermingled with merchant's and craftsmen's houses, churches, and monasteries in the area near the *Stubentor* (the Barrack Gate). Since the cohabitation of students, merchants, and monks was not always harmonious, in 1623 Emperor Ferdinand II (1578–1637) made the Jesuits (whom his predecessor, Emperor Ferdinand I, 1503–1569, summoned to Vienna) responsible for the university. The Jesuits consolidated the university by erecting, in 1624–1631, a whole complex of Baroque buildings, the Academic (Jesuit) College, which brought the different faculties together, if not under one roof, at least under one group of roofs. The complex centered on the *Jesuitenkirche* and the *Universitätshaus* right next to it, both on the *Sonnenfelsgasse* (Fig. 6.7). The former became one of the most opulently furnished churches in Vienna. The Universitätshaus, separated from the church by the narrow *Jesuitengasse* (Fig. 6.8) became the seat of the university administration, which included the *Rektorat* (the office of the *Rektor*, the head of the university) and the *Akademisches Senat*. It retained this function until 1884, when the whole university moved to the new building on the *Ringstrasse* and the complex of buildings in the inner city came to be referred to as the *Alte Universität*, the Old University. It must have been in the old *Universitätshaus* that Mendel registered and was admitted to university studies.

In the eighteenth century, under Maria Theresia and Joseph II, a century before the move, the Old University grew by the addition of new buildings. The two rulers also reformed the university. As part of the program of restricting the influence of the church on the schools of higher education, they lifted the Jesuit control of the university and in 1773 dissolved the order. As part of the university's expansion, they ordered the construction of a building opposite to the *Universitätshaus*. The showpiece of the building was the *Aula*, the Great Hall, on the second floor, with its richly decorated ceiling and walls (Fig. 6.9). Maria Theresia and her husband, the Emperor Franz Stephan von Lothringen inaugurated the Aula personally in 1756, and it then quickly became one of the most renowned salons in the town. During the 1848 revolution, however, the rebelling students used it as their center of command, and in the aftermath of the uprising, the *Aula* and the whole building fell into disgrace. For a while it was even used by the military as barracks before the Government restored its favor on it once more. The university, however, never got it back. In 1857, the unforgiving Franz Joseph made it instead the seat of the *Kaiserliche Akademie der Wissenschaften* (the Imperial Academy of Sciences), which Emperor Ferdinand founded in 1847. Its successor, the *Österreichische Akademie der Wissenschaften*, uses it to this day as its *Präsidium* (headquarters).

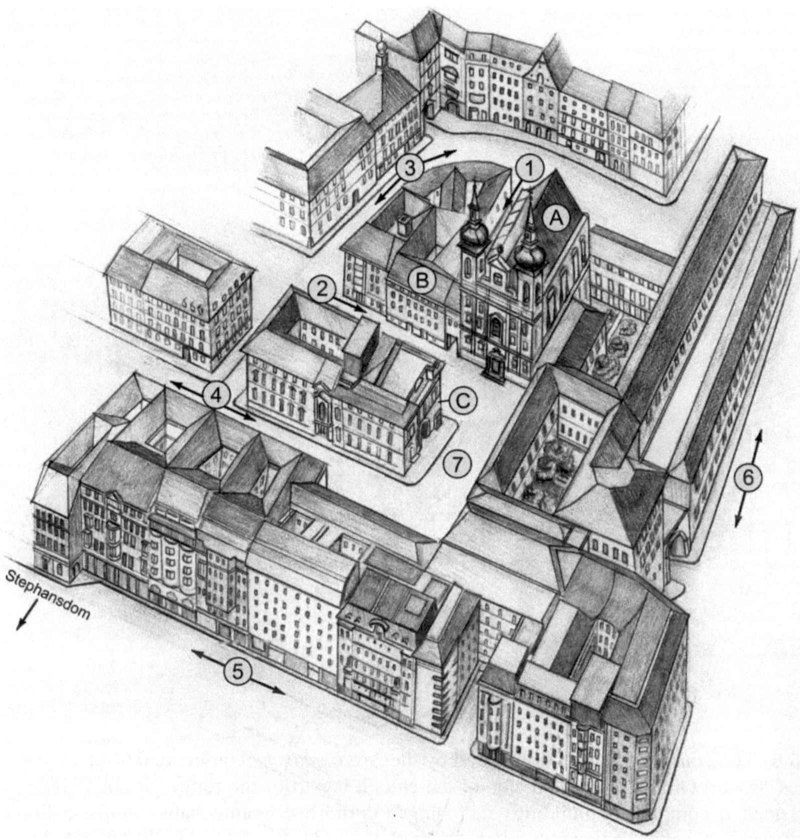

Fig. 6.7 The Old University of Vienna in the inner city. In the Middle Ages, buildings belonging to the university or serving as student hostels were scattered through the depicted area. In Mendel's time, however, the university began the process of moving out of the inner city. Mendel, though, still registered at the *Universitätshaus* (B) next to the *Jesuitenkirche* (A, the Jesuit Church). These two buildings were separated by a narrow street, the *Jesuitengasse* (1), the Jesuit Street. Opposite the *Universitätshaus*, across the *Sonnenfelsgasse* (2) named after Maria Theresia's legal adviser, Josef von Sonnenfels, stood a building (C) with the magnificent *Aula* (Great Hall) and lecture halls, in which Mendel might have heard some lectures before the building was given to the *Kaiserliche Akademie der Wissenscaften* in 1857. Some of the other streets in this area are *Schönlaterngasse* (3), *Bäckerstrasse* (4), *Wollzeile* (5), and *Postgasse* (6), as well as the square *Universitätsplatz*, now *Dr. Ignaz-Seipel-Platz* (7). (Based on Brook, S. B. *Vienna. Eyewitness Travel*. Dorling Kindersley. London 1994)

The governing body of the university was the *Consistorium*,[100] the university board, headed by the *Rektor* (president) and consisting of 18 additional members, the *Akademische Senat*. The latter consisted of the *Prorektor* (vice-president); *Kanzler* (chancellor); four *Dekanen* (deans), elected by the *Professorenkollegien* of each of the four faculties; four *Senioren* elected by each of the *Doktorenkollegien* of the four faculties; and four *Prodekanen* and four *Prosenioren* elected in a

Fig. 6.8 The *Universitätsplatz* dominated by the *Jesuitenkirche* consecrated to Sts. Ignatius and Francis Xavier. On the right-hand side of the church is part of the former Jesuit College, which represented a complex of buildings including dormitories, lecture halls, shops, a library, an observatory, a theater, and a wine cellar. On the left-hand side of the church, separated by a gate entrance to the *Jesuitenstrasse*, is a part of the *Universitätshaus*. In front of the *Universitätshaus* is a house, which was later replaced by one with the new great *Aula*

corresponding manner. The *Profesorenkollegium* (College of Professors) included all the professors/teaching members of a given faculty.[101] The *Doktorencollegium* (College of Doctors) comprised all the graduates (former students and alumni) with a doctoral degree from one of the four faculties, regardless whether they taught at the university or not. The professors could not be elected to become deans, and since the doctors always outnumbered them, they had little influence over the selection of the deans. Naturally, they were not happy with this arrangement. The professors were supervised by the Imperial Education Commission, which also selected and appointed the *Rektor*, the *Prorektor*, and the *Kanzler*. The university had four faculties: Law, Humanities, Theology, and Philosophy. The Philosophical Faculty, which was in fact the Philosophical Institute, was the weakest of the four. The commission prescribed the courses the students had to take and in what order, as well as the textbooks from which they had to study. Course attendance was compulsory and was recorded. At the end of each semester and each academic year, students had to pass general examinations. Like the professors, the students were

Fig. 6.9 The seat of the Austrian Academy of Sciences opposite to the former *Universitätshaus* of the Old University of Vienna. On the second floor of this building is the Great Aula. The building belonged to the university, but after it became the focal point of the student uprising in 1848, it was given to the academy

dissatisfied with the situation, and they showed it during the 1848 uprising. As a result some of the first reforms of the revolution forced onto the emperor concerned education. Surprisingly, they survived the postrevolutionary repression.

The reforms included abolition of the Imperial Educational Commission and the creation of the Ministry of Culture and Education in its stead. To the helm of the ministry, Emperor Franz Josef appointed Count Leopold ("Leo") von Thun-Hohenstein (1811–1888). The minister was by no means a liberal, but otherwise he was a reasonable man. He realized that the Austrian educational system was in a bad state and that if it was ever to catch up with the much more advanced systems in Prussia and other European countries, it would have to be thoroughly overhauled. He set out to do this with the two advisers he chose, Franz Exner (1802–1853), professor of philosophy, and Herrmann Bonitz (1814–1888), professor of philology. The trio instituted changes which came to be referred to as the "Thunian reforms."[101] They abolished censorship of scholarly works and lifted restrictions on lectures and textbooks. They granted more autonomy to the *Proffesorenkollegium*, which included the right to elect their own deans, and thus substantially reduced the influence of the *Doktorenkollegium*. They transferred the propaedeutic function of the Philosophical Institute, together with the professors, to the *Gymnasien*, and thoroughly revamped the Philosophical Faculty. They placed it on equal footing with the other three faculties and encouraged the recruitment of a new generation of research-oriented professors in an effort to turn the Philosophical Institute into a research institution. Of course, only professors who were not active participants of the 1848 revolution and were not disloyal to the Emperor stood a chance of being appointed to the new positions. This restriction notwithstanding, the effect of the reforms was remarkable. It was as if a fresh breeze began to blow through the stuffy corridors of the Austrian universities.

Getting Settled at Vienna

This, then, was the city to which Mendel traveled by train on the night of October 27, 1851. For a description of what it was like to use the Imperial-Royal Railways, we refer the reader to a testimony of an eyewitness.[102] Although he used the railway somewhat later than Mendel, things changed so slowly in the Austrian Empire that we may safely assume that his account applies also to the period we have now reached in our narrative. In Mendel's time, Vienna had three main railway stations, which are now called *Westbahnhof*, *Südbahnhof*, and *Franz-Josefs-Bahnhof*. Mendel used the *Kaiser-Ferdinand-Nordbahn* which connected Vienna with Brno, Olomouc, and Prague and which terminated at the *Franz-Josefs-Bahnhof* in the *Leopoldstadt* District of Vienna. Since the distance from the station to the inner city was too large to walk, especially with the burden of luggage, we can assume that Mendel took a *Fiaker*, a horse-drawn cab, to his destination. But what was Mendel's destination? Upon his arrival in Vienna, Mendel had two urgent matters to take care of: He had to find a place to stay and he had to matriculate at the university. It seems that he dealt with these tasks in the following order: He first found a short-term accommodation. The first few days he stayed in a house, most likely an inn or a hostel, on *Landstrasse* 79 (today's *Salmgasse*). The house no longer stands; it was torn down in the nineteenth century.[103] After securing a place to stay, Mendel then went to see Baumgartner and initiated the matriculation process, and while that was in progress, he searched for a long-term accommodation. The first step in the matriculation process was to deliver Napp's letter to Baumgartner and solicit his help. It seems that he was lucky and caught Baumgartner in his office at the ministry and that the minister had indeed intervened on Mendel's behalf in the matter of matriculation. He might have even directed him to the address, where he ultimately rented a room. Baumgartner was, at that time, a *Prorektor*, a vice-president, of the university and so a note from him should have cleared any hurdles, which might have stood in the way of Mendel's late matriculation. As it turned out, however, it was no longer possible to enter Mendel's name in the matriculation book[104,105] listing ordinary students (*ordentliche Hörer*) enrolled for the academic year 1851/1852 (the word derives from the Latin *matrix*, meaning a list). He was therefore admitted as an extraordinary student (*auserordentlicher Hörer*), whose name would not appear in the matriculation book, but rather would be listed in a separate document.[104] Even as an extraordinary student, Mendel had to produce a number of documents establishing his identity, origin, profession, and so on, as well as a letter stating the reasons for his late matriculation. For this part of the admission procedure, however, Mendel was well prepared, for he knew from his previous experiences how the k.k. bureaucracy worked. In the letter from November 5, 1851, he blamed the delay on unforeseen circumstances.[106,107] A part of the matriculation process was the selection of courses he wanted to take, the free selection being another achievement of the Thunian reforms. As one might have expected, he chose courses in physics and natural history offered by the Philosophical Faculty in the subjects he anticipated would prepare him for his teaching career. Somewhat unexpectedly, he signed up for more physics than natural history courses. According to the *Verzeichniss der Vorlesungen, welche der Studierende zu hören beabsichtigt* (index of lectures which the students intends to hear),[102] he took physics in all four semesters (in the first semester exclusively and in the fourth nearly so) and natural history largely in the second and the third semester (Tables 6.1 and 6.2). (The academic year had two semesters, the winter semester from October to April and the summer semester from April to July.) Recently, an attempt has been made to tie Mendel's status of an extraordinary student (*auserordentlicher Hörer*) to his signature on the 1848 petition submitted to the National Assembly.[108] The suggestion is that Mendel could not enroll as an ordinary student at the University of Vienna because of the petition and so the sympathizers with the 1848 revolution (Baumgartner and Doppler) made him, in effect, a "clandestine" student of physics. This interpretation, however, is contradicted by all the historical facts concerning his enrollment,[109] which we described above. His late arrival was Napp's fault,[109] which Napp then tried make good by involving Baumbartner (and indirectly also Doppler) in the case. There was nothing clandestine about Mendel's becoming a student of physics at the university. The documents pertaining to the case are all now accessible in transcription and translation[109] for all to see and to dispel any fantasies about a clandestine student punished for his revolutionary activities. The university registration document also contains, in addition to the *Verzeichnis*, a *Nationale*,[110] a sort of identification paper with Mendel's personal data. From this document we learn where Mendel found his long-term accommodation. The address is given as *Landstrasse* 358, in the *Landstrasse* District (Fig. 6.10). It was a house at the corner of *Landstrasse* and *Invalidenstrasse*; in 1910 the city authorities changed its address to *Invalidenstrasse 13*. It was, in fact, a complex of apartment houses enclosing a large, square courtyard. The main entrance to the house was, in Mendel's time, at the street level, but later as the level was raised the entrance ended up below it (Fig. 6.11). The house belonged to the religious order of St. Elizabeth, which had a nunnery, church, and hospice in the same area.[43] The nuns rented apartments and single rooms to boost their income. Whether Mendel rented a separate room or sublet a room from another tenant is not known. Also unknown is how he found the accommodation. It is rather unlikely that upon his arrival in Vienna he ambulated from one monastery to another inquiring about a free room. More probable, he acted on a tip, which might even have come from Baumgartner. From Mendel's point of

Table 6.1 Courses that Mendel took at the University of Vienna

Semester	Course	Lecturer	Place	Location	Frequency/week	Time (hours)
Winter 1851/1852	Demonstrative experimental physics	Christian Doppler	Physikalisches Institut	Hauptstrasse 104, Erdberg	5×: Mo, Tu, We, Th, Fr	10–12
Summer 1852	Demonstrative experimental physics	Christian Doppler	Physikalishes Institut	Hauptstrasse 104, Erdberg	5×: Mo, Tu, We, Th, Fr	11–13
	Zoological systematics	Rudolf Kner	Akademische Schulgebäude		2×: Tu, Th	12–13
	Zoological practicum	Rudolf Kner	Akademische Schulgebäude		3×: Mo, We, Fr	12–13
	Morphology and systematic of phanerogamic plants	Eduard Fenzl	Musealgebäude des Universitäts Gartens		5×: Mo, Tu, We, Th, Fr	5.45–6.45
	Exercises in plant analysis and description	Eduard Fenzl	Musealgebäude des Universitäts Gartens		2×: We, Fr; second half of May only	16–19
Winter 1852/1853	Demonstrative experimental physics	Andreas von Ettingshausen	Physikaliches Institut	Hauptstrasse 104, Erdberg	5×: Mo, Tu, We, Th, Fr	10–12
	Zoology, especially for students of medicine and pharmacology	Rudolf Kner	Akademische Schulgebäude		5×: Mo, Tu, We, Th, Fr	12–13
	General and medical pharmaceutical chemistry	Joseph Redtenbacher	Chemisches Hörsaal, Theresianum	Favoritenstrasse 15	5×: Mo, Tu, We, Th, Fr	8–9
	Methods of analytical chemistry	Joseph Redtenbacher	Chemisches Hörsaal, Theresianum	Favoritenstrasse 15	5×: Mo, Tu, We, Th, Fr	9–10
	Calculation, arrangement, and use of logarithmic-trigonometric tables	Franz Moth	Saal No.31, Theresianum	Favoritenstrasse 15	1×: Sa	9–10
	Anatomy and physiology of plants	Franz Unger	Alte Universitätsgebäude		3×: Mo, We, Fr	18–19:30
	Practical exercises in the use of the microscope and setting up physiological experiments	Franz Unger	Musealgebäude des Universitäts Gartens		1×: Su	11–13

	General paleontology[a]	Friedrich Zekeli	Geologische Reichsanstalt		4 h	
	Guide shells[a]	Friedrich Zekeli	Geologische Reichsanstalt		2 h	
Summer 1853	Preparation and use of physical instruments	Andreas von Ettingshausen	Physikalisches Institut	Hauptstrasse 104, Erdberg	3×: Mo, We, Fr	16–17
	Higher mathematics	Andreas von Ettingshausen	Physikalisches Institut	Hauptstrasse 104, Erdberg	2×: Tu, Sa	16–17:30
	General and medical pharmac. chemistry	Joseph Redtenbacher	Chemisches Hörsaal, Theresianum	Favoritenstrasse 15	5×: Mo, Tu, We, Th, Fr	7:30–8:30

[a]Not in the *Vorlesungsverzeichnis* (index of lectures) but in the *Nationale* (identification papers)

Table 6.2 Summary of courses Mendel took at the University of Vienna in the different subjects (Modified from Orel and Kuptsov 1983; see Table 6.1)

	Hours per week in the academic year/semester					
	1851/1852		1852/1853			
Subject	I	II	III	IV	Total	% total
Physics[a]	10	10	10	3	33	36.5
Mathematics	–	–	1	3	4	4.4
Chemistry	–	–	10	10	20	22.1
Zoology[a]	–	5	5	–	10	11.1
Paleontology	–	–	6	–	6	6.6
Botany[a]	–	11	6.5	–	17.5	19.3

[a]Includes practical exercises

view, it was a good solution to his problem, for the house was at a walking distance from the various institutions at which he attended lectures. Whether the solution satisfied Napp was doubtful. Although the house belonged to a religious order, it was by no means part of a convent and so it did not fulfill the Bishop's stipulation. Mendel, however, could fend off any criticism by pointing out that under the circumstances, it was the only option he had. Having thus accomplished the two most important chores of his first few days in Vienna, he could turn his attention to the actual purpose of his presence in the city—his studies.

Student of Physics

As we mentioned earlier, Mendel fully devoted his first semester at the university to physics. The first reason behind this decision was his preference for this subject, enforced by experience at the qualifying examination one year earlier. The second reason was purely formal and coincidental: By starting with physics, he did not miss any lectures despite his late arrival because in that particular semester the physics course happened to be delayed by one month. The circumstance responsible for the delay had to do with the Thunian reforms. The integration of the Philosophical Institute into the *Gymnasien* and the emphasis on teaching physics and natural history in their seventh and eighth grades led to a heavy demand for professors competent in teaching those subjects and to a pressure on the universities to produce them.[94] While in several other European countries the emphasis at the universities had been shifting from humanities to natural sciences, at the University of Vienna, these subjects were still poorly supported, inadequately staffed, and generally neglected. In physics in particular, in other European universities, professors were making one important discovery after the other, but the Vienna University had nothing of that sort to report. This was the message Leo Thun conveyed to the emperor on December 1, 1849, when asked to inform him about the status of physics at the imperial universities. He pointed out that in the imperial city itself, physics teaching had to content itself with a crowded room in a building occupied by the military and with

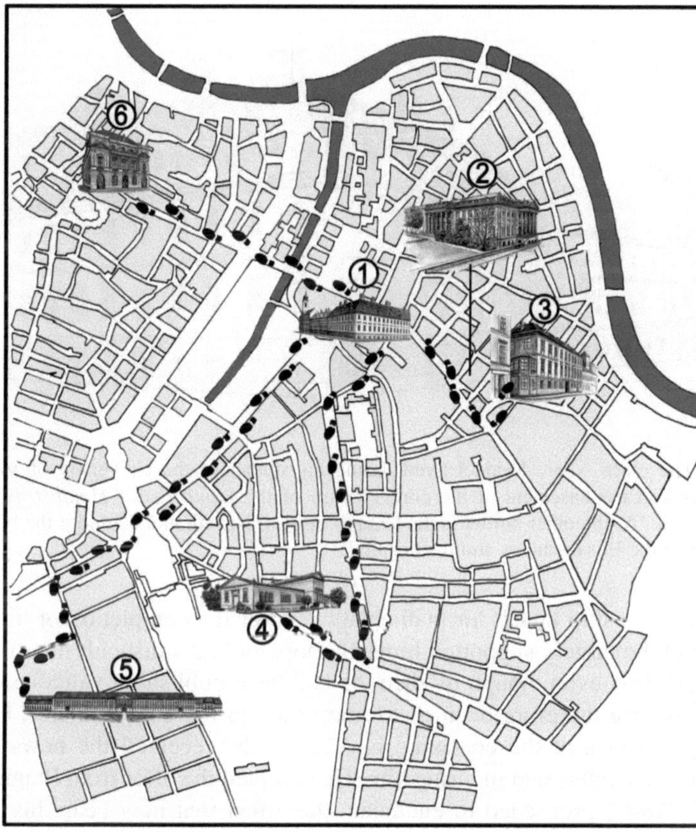

Fig. 6.10 Mendel's Vienna: The location of the buildings in which he attended lectures and practical exercises given by the university professors. The highlighted streets indicate probable routes Mendel took when he walked to the lectures from the place, where he lived

demonstrations performed with obsolete instruments. He urged the emperor to found a new physics institute, have it furnished with modern equipment, and appoint a first-rate physicist, specifically Doppler, to lead it. The emperor listened to and acted on Thun's recommendations surprisingly fast.[94] On January 17, 1850, he approved founding the Physics Institute and appointed Doppler as its head.

Johann Christian Doppler (1803–1853; Fig. 6.12)[111] was the son of a stonemason in Salzburg.[112] Junior showed a certain talent for his father's profession, but since he was of a delicate health, the father decided to make a businessman out of him. This intent came to nothing when a local teacher discovered that junior had an

Fig. 6.11 The place where Mendel lived during his studies at the University of Vienna. His apartment was in the basement of the corner house of the *Landstrasser Hauptstrasse* and the *Invalidenstrasse*. At the other corner is the Church of St. Elisabeth and behind it the buildings of the Convent of the Elisabethines and the hospital

even more promising knack for mathematics. After the completion of his studies, during which he largely supported himself, Doppler had a difficult time finding an adequate position even though by then, he had been publishing valuable scientific contributions. He therefore decided to immigrate to the United States, but while applying for a visa at the consulate in Munich, he received the news that two institutions were interested in hiring him. He accepted the offer from Prague, where he then (in 1842) proceeded to enunciate the effect that now bears his name.[113] After a couple of other positions at other places, he accepted the offer of the professorship at the University of Vienna and of the directorship of the Physics Institute. Doppler plunged with great energy into the organization of the institute, but discovered quickly that establishing a well-functioning physics research and teaching center at the *Unversitätshaus*, in the limited space and with the outmoded equipment available, was nearly impossible. Eventually, however, the university administration gave in to his nagging and rented for him two floors and a large garden of a private house in the *Landstrasse* District, at the corner of the *Erdberger Hauptstrasse* and *Parkgasse* (Fig 6.13).[114] The rooms did not become available until September 29, 1851, and so, after the completion of their renovation and furnishing, the Institute of Physics opened five weeks after the beginning of the winter semester, at the same time as Mendel arrived in Vienna.[94,115] The students, for some reason called the *Eléven*,[116] were formally divided into ordinary and extraordinary, but Doppler seemed to treat both groups equally. According to the institute statutes,[117] which Doppler prepared and the ministry approved, the number of the ordinary *Eléven* should have been limited to two dozens, whereas the number of extraordinary students remained unspecified. The total number of *Eléven* in any one year might have been about 40. The education was tuition free and the 12 ordinary *Eléven* could even apply for a one-year stipend (with the possibility of a

Fig. 6.12 Mendel's professors of physics. Johann Christian Doppler (1803–1853). Andreas von Ettingshausen (1796–1878)

half a year extension) of 40 florins per semester, provided they were talented and needy. The stated goal of the Physics Institute was to educate physics teachers for *Gymnasien* and to prepare them not only in terms of knowledge but also in the ability to demonstrate the nature of physical phenomena by simple experiments. Doppler planned the physics course for four semesters, of which the first three would be devoted to "demonstrative experimental physics" and the fourth semester to the "preparation and use of physical instruments." It was a full-time occupation for both the students and the professor with his assistant, Franz Pekarek, who was, like Mendel, a substitute teacher at Znojmo during the winter semester of 1849/1850. The professor lectured for two hours every day from 10 AM to 12 PM, five times a week, and accompanied his discourse with demonstrations. The students then spent the rest of the day experimenting in the laboratories, researching the scientific literature in the library, evaluating or planning experiments, and writing reports. Doppler designed the last semester so as to meet specifically the needs of the future *Gymnasien* teachers, who may not have at their disposal all the instruments necessary for practical demonstrations. He showed them how to improvise in such cases by simple means. Doppler's method of instruction was quite revolutionary at Austrian universities in that it led students to appreciate the importance of experimentation in research and teaching and of independent thinking. These messages were certainly not lost on Mendel.

Unfortunately, Doppler's poor health did not allow him to realize his plans. At the beginning of the winter semester 1852/1853, he fell so ill that he had to ask the ministry to suspend him from his duties. Toward the end of the year, the chronic lung disease, which tormented him since his Prague years, progressed to an acute stage.

Fig. 6.13 The Physical Institute at the University of Vienna at the corner of *Erdbergstrasse* and *Parkgasse*

Hoping that a change of climate might stave off its progression, he moved to Venice, but by then it was too late for him. He died there on March 17, 1853. The ministry, apparently not expecting him to recover any more, removed him from the directorship on November 1, 1852, already. This news certainly could not have brightened his last days. In his stead the ministry appointed Andreas von Ettingshausen (1796–1878; Fig. 6.12)[37,118]

As the son of a major general, Ettingshausen was expected to embark on a military career and so he received initially education at various military schools, including a bombardment institute. When, however, it began to look as if peace would last in Europe for a while and the prospect of a rapid advancement dimmed for the young Ettingshausen, he changed his orientation to mathematics and physics with the goal of becoming a professor of these two disciplines. He advanced fairly smoothly toward this goal and in 1834 succeeded Baumgartner in the chair of physics at the University of Vienna. In this position he made a name for himself as a promoter of daguerreotype in the imperial capital. In 1848 he switched over to the Engineering Academy, and four years later, when the latter was turned into a military school, he switched again to the Polytechnic Institute, only to take over a few months later the directorship of the Physics Institute at the university. Although the ministry rushed the appointment so as to avoid an interruption in instruction, Ettingshausen may not have begun teaching until the middle of November 1852, about six weeks after the beginning of the winter semester.

It is doubtful whether Mendel was able to hear any of Ettingshausen's lectures in that semester, not only for this reason, but also because their timing clashed with the natural history courses. He did, however, sit in on the entire physics course in the summer semester of 1853, albeit with certain limitations having to do with the institute's cabinet. Although it was Ettingshausen who, during his earlier employment at the university, essentially furnished the cabinet, he complained to the ministry about its state in a not very collegial manner as far as his deceased predecessor was concerned.[94] He insisted on a full audit of the cabinet, which meant that for most of the semester it remained closed in the afternoons. Because of all these circumstances, we may assume that Doppler's influence on Mendel was much more pronounced than Ettingshausen's.

What, then, did Mendel learn at the Physics Institute? Apart from gaining solid knowledge of theoretical and experimental physics, he got a good grasp of the method of scientific research. From Doppler he learned how to approach unresolved scientific problems through experimentation. Doppler taught him also that an experiment can only then resolve a problem, if it is designed to answer clearly formulated, simple questions and if it is carefully planned, meticulously prepared, and precisely executed. The planning has to take into account all conceivable complicating factors, which might muddle the outcome and confound the answers. Each of these factors must be held in check by an appropriate set of controls. The preparation must include steps to test the reproducibility of the outcome. One of these steps must be checking the input material whether it is suitable for the intended purpose. The execution of the experiment requires skills, which must be gained by practicing the procedures involved until they become a routine. The evaluation of the results must be based on theoretical concepts underlying the experiment. It might have been the physics course that made Mendel aware of the theoretical concepts he had to master for the experiments that might have by then already been germinating in his mind, concepts such as the theory of probability, combinatorial analysis, and statistics. Both Doppler and Ettingshausen covered these topics in their lectures. In short, it might have been the physics course that gave Mendel the idea of addressing biological questions as if they were physics problems.

Natural History at the University of Vienna

Having acclimatized himself to the university milieu in the winter semester of 1851/1852, Mendel began expanding his learning horizon by including natural history courses in his curriculum in the summer semester of 1852 (Table 6.1). It was a very different natural history than the kind he encountered before—a natural history fragmented into several disciplines, each taught by a different professor in a different building (Fig. 6.10). Behind the scattering of courses all over Vienna was insufficient space at the Old University to accommodate the expanding disciplines. But it must also be said that after 1848 the imperial government was only happy to scatter students over Vienna, fearing that their concentration at one place might spark new unrest. Figure 6.10 shows the extent of the scatter and also the distances Mendel had to cover to attend the courses for which he signed up. Closest to Mendel's place of residence, the house on *Landstrasser Hauptstrasse 358*, practically around the corner from it,[94] was the Physics Institute on *Landstrasser Hauptstrasse* 104 in the Erdberger District (Fig. 6.10). At about the same distance was the *k.k. geologische Reichsanstalt* (Imperial-Royal Geological Institute) on *Rasumofskygasse*, where Mendel attended courses on paleontology offered by Friedrich Lucas Zekeli (1823–1881). The Geological Institute (Fig. 6.14) was (and still is) located in the former Razumovsky Palace, named after its previous owner, the Russian Ambassador to Vienna, Count Andrey Kyrillovich Razumovsky (1752–1836).[119] The palace (and garden) was located on the same street (Landstrasse) as the one on which Mendel lived. Slightly farther away (about one kilometer), but in the opposite direction, was the *Alte Universität* (the *Universitätshaus* with the *Aula*) in the inner city. There Mendel sat on Kner's zoology lectures and on Franz Unger's Plant Anatomy and Physiology course. Farther still (about 1.5 kilometer), was the museum building in Vienna's Botanical Garden (Fig. 6.10), where Mendel attended Eduard Fenzl's lectures on Plant Morphology and Systematics (Fig. 6.15). In the same building, both Fenzl and Unger also conducted their practical exercises, the former in Plant Analysis and Description and the latter in Microscopy and Performance of Physiological Experiments. The botanical garden was adjacent to the Belvedere[120] Gardens with their two palaces, the Upper and the Lower Belvedere, in the Landstrasse District. Founded by Maria Theresia in 1754 on the urging of her personal physician, the *Hortus medicus* (Medical Garden) served originally the medical students to acquaint themselves with medicinal plants. With time, however, it developed into a botanical garden serving students of natural history. Farthest away from Mendel's lodging was the *Theresianische Akademie* or *Theresianum* (Fig. 6.10), erected in the early seventeenth century on the outskirts of Vienna. The original building served as the summer palace of the imperial family. Favored by a succession of emperors, it was nicknamed the *Favorita,* and to this day the street that runs by it in the Wieden District is called the *Favoritenstrasse*. Maria Theresia, however, did not like the place because her father died there, and so she moved instead to the Schönbrunn Palace and in 1764 gave the *Favorita* to the Jesuits to run it as a school and dormitory for young men from impoverished aristocratic families to be educated as government officials. The building has ever since been used for educational purposes. In Mendel's time several university professors used some of the rooms there as lecture halls. There, Mendel took chemistry courses taught by Josef Redtenbacher there as well as instructions on the use of logarithmic and trigonometric tables by Franz Moth.

The dispersion of courses presented a logistics problem for Mendel because some of the courses overlapped partially or followed each other tightly. Mondays in the winter semester were the worst in this regard. According to his schedule, he was to attend Redtenbacher's inorganic chemistry from 8 to 9 AM and analytical chemistry from 9 to 10 at the Theresianum, from 10 to 12 Ettingshausen's physics course at the Physics Institute in a different district of Vienna, and from 12 to 1 PM Kner's zoology at the Old University; then for the rest of the afternoon, he was to participate in the practical exercises back at the Physics Institute, only to be back at the Old University for Unger's botany lecture from 6 to 7 PM. Even if Mendel had

Fig. 6.14 The Razumovsky Palace, the seat of the Geological Institute of the University of Vienna

Fig. 6.15 The Botanical Institute in the Botanical Garden of Vienna. Based on a photograph from c.1903. At that time the building, then called the Museum, still had the same appearance as in Mendel's time. It was destroyed by a bomb in 1944. (Photograph was provided by Dr. Robert Stangl.)

dashed from one class to the next (which is hard to imagine since he was anything but the athletic type), he still could not have made it to all of his classes in time. How he solved the problem we don't know, but we do know that, for example, he may not have attended a single zoology lecture given by Kner.

Student of Botany

Botany had two professors at the University of Vienna. This innovation, like the creation of the Physics Institute, was part of the reform package prompted by the 1848 revolution. It was meant to follow the example of Prussia and of other European countries in turning natural science departments into research institutions. Botany was traditionally a descriptive science and at the University of Vienna the tradition had deep roots. The last representative of the traditional, pre-reform botany was Stephan Ladislaus Endlicher (1804–1849), whose magnum opus, the multivolume *Genera plantarum* (published from 1836 to 1850) was a catalogue of the known plant genera arranged into a natural system, a classification taking into account natural similarities between species, rather than some arbitrary, artificial characters such as the number of anthers. When Endlicher committed suicide in 1849, the university still followed the custom of choosing his protégé as his successor. He was the systematist Eduard Fenzl (1808–1879; Fig. 6.16a),[121] who was the son of a patrimonial magistrate (a judicial officer employed by a prince) in Krummnussbaum on the Danube in Lower Austria, halfway between Vienna and Linz. He received his basic education from his father at home and then at the *Gymnasium*. Noting his interest in plants, the father sent him to Vienna to study medicine. At the University of Vienna, Fenzl came in contact with Endlicher, Jacquin, Unger, and other botanists. A few days after taking his doctorate, Joseph Freiherr von Jacquin (1766–1839), whose attention Fenzl drew to himself by his brilliant performance at the botanical examination, hired him as his assistant. The acquaintance with *Herr Baron* not only ushered Fenzl into the "high-society" circle of scientists but also made him a potential candidate for a place in the line of succession leading to university professorship and directorship of the botanical garden. *Herr Baron* himself inherited both positions from his father, Nicolaus Joseph Freiherr von Jacquin (1727–1817), who held them from 1768 to 1796. The son took over the positions on his father's retirement and held them until he died in 1839. It was then Endlicher's turn, since he functioned since 1836 already as *Kustos an der botanischen Abteilung des Hof-Naturalien-Kabinettes* (curator of the Botanical Division of the Court Natural History Museum)[122] and in this function hired Fenzl as his *Kustosadjunkt* (assistant). When Endlicher advanced to the directorship, Fenzl rose to the curator status, and when Endlicher died, Fenzl was promoted to the directorship of the botanical garden and professorship at the University of Vienna. It was all quite orderly and in accordance with tradition. The successor had to wait for his predecessor to retire or die, whichever came first, and to get onto the waiting list in the first place, the candidate had to be someone who would follow in his predecessor's footsteps. And so plant systematist Fenzl succeeded plant systematist Endlicher. Was Fenzl at least a good systematist? His colleagues apparently thought so: Endlicher, for example, named three new plant species after him and others wrote salutary articles and even a booklet about him while he was still alive.[121] Posterity, though, seems to judge otherwise: His name is not mentioned in the histories of botany[123] and on the Web it comes up only in connection with Mendel's. Even the Ministry apparently did not expect Fenzl to modernize botany at the university, but it needed him to run the museum and the botanical garden and for these functions he seemed eminently suited. The ministry made him also a university professor, but under the condition that this appointment would not interfere with his curatorial and directorial duties. This stipulation might explain why he held his lectures at such ungodly hours as from 6 to 7 AM and his practical exercises in the second half of May only.[94] As expected, he concentrated largely on his administrative responsibilities and social functions, limited teaching to a minimum, and gave up research altogether. The Fenzl appointment placed the museum and the botanical garden into competent hands, but it did not contribute to the modernization of botany

Fig. 6.16 Mendel's professors of botany. (**a**) Eduard Fencl (1801–1879). (**b**) Franz Xaver Unger (1800–1870)

at the university. To achieve the latter, the Ministry took the unprecedented step of creating a second botany chair and appointed Unger to it.

Franz Xaver Unger (1800–1870; Fig. 6.16b) was one of the most interesting persons[124] Mendel came in contact with in Vienna. His CV is framed by two rather extraordinary circumstances: a large family at its beginning and a mystery at its end. He was the first-born son in his father's second marriage into which Franz's mother brought seven children and then went on to bear nine more before she died at childbirth. There was, however, no shortage of space at Ungers' home to rear the 16 children, for their mother inherited a good-size property, the Amthof estate near the village of Leutschbach in the south Austrian province of Steiermark or Styria.[37,125] The mystery surrounding Unger's final exit was in the circumstances under which his body was found. His wife discovered it one morning with minor injuries on the head, wounds on other parts of the body, and blood spots on the floor. The press had a field day speculating who the murderer might have been (by then, Unger was a well-known personality in Austria), but the police declared his death to have occurred from natural causes. Between the beginning and the end was a life of travel, adventure, work, and confrontations. It included seven months in jail, when upon his return from a trip through the German countries, the Austrian police discovered that this apparent vagabond traveled without the proper papers. This experience could not have been as bad as it might seem, since he was able to persuade his jailor to escort him on specimen-collecting forays into nature so that he could finish his dissertation in the prison cell. He was a man of many talents and even more interests. As a youth he was adept at rhyming and as a retiree he picked up painting. In between he was a physician, paleontologist, microscopist, evolutionist, ecologist, anatomist, and science writer, but above all these a superb botanist. There is a long list of species named after him and an equally long list of societies and institutions that made him their honorary member. To all German-speaking botanists he was the author of the *Grundzüge der Anatomie und Physiologie der Pflanzen* (Essentials of Plant Anatomy and Physiology) as well as of other textbooks.[126] To Vienna's public he was familiar as the author of a column on scientific subjects in the daily *Wiener Zeitung* (The Vienna News). And all his students remembered him for his lectures. Not because he was a brilliant speaker, but because he spoke with enthusiasm and clarity, had interesting information to convey, and had a way of inspiring and captivating his audiences.

Unger received his elementary and secondary schooling at the *Gymnasium* associated with the Benedictine Abbey at Admont, Styria, where he also graduated from the Philosophical Institute. On his father's wish he then began to study law, but had little taste for it and so turned to medicine instead, first at Vienna and later at Prague. After taking his medical degree in 1827, he practiced medicine as a *Landesgerichtsrat* (counsel for the provincial court) at Kitzbühl in the Austrian Tyrol, until his botanical works landed him a professorial position at the Johanneum[127] at Graz in 1835. In 1849 the Ministry of Education appointed him professor of plant anatomy and physiology, a newly created botany chair at the University of Vienna. He remained in this position until his retirement in 1866. Unger's scientific contributions cover several disciplines. In cytology (the study of cell structure and function), he examined the growth of plant stem and root tips and concluded that new cells originate exclusively from existing cells. His studies on the distribution of plants in relation to the soil on which they grow made him one of the founders of ecology, the science of the interaction between organisms and their physical and biological environment. In paleobotany, the study of fossil plants, he demonstrated that the earliest geological periods were entirely devoid of plant life and that in the subsequent successive periods emerged plant species of ever-increasing complexity. He interpreted these findings as an indication that plant species had developed on a geological time scale in a manner akin to the development of an embryo into an adult form. Unger's vivid descriptions of the changes the plant world had undergone in the past so fascinated his students and other listeners that they urged him to attempt a graphic rendering. Ultimately he did just that in collaboration with an artist, Joseph Kuwasseg (1799–1859), a prominent landscape painter in Graz. Under Unger's supervision, Kuwasseg produced 14 drawings, each of them representing a landscape reconstructed from the fossil record for a given epoch of earth's history. An artisan, Leopold Rottman, then prepared large lithographs of the drawings and the set, accompanied by Unger's explanatory text in German and French, was published as a folio atlas in 1851 under the title *Die Urwelt in ihren verschiedenen Bildungsperioden*[128] (The Primeval World in its Different Periods of Formation). The atlas was the first of its kind. Previously, paleontologists published individual reconstructions of some of the periods, which they themselves drew. The novelty of Unger's atlas was in that it covered the entire earth's history and that a professional artist executed the drawings. Both the scientific and general public welcomed enthusiastically this magnificent work, which to this day has not lost any of its charm. There were, however, some dissenting voices to be heard and the loudest of them was that of Sebastian Brunner, who became the proverbial flee in Unger's fur coat.

Doctor of philosophy and theology, Sebastian Brunner (1814–1893)[32,37] was a Catholic priest, self-appointed spokesman for the right wing of the Catholic Church in Austria, fighter against the remnants of Josephinism in the government, and a guardian of Catholic influence at the university. A prolific writer, he was the author of books on theology, literature, and history; he wrote poetry, satire, essays, criticisms, and homilies; and he was the editor and publisher of the *Wiener Kirchenzeitung* (The Viennese Church News). He followed closely activities at the university, standing watch for signs of deviations from orthodoxy.[101] Unger came under his scrutiny when he began publishing his weekly column on plant life in the *Wiener Zeitung* on May 28, 1851. At first, Brunner did not know who the author of the essays was, because they were published anonymously, but later, piqued by their heretical undertone, he found out. In the concluding paragraph of the last installment, Unger waxed poetic about the force he held responsible for the development (today one would say "evolution") of plants from primitive to more complex forms described in the preceding essays. Brunner might not have been comfortable with the idea of "development," but it alone might not have provoked his attack. So, he was prepared, to paraphrase Matthew (22:21), *to render unto science the things which were scientific; and unto religion the things that were God's*.[129] In the essay, however, Unger transcended the religion-science boundary and stepped onto a territory Brunner considered his own turf, namely, theology, and this the theologian could not tolerate. Unger implied that a teleological, spiritual force, whose ultimate predestined purpose was to bring forth human beings, drove the development. Brunner's position was that if there were such a thing as "development," humans would have to be excluded from it. He might have been willing to go as far as admitting the production of the human body by the process of "development," but being a strict dualist, he insisted that the human soul came directly from the triune, the personal God of Christianity. Unger's closing paragraph, however, seemed to have been taken from the writings of one of the German *Naturphilosophen* (see Vol. 2 Chap. 1) and could have been interpreted as championing the existence of an impersonal force pervading all nature. This, to Brunner's mind, smacked of pantheism, paganism, and atheism, the three "isms" Roman Catholic theologians loathed.

The *Wiener Zeitung* published Unger's essay on October 18, 1851; one week later, on October 25, the *Wiener Kirchenzeitung* carried an editorial titled *Unsere Hochschulen* (Our Universities), in which Brunner framed his attack as a critique of what he perceived to be dangerous trends at the university. Brunner presented Unger as an example of Austrian Catholic educational institutions becoming hotbeds of agnosticism and paganism masquerading as science. He also attacked the supposedly Catholic *Wiener Zeitung* for providing forum for such views. Unger did not heed the attack and in the winter of the same year published *Die Urwelt*, depicting the "development" of life on earth. The worked-up Brunner saw red and reacted on April 17, 1852 in the *Wiener Kirchenzeitung* in an article entitled *Die Fabel der Schöpfung* (The Fable of Creation). In it he reiterated his position on the creation of the human species and then proceeded to attack Unger personally. The attack must be viewed in connection with the so-called Bonitz controversy, which then reverberated through the university halls. In the summer of 1851, the College of Professors of the Philosophical Faculty elected as its dean Thun's advisor, the Protestant Hermann Bonitz (1814–1888). The Catholic hardliners in the *Consistorium*, including the dean of the College of Doctors, contested the election on the grounds that it provided foothold for a non-Catholic influence on the university affairs and that it degraded the *Consistorium* to a mere extended arm of the ministry. Brunner, a member of the *Consistorium*, whipped up the controversy in a series of articles in the *Wiener Kirchenzeitung* that achieved the ultimate goal, as the governing body of the university annulled the election on technical grounds. In the *Die Fabel* article, Brunner linked Unger to the controversy by claiming that the botanist was the dean of the College of Professors: After a Protestant dean, he thundered, the college is now headed by a man who denies creation and the Creator Himself! In reality, Unger was never a dean. When it was pointed out to Brunner that not Unger but Fenzl was the dean,[130] the theologian published a *Besichtigung* (literarily "inspection"), a sort of clarification in the sense: Oh, well, one botanist or another, what's the difference?

Brunner continued his attacks on Unger, at one point even demanding his dismissal from the university. He reached the lowest level of his vituperations in an article entitled *Isispriester und Philistine* (A Priest of Isis and a Philistine)[131] published in the *Wiener Kirchenzeitung* on January 29, 1856. In this piece he attacked Unger's scientific credentials and compared him to a man who transplants nicely smelling plants into a reeking dung bed. This time, however, he went too far. There came an outcry from the press, including Catholic journalists, against this sort of defamatory rhetoric, and at the university some 400 medical students signed a petition asking Minister Thun to shut the priest's foul mouth. The students pointed out that in his lectures Unger had always restricted himself to scientific issues and never ventured into anything even distantly related to religion. Unger, who until then met all the attacks with dignified silence, filed a lawsuit for slander against the priest and published a "clarification" in various newspapers on March 4, 1856, in which he denied the accusations of pantheism and materialism. Although the court dismissed the lawsuit as groundless and Thun's attempt to mediate in the controversy failed, Brunner, claiming victory, stopped the attacks shortly afterward.[101]

Mendel must have known about the Brunner-Unger controversy. His reaction to it is not known, but it is hard to imagine that he stood on Brunner's side. More likely, the offensive behavior of a fellow priest must have embarrassed him. He himself could not have had any difficulty with accepting Unger's evolutionary concepts and his interpretation of earth's history. We can assume that he enjoyed Unger's lectures and, like many other students, was inspired by them. Unger, like Doppler, was a strong believer in the importance of an experimental approach to science, and Mendel must have been eager to acquire a microscopist's skills. Together with

Doppler and Ettingshausen, Unger must have exerted the strongest influence on Mendel at the University of Vienna. Whether Mendel got to know Unger closer than other student is rather doubtful, since any attempts in this direction might have been dampened by the Brunner affair. Unger might not have been quite comfortable with the presence of a priest in his class, but that this uneasiness went as far as suspecting Mendel of spying for the church, as some biographers suggest,[94] seems rather improbable in view of what we know about Unger's character.

In addition to influencing Mendel in a general way by revealing to him the experimental nature of biological sciences, Unger might have given him also specific stimuli to the concretization of the experiments crystallizing in the student's head. These, however, could be inferred only obliquely, since Mendel does not acknowledge his debt to Unger in any of his publications and not once does he even mention his teacher's name. A footnote in Mendel's main work refers anonymously to the *Ansicht berühmter Physiologen* (opinion of famous physiologists) and concerns the nature of the fertilization process in flowering plants.[132] The process was a controversial issue on which Unger stood on the side that ultimately prevailed and Fenzl on the losing side. Mendel's omission is puzzling. Was he, perhaps, afraid of acknowledging openly that he was in the same league as the pantheist Unger? This possibility is not as forced as it might appear. In Brno, where the scientific and clerical communities intermingled, the Brunner-Unger controversy must have been followed with great interest. Schaffgotsch's spies would undoubtedly have reported to the Bishop that a friar from the St. Thomas Abbey cited approvingly the pantheist Unger, had this been the case. Mendel must have been well aware of the tension between the bishopric and the abbey, which culminated in the Bishop's attempt to have the latter closed down. He might have therefore been concerned about not providing any additional ammunition for the bishop's guns. It might have been this concern that was behind his being overly cautious in the formulation of the footnote. A few examples of Unger's specific influences on Mendel follow.

One of these influences was related to the controversy concerning the nature of the fertilization process in flowering plants mentioned above. The understanding of the process was essential for Mendel's interpretation of his hybridization experiments, which he initiated after his return from Vienna. At issue was the question whether only one parent (the donor of the pollen) or both parents contributed to the genetic makeup of the embryo (see Vol. 2 Chap. 3). In his textbooks and hence presumably also in his lectures, Unger defended vehemently the latter alternative. According to him the embryo grew out from a cell that resulted from the fusion of an egg cell in the ovary with a material from the terminal part of the pollen tube. Both the egg cell and the pollen tube contributed equally to the makeup of the embryo. A few years later Mendel realized that only with this view did his hybridization experiments make sense. Another example of Unger's influence on Mendel is the terminology he used in his report on those experiments.[132] At the time he wrote the report describing them, the controversy about the fertilization process had not yet been resolved, and consequently different authors used different names for the cells and parts of the plant reproductive system. Mendel, however, used largely Unger's nomenclature. Unger might have influenced Mendel also in the choice of a method. In his writings and lectures, Unger promulgated artificial fertilization as a method for the production of new varieties of cultivated plants. Although Mendel was already familiar with the method from Diebl's lectures in Brno, Unger's vivid description and perhaps even demonstration of artificial pollination might have contributed to his resolve to use it. Finally, in 1849 Karl Friedrich von Gärtner (1772–1850) published a monograph, which summarized and evaluated critically all the experiments on plant and animal hybridization.[133] Unger drew the students' attention to it and recommended it warmly. Mendel bought a copy and studied it thoroughly, as the many markings and marginalia he made in it indicate:[134] It became his bible.

Fenzl's influence on Mendel, on the other hand, was probably only slight. One reason was that Fenzl was not an inspiring teacher and, another, that plant systematics was not one of Mendel's favorite subjects. Mendel learned plant identification and classification from Klácel and the botanists of the Brno circle, but his actual interest in plants was oriented in a different direction.

The Lackadaisical Student of Zoology

It was only Rudolf Kner's third year at the University of Vienna when Mendel began his studies there. After a ten-year stint in the provinces (Lvov in the Ukraine), Kner took over the zoology chair in the imperial city in 1849. He offered three courses: one in general zoology for students of medicine and pharmacology (five hours per week); another in zoological systematics (two hours per week); and the third, complementary to the second, on practical zoological exercises (three hours weekly). Mendel signed up for all three courses, the first in the winter semester of 1852/1853 and the second and the third in the summer semester of 1852, but took none of them and the registration office crossed out all three in Mendel's course record. Kner offered the same three courses again in 1853, but this time Mendel did not even bother to sign up for them. Moreover, besides Kner's courses, Johann Friese (1792–1866), professor of natural history at the Philosophical Faculty, offered a zoology course every other semester, but again Mendel did not take any of them. About Mendel's reasons for this curious neglect of zoology, one can only speculate. The simplest reason could have been that the zoology courses collided with physics courses, which Mendel preferred. This was true, but the overlap was only partial and Mendel could have taken at least some of the zoology courses, either Kner's or Friese's. Another possibility is that the courses did not take place because of Kner's absence from the university, either because of a trip or illness. There is, however, no evidence to support this explanation. A third possibility remains that Mendel simply did not like Kner and when he discovered that he could get training in zoology without attending Kner's courses, he grabbed it. This opportunity offered itself when Mendel met Vincenz Kollar (1797–1860; Fig. 6.17)[37,135] the head of the zoological division of the *Hofnaturalien Kabinett* (the museum).

How and when Mendel and Kollar met is not known, but once they did, they found a mutual liking in each other, despite the 25-year age difference between them. They discovered that they had at least three things in common. The first of

Fig. 6.17 Mendel's professors of zoology: Vincenz Kollar (1797–1860). For a portrait of Rudolf Kner (1810–1869) see Fig. 6.4

these was their origin from the same corner of central Europe, even if their birthplaces were separated by a frontier between two quarreling (and sometimes warring) countries. The corner was Silesia, where Kollar's birthplace, the village Krzanowice, was in the Prussian part, whereas Mendel's Hynčice was in the Austrian part. Krzanowice is now in Poland and Hynčice in the Czech Republic. The second thing Kollar and Mendel shared were memories of a hard childhood on small farms, of tough student years with only meager or no support from their peasant parents and of teachers who recognized their talents. Unlike Mendel, however, Kollar managed to pull through without having to pledge his soul to God. Kollar studied at the *Gymnasium* in Glubczyce (in the past called Lubczyce, Hlubčice, or Leobschütz) and later at the University of Vienna. In the imperial city his hobby (insect collecting) brought him to the entomology section of the Court Museum and to the attention of the section head, Franz Ziegler. Impressed by the young man's zeal, Ziegler managed to secure a small stipend for Kollar that enabled him to work on the collection as a helper. He worked passionately, neglecting his medical studies. Ultimately he gave them up altogether and took a doctoral degree in entomology. He stayed at the Court Museum for the rest of his active life, working his way up through all the hierarchical levels of a museum's employment, eventually replacing Ziegler upon his retirement. By then Kollar was an internationally renowned entomologist, an author of many scientific publications and several compendia. Like Mendel—and this is the third thing the two had in common—he did not forget his roots and throughout his life retained keen interest in the agricultural aspect of his discipline. His specialty became insect pests inflicting damage in farming and forestry. The emperor acknowledged the civil

Fig. 6.18 The *Theresianum* in which Professor Joseph Redtenbacher taught chemistry until the construction of a separate Institute of Chemistry of the Vienna University was completed

servant' accomplishments, honoring him, two years before Kollar's death, with the title *Geheimer Regierungsrath* and decorating him with the *Ritterkreuz des Franz-Josephs Ordens*. Kollar was a kind, unassuming man with whom Mendel liked to associate. In his first scientific publication, Mendel acknowledged him as his *hochgeehrter Lehrer* (highly respected teacher). Mendel spent as much time as his schedule allowed him at the museum and learned a great deal of zoology from Kollar and his associates. Kollar also provided stimuli to Mendel's first two scientific reports and sponsored him for membership in the *Zoologish-botanisches Verein* (of both these development later). Mendel remained in contact with Kollar until the latter's death in 1860.

Student of Chemistry

Prior to 1848, chemistry, like physics, vegetated at the University of Vienna. The little of it that was taught to medical students was in reality antiquated pharmaceutics. In the first half of the nineteenth century, when Joseph Louis Gay-Lussac (1778–1850) at the École Polytechnique in Paris, Jöns Jakob Berzelius (1779–1848) at the Karolinska Institutet in Stockholm, Justus Liebig (1803–1873) at the University of Giessen in the German state of Hessen, and others at other places in Europe were exploding the discipline; at the University of Vienna, meanwhile, there was still no chemistry research being done. But as with physics and botanical physiology, the 1848 revolution changed the situation. In 1849 the Ministry of Education transferred chemistry from the Medical to the Philosophical Faculty and appointed Joseph Redtenbacher professor of the first chemistry laboratory. The chemistry laboratory eventually became the Chemistry Institute, when it moved from the Theresianum (Fig. 6.18) to a separate building. Later still, when students of medicine, pharmacology, botany, and zoology crowded the chemistry course (more than 500 in some semesters), the ministry was forced to open a second chemistry laboratory and divide inorganic and organic chemistry between the two.

Joseph Redtenbacher (1810–1870; Fig. 6.19)[37,136] son of a merchant at Kirchdorf an der Krems in Upper Austria, studied medicine at the University of Vienna. After graduation, a stipend enabled him to visit different places of learning in Europe, including the University of Giessen, then the Mecca of many a budding chemist. There, Justus Liebig developed a new type of a teaching institution, at which students learned primarily in the laboratory. The beginners experimented under the close supervision of the professor and his assistants, whereas the more

Fig. 6.19 Joseph Redtenbacher (1810–1870), Mendel's professor of chemistry at the University of Vienna

advanced students worked more or less independently on assigned research projects. Liebig himself was a superb experimentalist and his research accomplishments were legendary. Not only was he one of the founders of organic and analytical chemistry, but he was *the* founder of food and agricultural chemistry. Convinced that chemical reactions in plants convert inorganic substances from the soil and air into organic matter, he pioneered the study of soil composition with the goal of improving it. He introduced the practice of selectively adding to the soil chemical substances (artificial fertilizers) which the plants had been deficient in and which they needed in order to grow. From Liebig's laboratory came many outstanding chemists, and the stream of students to it continued for as long as its doors remained open. Redtenbacher spent several months in it and then, in 1840, accepted a university position at Prague, where he established a teaching laboratory modeled on Liebig's. In 1849, he then moved to Vienna to reform chemistry there.

In Vienna, Redtenbacher offered three courses: Inorganic Chemistry, Organic Chemistry, and Methods in Analytical Chemistry, five hours per week each. Mendel took all three courses, the first and third in the winter semester of 1852/1853 and the second in the summer semester of 1853 (Table 6.1). His decision to take these courses is somewhat surprising, and the overall combination of courses he signed up for was not that of a typical Vienna University student.[137] But then, Mendel was anything but a typical student. Since he intended to teach physics and natural history, he did not need to take chemistry, but rather should have taken zoology instead. Why, then, did he include chemistry in his curriculum? The most obvious answer could be that he intended to be a teacher *and* researcher and felt that for the latter he had to have a chemistry background. Another possibility is that as a peasant's son interested in the application of natural sciences to agriculture, he was anxious to learn what the new agricultural chemistry had to offer toward easing his father's toils.[94] Another incentive might have been his natural curiosity about chemistry as a science that had much to offer toward the understanding of life's processes. He owned a textbook of chemistry[138] covering the latest advances in that discipline and so, in his studies of the subject, he might have gone far beyond what Redtenbacher presented in his lectures.

The impact of the chemistry courses on Mendel can only be conjectured. Mendel's main opus, his other publications, as well as some other documents betray him as a person comfortable with a physicochemical interpretation of natural phenomena. Klácel's *Naturphilosophie* musings and Unger's allusions to the operation of spiritual forces in natural objects left him cold. Although he never said so in so many words, between the lines of his writings, it is possible to perceive a view of life based on interactions of atoms and molecules and on chemical reactions resulting from them. It must have been this—one is tempted to say "materialistic," but that would be going too far—view of life that makes his classical paper to appear so modern. The physics and chemistry courses he took at the University of Vienna might not have been responsible for the acquisition of this worldview, but they certainly must have affirmed him in it.

Various attempts have been made to identify specific stimuli the chemistry courses might have given Mendel in regard to his interpretation of the hybridization experiments. One suggestion[139] attributes the origin of Mendel's concept of particulate inheritance to the theory of chemical radicals. The essence of the theory is that the common currency used by the trading partners in chemical reactions is a molecular fragment called *radical*[140]—an atom or a group of atoms bound together and functioning as units of exchange. Redtenbacher, as a student of Liebig, who was a major contributor to the theory, must have expounded on it in his lectures and so had given Mendel the idea that radical-like substances might function as units of inheritance. Mendel might have thought that, analogously to chemical reactions, in which the input substances give rise to new a compound by exchanging radicals, in heredity the exchange of semipermanent radical-like substances might be responsible for the appearance of new character combinations in the progeny of two parents. This attribution to Mendel, however, appears to us farfetched in view of the fact that there is not the slightest indication in Mendel's writings of his thinking along these lines. What's more, the hypothesis, when thought through, would have led Mendel to conclusions that would be incompatible with the results of his experiments. Another suggestion[141] has been inspired by Mendel's use of the word *Element*[M6] in reference to the agents responsible for the development of characters and sometimes also for the characters themselves. Might he not have meant by the word, so the suggestion goes, actual chemical elements? This possibility is, however, even less tenable than the first one. If he learned anything from the chemistry courses, he could not have thought of the *Element* in terms of chemical elements such as hydrogen or oxygen, consisting of single kinds of atoms. Moreover, the ambiguity with which Mendel uses the word (sometimes in reference to what today is called *gene* and at other times in reference to a *character*) indicates that he availed himself of *Element* in its common usage as another word for a "component" or "factor."

Student of Paleontology and Mathematics

In addition to the main courses, Mendel also sat in on three fringe lecture series, two on paleontology and one on specialized mathematical topics. Fossils, the subject of the paleontology lectures, have been known since antiquity, but generally recognized for what they were only at the end of the eighteenth and beginning of the nineteenth centuries. The word "fossil" was originally applied to any curious object dug out of the earth, whether organic or inorganic in derivation (it comes from the Latin *fodere*, to dig). Only in the first part of the nineteenth century did it begin to assume its present-day meaning of any remnant, impression, or track of past life forms preserved in the earth's crust since prehistoric times. It was also only in the first half of the nineteenth century that the existence of fossils came to be connected with the concept of evolution. It was then that geologists noticed the succession of fossils from the simplest to the progressively more complex in layers of sedimentary rocks deposited in the order from the oldest to the youngest. They deduced from this observation that life on earth must have evolved over an enormous time span and that fossils characteristic of individual geological periods, the so-called index fossils, could be used to establish correspondence of sedimentary rocks in different parts of the globe and to divide geological time into intervals. These revelations and concept contradicted the account of the origin of life as described in the Bible, specifically in the Book of Genesis. The perspicacious part of the clergy and the laity learned to live with this development by admitting that the Biblical account must not be taken literarily. The bigots, however, dug their heels in and insisted that not the Bible but science got it wrong. In most of Europe, this particular antagonism between religion and science resolved itself with time and reason prevailed, but in some countries widespread bigotry continues to this day. In Mendel's time in Vienna narrow-mindedness revealed itself through the Brunner-Unger controversy. It was therefore an act of courage on Mendel's part to take the paleontology course. By taking it Mendel demonstrated once more that he had no problem with accepting the view that geology and the existence of fossils intimated. Actually, by signing up for the course, Mendel was doubly courageous, because he thus showed that not only did he not have any qualms about learning this subject but also that he was not bothered by the fact that the instructor was a Protestant preacher.

The instructor was Lukas Friedrich Zekeli (1823–1881), son of a Protestant preacher and he himself an ordained preacher.[142] He was born in Sighisoara in Transylvania, then a province of the Austrian Empire, but now part of Rumania. He studied theology at the University of Halle an der Saale, then a center of Protestant education in the German state Sachsen-Anhalt. On the side he attended lectures on natural history and developed an interest in paleontology. He then continued his studies at the University of Vienna, taking courses in anatomy, botany, and mineralogy. In 1844 his financial situation forced him to return to Sighisoara and accept a teaching position at the *Gymnasium*. He then became a preacher, but when the civil war broke out in the wake of the 1848 revolution and the city was overrun first by Magyar and then Russian troops, he was removed from the post, resumed teaching, and devoted his free time to paleontology. In 1850 he was back in Vienna, where he became an assistant at the Geological Institute and in this function participated in the geological survey of the Alps. After completing his dissertation and taking a doctoral degree from the University of Halle, he began teaching the courses that Mendel took, one in general paleontology and the other on index fossils. Mendel must have been an assiduous participant since Zekeli signed his course register with the note "extremely diligent and attentive right up to the end of the term."[43,142] Unable to secure professorship at the university, presumably because of his religion and also because of a scientific disagreement he had with an old-guard professor, Zekeli moved to Berlin and there he taught at the *Gymnasium* until his death.

Before the advent of electronic calculators, logarithmic and trigonometric tables were an indispensable tool to all who had to do a lot of multiplication, division, raising numbers to their power, or finding roots of numbers. Invented in the early seventeenth century, logarithms were at first rather awkward to use, but after the introduction of the much simpler common logarithms and of the logarithmic tables, their popularity rose rapidly and so universities often offered courses on their use. At the University of Vienna, the teacher of the course was Franz Xaver Moth (1802–1879), native of Žlutice in western Bohemia. Moth graduated from the Prague University, where he then became a substitute professor of higher mathematics. Subsequently he moved to Lower Austria, first as a professor at a secondary school and then professor of elementary mathematics at the University of Vienna. Mendel might have taken Moth's courses knowing that in the not too distant future he would be doing a great deal of computations.

Membership in a Society and the First Publications

As in other large cities, in nineteenth century Vienna, scientists[144] tended to organize themselves into societies to facilitate communication among themselves, disseminate knowledge to the public, and put pressure on the government on issues such as education. In 1851 a group of Viennese scientists decided that the time was ripe for founding a *k.k. zoologisch-botanischer Verein in Wien* (Imperial Royal Zoological and Botanical Society in Vienna). Including the "k.k." in the title was the required libation to the imperial-royal gods and a display of the seal of government approval. The group of founding members included Kollar, Unger, Fenzl, Kner, and other prominent experts in the two disciplines. Kollar, who was highly active in the organization of the society, gave one room in the museum to its disposal. There the *Verein* met regularly in the late afternoon every first Thursday of the month to hear presentations of its members. At least some of the presentations were open also to nonmembers, while other parts of the meetings were conducted behind closed doors. The *Verein* published the lectures, reports, and miscellaneous news from its life in its own *Abhandlungen* (proceedings). Mendel began attending the society's meetings from summer 1852, possibly on Fenzl's instigation. On January 5, 1853, the society admitted him among its members and on June 1 of the same year he stood for the first time in front of the plenum and read his first scientific report. Here is how it came about.

In the summer of 1852, while on vacation in Brno, Mendel noticed in the abbey's vegetable garden a plot of radishes attacked by a pest. The radish (*Raphanus sativus*) is grown for its thickened stem, but since it is an annual plant, its seeds must be harvested for the next year's planting. On this particular plot, caterpillars, some of which could still be seen feeding on the plants, destroyed the seed-containing pods. Mendel collected some of the caterpillars, and when he returned to Vienna, he took them to Kollar for identification. Kollar recommended letting the caterpillars go through the rest of their life cycle and wait for the butterflies to appear. And this was what Mendel did. Kollar then identified the butterfly as *Botys (Scopula) margaritalis* Hübn 1899 from the family Pyralidae (Fig. 6.20a). The species was known to cause damage to field mustard (rape), *Brassica rapa*; Mendel's was apparently the first observation of its attack on radishes. Kollar therefore encourage Mendel to prepare a report on his observation, read it at the next meeting of the *Verein,* and publish it in the *Abhandlungen.*[145]

Mendel's second report was also about a pest, this time a pest of a plant that one day would become emblematic of Mendel—the garden pea, *Pisum sativum*. The report had the form of a letter he sent to Kollar from Brno in 1854. Kollar read it at the monthly meeting of the society on April 5, 1854, and then published it in the *Abhandlungen*[146] together with his own commentary. In the letter Mendel reported on the devastation of the pea crop in 1853 in the fields around Brno caused by the pea weevil, *Bruchus pisi* (Fig. 6.20b). Much of the crop that was salvaged and brought to the market could not be sold because it did not pass the market inspection. Mendel then describes how he examined in early January of 1854 the infested seeds, but superficially could not find any difference between the infested and noninfested seeds. On a closer examination, however, he noticed a tiny prick, as if punctured with a needle, and opposite to it, on the other side of the seed, a small, circular darker spot. When he opened the spotted seed, he found inside a larva, which had gnawed itself at one point nearly to the surface where the darker spot was located. From these observations he deduced that the larva entered the seed at the site of the prick and the beetle exited it at the site of the spot, and described correctly the entire life cycle of the beetle, which was until then still somewhat controversial. The larva pupated while still in the seed and in the late spring turned into a small beetle, which then left the seed and climbed on the pea plants growing nearby. It fed on the flowers, grew into an adult, and mated. The females laid eggs on the forming pea pods, at the site where the pods curved over the seed. In the late summer and autumn, the larvae that developed from the eggs gnawed themselves through the pod and through a tiny hole entered the seed to repeat the cycle.

The two papers represent only a very minor contribution to science. The first report merely extends the host range of a parasite that is not a major threat to crops, whereas the second, at best, contributes to the clarification of an obscure point in the life cycle of an important pest. The significance of the reports is, however, not in what they contribute to science, but in what they reveal about Mendel. They betray, at this early stage of Mendel's development into a scientist, three characteristics that some ten years later would shine through his *magnum opus*: a gift for acute observation, strongly developed power of reasoning, and remarkable clarity of

Fig. 6.20 (**a**) The imago and caterpillar of the moth *Botys (Scopula) margaritalis* Hübn 1899 feeding on the field mustard (*Brassica rapa*), the subject of Mendel's first scientific publication. Drawing taken from the Internet. (**b**) The pea weevil (*Bruchus pisi*), which devastated the crop of the garden pea in Brno and was the subject of Mendel's second scientific report

expression. The next time he used these gifts, it would not be on minor issues, but on solving one of the most fundamental problems of life.

Private Life in Vienna

Historians perceive the nineteenth century as the golden age of Vienna. What they mean by this is that in the first quarter of the century, there assembled in Vienna a pleiad of great composers (Haydn, Mozart, Beethoven, Schubert) and that in the last quarter, Vienna glittered again with prominent names, this time in diverse cultural enterprises: music (Brahms, Bruckner, Mahler, Schönberg), visual arts (Klimt, Kokoschka), literature (Kraus, Hofmannsthal), philosophy (Mach, Wittgenstein), and psychology (Freud). This perception, however, ignores several facts. First, for the period between these two peaks Vienna could not boast about many great names, with the exception of popular music (the Strauss family). Second, throughout the entire century, the Viennese public had not demonstrated more refined tastes in culture than the populaces of other great cities, if anything just the opposite might have been the case. And third, the "gold" concerned only a small fraction of the society, namely, those who were well off.[147] The great majority of Viennese was *not* well off; in fact it lived quite miserably. The wealth and glitter concentrated in the *Hofburg*, where the imperial family lived, and in the circle around it, in which the 200 palaces of the aristocracy were located. It then diminished centrifugally from the center to the periphery. After the center of the wellborn, followed the zone of the nouveaux riches bourgeoisie (high civil servants, financiers, great landlords, industrialists), bounded by the *Ringstrasse*. Next was the zone of the middle class proper (managers, public servants, businessmen, doctors, professors) bounded by the *Gürtel*. Outside of the *Gürtel* lived the bulk of the Viennese (small merchants, skilled or unskilled laborers, and the proletariat), most of them under deplorable conditions, crammed in large apartment buildings, frequently ten or more persons per room. And for these rat holes with communal lavatories and water sources, the landlords charged the tenants rent, which amounted to one quarter of a man's wage. Consequently, tuberculosis, intestinal infections, and many other diseases were rampant. Thus, for most Viennese the golden age was crowded, noisy, malodorous, and unhealthy. To escape the dread at home and at work, the poor people sought distractions in taverns, dance halls, cheap theaters, and entertainment parks, which were all equally crowded, noisy, and smelly. They had no access to the salons, expensive theaters, ballrooms, concert halls, and coffee houses, where the rich sought their diversions, and where the golden age played itself out.

How did Mendel fit into this picture? He was not rich, but neither was he destitute. He had a room for himself and so in this respect was better off than most Viennese. He probably had to share a lavatory and tap water with other tenants on the same floor of the apartment building, but that was no different condition from what he had in the abbey. His apartment was just outside of the *Gürtel*, in an area which the rich considered still acceptable, judging from the fact that some of them had their palaces there. All the entertainment places of the poor and some of those of the rich were theoretically accessible to him, but it is not very likely that he visited any of the dance halls nor that he was a frequent guest in one of the many taverns. On the other hand, on Sundays he might have joined the crowds in the amusements parks or the promenades, and occasionally, the purse allowing, he might have purchased a theater ticket to see a play or an opera, with or without a chaperon. He might have dined in one of the neighborhood restaurants, feasting on *Leberknödelsuppe* (liver dumpling soup), *Wiener Schnitzel* (breaded veal or pork), *Gefühlte Paprika* (green pepper stuffed with ground beef and rice), or *Tafelspitz mit G'röste* (thickly sliced, boiled beef with fried, grated potatoes), the favored dish of the emperor. To quench the craving of his sweet tooth, he might have stopped at one of the *Konditereien* to taste one of the many varieties of *Torten* (gateaux, cakes), perhaps even the famous *Sachertorte*.[148] To inform himself about the state of the empire and the world, he might have spent an evening now and then at one of the cafe houses to sip coffee and read the selection of newspapers and magazines the house offered to its guests. But in general he had to be thrifty to come out with his meager allowance. Out of thriftiness, he had, as in Znojmo, his laundry washed at the abbey. On the other hand, he managed to save a few *Groschen* to purchase a lottery ticket: Was he dreaming of winning big to bail himself out of the monastery life? He never won anything; apparently God wanted him to stay in His service.

A letter to Rambousek[149] in the summer of 1852 reveals him in good humor, joking almost frivolously about the *Strapazen* (small problems) of his daily life and his upcoming vacation. In it he informs his friend that a shipment of washed laundry got lost. It contained clean shirts and now he has none to put on. Of the dozen he took with him to Vienna, all are now worn down and have holes. He therefore asks Rambousek to tell Frau Smekal to buy, on Mendel's account, linen for three shirts and have them made as soon as possible, because the situation is quite urgent. In a few days he will be returning to Brno to participate in the spiritual exercises[150] ordered by the abbot for the summer. Without the new shirts, Mendel says, he would have nothing to wear, and adds: Would it not be a disgrace if he emerged from the pious exercises as a new man but in a shirt full of holes? "I would be ashamed of myself if I had to (Apocalypse: *Stantes amicti stolis albis*)[151] parade in a worn out robe! Herr Prelate [Napp] has notified me that I am to attend the exercises, which will take place in the last week hujus.[152] Since it is generally known that the semester at the university ends on July 20 and since it would do no good to piss against the wind, I have fixed July 24 as the day of my departure. I will arrive in Brno at noon of that day. P. Mathew [Klácel], I presume, still dwells in the primeval forests of Česká Třebová. The lucky devil. . .Should I win tomorrow 25,000 florins, Frau Smekal would receive an (encoded) telegram. Ask her in the evening. I look forward (?) to seeing you soon."

The letter is interesting for several reasons. It reveals a different Mendel than other documents and most biographers present him as. It shows him as possessing a good sense of humor colored with a sarcastic undertone. When communicating with a friend, he has no inhibition about using an earthy phrase such as "pissing [*brunzen*] against the wind," which in those times, as well as today, might have been regarded as vulgar, especially coming from a priest. It is refreshing to know that he was not exactly the pious monster the Znojmo's professorial staff painted in their reference letter. It may surprise some how lightly, almost flippantly, Mendel writes about the spiritual exercises. He postpones his return to the abbey, as much as he dares, to escape at least part of the tedium; he envies Klácel for escaping them entirely; and, by quoting the New Testament, he makes fun of the ritual by imagining himself parading in a perforated robe. Apparently he was not the only friar who hated the mind-numbing ritual. Napp did what he could to spare his flock from the worst excesses of the monasterial routine, but with Schaffgotsch on his back, he had to show some activity and the *Exercitiae* were part of this effort. Surprising is also the familiar tone of Mendel's references to Frau Smekal.[153] True, she was a servant and an elderly married woman (49 years old in 1852), but a *woman* nevertheless, and the friars were supposed to stay away from the fair sex of any social status and any age. Apparently, Mendel and the other friars liked her and liked to tease her. Perhaps they found in her a substitute for their mothers, from whom they had been separated at an early age. As for Mendel's life in Vienna, the letter reveals that he did not walk around as impeccably dressed as his status might have called for, but in that he was hardly alone.

Visit in Hynčice

Mendel would have preferred to stay in Vienna continuously for the entire two years, but Napp insisted on his returns to the abbey for the two months of summer vacation and one to two weeks of Christmas and Easter holidays. At the end of the 1852 summer vacation, Mendel returned to Vienna on September 30 to register for the winter semester, but on October 9 he left again, to travel to Hynčice for his sister's wedding. Presumably, he took the train all the way to Mankovice, a village that was only a short distance from Hynčice. The wedding took place on October 12: The 23-year-old Theresia married Leopold Schindler, a *Bauer* from Hynčice No. 3. The wedding ceremony conducted by Pfarer Kahlig at the church in Dolní Vražné[154] was followed by a wedding feast, which the custom dictated had to be held in the house of the bride's parents. Since 1842, however, the house was in the possession of Mendel's other sister, Veronika, and her husband Alois Sturm. Presuming that Theresia continued to live in the house after Veronika's marriage, the wedding feast would have been held there. Whether Anton and Rosina Mendel, Gregor's parents, also lived in the house in a separate room as retired *Ausgedinge* is unclear. According to some accounts, after Veronika's marriage they moved to another village. Anton Mendel, however, signed the letter to his son, in which he asked him for his consent regarding Theresia's marriage, as *Bauerausgedinger in Nr. 58 in Heinzendorf* (retired farmer at No. 58 Hynčice); the signature could therefore be interpreted as an indication that he still lived in that house at the time of Theresia's wedding. Whatever the arrangement might have been, the wedding feast must have been a copious affair with plenty to eat and drink, with music, singing, dancing, and general merrymaking. Most of the villagers took part in it, either as invited guests or as neighbors who just stopped by to congratulate the newlyweds and to pay respects to the *Hochwürdigen* (the highly respected one), as they now addressed Gregor, whom many of them knew as the boy named Johann Mendel. They recalled how he suddenly "disappeared" from the village, only to return now, at the age of 30, as a learned and revered man. They must have been pleasantly surprised to discover that the transformation of Johann into Gregor had no effect on his character: He had remained as respectful, modest, and simple as they remembered him from his younger years. The learned man and priest was not even ashamed to revert to the local dialect when speaking with them. The feast continued deep into the night, long after the groom led the bride to his house, where the feasting would resume the next day, though on a more modest scale. Mendel spent the rest of his stay in Hynčice talking to his parents and the rest of the family. They were all interested in hearing about his life in Vienna and Brno. He could not have helped noticing how his parents were aging rapidly, especially his father who had never fully recovered from the accident years ago. On the other hand, he must have been pleased to discover that Veronika and Alois Sturm were managing the farm well and that Theresia found herself a fine man. Leopold Schindler had made a good impression on him, an impression that would later be reaffirmed through further visits and correspondences. He remained in close contact with the Schindler family throughout the rest of his life and even took their boys under his protection when they were studying in Brno. After visiting some of his favored places in the area, on October 20, he returned to Vienna, relaxed and enriched in memories. Before he realized it, he was finished with his studies. Mendel's stay at the university was over and he had to head back to Brno for good. He returned from Vienna wiser, more self-confident, and full of plans for research he wanted to embark on.

Failed Examination: Act Two

Those familiar with Mendel's predicament expected him to have a second go at the teacher's qualifying examination immediately after the conclusion of his studies, when all the learned material was still fresh in his head. Mendel, however, seemed in no hurry to do so. Instead he simply slipped back into the routine of the monastic life, helping out with pastoral work (possibly as a catechist at Brno's elementary schools[94]), taking care of the garden, and studying. It was during this time that he made the observations on the pea weevil and prepared the report, which Kollar then read at the Zoological and Botanical Society meeting. Other than this, however, we have no reliable information about Mendel's activities during the interval from the end of July 1853 to the end of May 1854. It seems that he turned his back on the Examination Committee and did not reapply for readmission to the tests during those months. What might have been the reason for this unexpected behavior? Of the various possible explanations, the most sensible one to us is that Mendel was very nervous about facing the examining board again and so he sought a way of becoming a qualified teacher without having to go through that grueling experience again. His exposure to the scientific milieu in Vienna wetted his appetite for research, and combining research with teaching was what he had in mind for his career anyway. He might have therefore decided to focus on research that would lead to a dissertation and a doctorate, which would have the added benefit of automatically qualifying him for teaching at the secondary school level. Hence, the 1853/1854 interval could have been a period of Mendel's searching for a suitable research topic or even preparation for research on which he had decided already. If these were Mendel's plans, the Fates crossed them out on May 1854, when Brno's newly founded school, the *Oberrealschule*, urgently needed a physics teacher. The school's director decided that Mendel was his man and he got him. We describe how this happened in the next section, but first we must deal with the consequences of this new development.

Following his appointment as a substitute teacher, the pressure on Mendel grew to pass the qualifying examination and become a full-fledged professor. Nobody more than he himself must have desired to end this uncertainty about his status and bring stability into his situation. Since the appointment crossed his plans to become a certified teacher through a doctorate, the only venue left for him toward attaining this goal was now to go through the strenuous examination process again. Wiser from the first attempt, more knowledgeable from his university studies, and better aware of what might be expected of him, he began filling the gaps in his familiarity with the two subjects he had now been teaching.

The circumstances of Mendel's second attempt to pass the examination are blurred because only a handful of not very informative documents pertaining to the attempt have survived.[54,66] The critical documents were either destroyed or lost. As commonly happens in cases when facts are sparse, speculations burgeon, and few biographers have desisted the temptation to spice up the life of their subject with tales of intrigue and cover-ups. In reality, however, the few facts that *are* known about the case are sufficient to ground any such flights of fancy.[156] The facts come from such inglorious sources as an accounting book, police records, and an admission register; important evidence is, however, also provided by one private letter. As explained earlier, the abbey employed an accountant, who kept a careful record of all the incomes and expenditures, no matter how trifling they might have been. Curiously, while many documents about Mendel's life have been lost, the accounting book survived. And so one can learn from it that on April 9, 1855, the accountant/cashier dispersed 20 florins to Pater Gregor for a trip to Vienna regarding his teacher examination.[157] Presumably, he went to *arrange* the examination, but why he had to do it in person rather than by mail is not immediately obvious. Clear is, however, that at the beginning of April of that year, Mendel applied a second time for admission to the examination. Presumably, he had to produce most of the same documents he collected for the 1850 examination, plus some new ones, for example, a reference letter from the director of the *Oberrealschule*, Joseph Auspitz. Mendel might have also managed to set up a date for the examination, although this possibility was rather unlikely because it needed the consent of all the members of the Examination Committee, which required time. So, as in the case of the first examination, he was probably informed about the date by mail. Whether Mendel had to go again through all three phases of the examination is unclear. Since he passed the homework written test in the first examination, one might think that the university would have waived this part of the procedure. There is, however an indication, as we shall learn shortly, that this was not the case and that Mendel had to write essays on new sets of questions. This would explain, why he was not admitted to the other two phases of the examination until more than one year later. Furthermore, this time around Mendel changed the application. Instead of wanting to qualify for teaching natural history in all the grades and physics in the lower grades of the *Gymnasium*, he now reversed the order and wanted to be certified for teaching *physics* in all the grades and natural history in the lower grades only. It might have been this change that led the university administration to insist on Mendel's taking all three parts of the examination anew. There might have been several reasons for Mendel's change of mind (e.g., his better performance in physics than in natural history at his first examination or his greater exposure to physics at the University of Vienna), but the main reason probably was that he felt more at home in the former than in the latter subject. If he did write new essays in the two subjects, the committee must have graded them as satisfactory, for he was allowed to take the second part of the examination.

The second piece of hard evidence provided by the abbey's accounting book places the date of the second phase of the examination to the beginning of May 1856, since an entry dated May 2, 1856, indicates that Mendel received again 20 florins for a "trip to Vienna to the *Klausurprüfung*."[157] When exactly the written test took place is uncertain. Iltis[43,158] dates it to May 5 on the basis of an examination protocol, which he seemed to have had access to, but which is now unaccounted for.[158] Although most biographers accept Iltis' date, some argue that the examination took place later in that year.[54,159] We come to the latter possibility momentarily, but first let us cite additional evidence for the May 5 version. It comes from a letter Klácel wrote to Bratránek, the letter we alluded to earlier, when we discussed Mendel's illness. In the present context the letter acquires additional significance and so we quote from it the entire passage pertaining to Mendel and his examination:[160] *I will not be able to come for the holidays—many hurdles had come in the way. Problems arose last week already: P. Gregor was called to an examination at Vienna. He left and there was no chance that he would be back for the holidays, and since there was a shortage [of priests] for services, I had to stay home. P. Gregor was unlucky. Although he drew easy questions, he fell ill during the first Klausurprüfung and as a consequence was unable to write. He seems to have problems with his nerves generally since he endured several such insidious attacks already and they say that in his youth he suffered from epilepsy. The day passed and nothing was achieved. One has to feel sorry for him, since his homework* etc. *was graded as excellent. But formalities are formalities; in this case it was not possible to continue* [with the exam]. *Afraid that further attacks might follow, he returned home without accomplishing anything. I am very sorry for him, especially since he is anyhow discontented and so will eat himself even more.*

The letter, though not known to Iltis, supports his version of the events. Early in May 1856 Mendel did indeed report for the *Klausurprüfung*, began it, but seized by an attack of agraphia (the loss of the ability to write), gave it up, and returned to Brno seriously ill. His illness was so serious that Napp, fearing for Mendel's life, informed the patient's father. This sequence of events is indicated by another entry in the abbey's accounting book. Under the date May 31, 1856, the entry reads *Vater u. Bruder P. Gregors* (father and brother of Pater Gregor), indicating that the two stayed at the abbey as guests without charge. Since Gregor Mendel did not have a brother, the second visitor must have been Anton Mendel's brother, Johann Mendel, Gregor's uncle. Anton Mendel was not in good health himself, so there must have been a very grave reason for his undertaking the long, exhausting trip to Brno.

The question then is: What triggered Mendel's breakdown? Iltis[43] records a story that was supposedly making rounds in Brno after Mendel's return from Vienna. Iltis' source of the story was Adolph Nowotny,[161] one of Mendel's colleagues at the *Oberrealschule*. The rumor was that Mendel had a confrontation with one of the examiners, presumably the botanist, regarding a controversial scientific issue. The encounter so upset Mendel that he suffered a nervous breakdown. In Brno people might have been inclined to believe the story as another example of Viennese arrogance toward everything coming from the provinces. If so, it made Mendel into a kind of martyr-hero and his failure to pass the examination therefore did not damage but rather boosted his reputation. Much later some biographers[132] embellished the story by adding piquant details to it regarding the identity of the examiner and the subject of the controversy. The cause of disagreement was purportedly the origin of the plant embryo. Mendel apparently insisted that elements from both the egg and the pollen contributed equally to the makeup of the embryo, whereas the examiner asserted that the pollen tube had a mere nursing function. This latter contention identified Fenzl as the examiner, because it was he, of the two botanists at the university, who held this view. Furthermore, the contention fit Fenzl's easily irritable and often abusive character in his relation to students. But there was even more to the story:[132] Wasn't Fenzl the grandfather of Erich von Tschermak-Seysenegg, one of the rediscoverers of Mendel's work some 50 years later, and wasn't Tschermak-Seysenegg professor at the University of Vienna?

Couldn't it be that the whole examination affair was rather embarrassing to Tschermak-Seysenegg since it turned out that Mendel was right, and so the former made the documents pertaining to it disappear?

It could be, but *was* it like that? It is a fact that in the middle of the nineteenth century, botanists had still not been united in regard to the pollen's contribution to the embryo and thus to the developing plant. At the international level, the opposing sides were represented by the Italian astronomer, mathematician, and microscopist Giovanni Battista Amici (1786–1881) and the German botanist Matthias Jacob Schleiden (1804–1881). Amici was the first to describe (in 1823) the pollen tube growing out of the pollen grain and toward the egg cell in the ovary. He suggested that the plant embryo arose from the union of elements in the tip of the tube with the egg cell. Schleiden, on the other hand, maintained that the pollen tube was a mere incubator or a "nurse" of the egg cell. In Vienna the two opposing views were prominently represented by Unger and Fenzl, respectively. Furthermore, it is also true that Fenzl was a combative type, whereas Unger was a more peace-loving person. Finally, although on the surface Mendel appeared docile and timorous, when provoked he could turn into a stubborn fighter. So, in theory, the potential for a confrontation with Fenzl was present at the examination. Nevertheless, there are incongruities in the whole story that cast a serious doubt about its verity. Indeed, the story crumbles under serious scrutiny. Not only are there no hard facts to support it, but it also does not make much sense. To begin with, in the abbey's accounting books, there is also an entry from the beginning of January 1856 documenting the purchase of an unspecified but rather expensive (2 florins and 32 Kreizers) medicine for Pater Gregor from a pharmacy (Eder) from which the friars normally had not been getting their medicaments (they had been getting them cheaper from the pharmacy of the Brothers of Mercy). This entry indicates that from the beginning of 1856, the year of the approaching examination, Mendel had not been feeling well and the purchase of a specialized medicine suggests that his illness might have been neurological. He must have been working hard preparing for the examination and giving his utmost as far as his duties at the *Oberrealschule* were concerned. By the time he went to Vienna to sit the examination, he was probably well prepared, but totally overworked. His labile nervous system did not need much to short-circuit. There was no need for a clash with Fenzl to trigger a nervous breakdown, as little as being locked up in the same room where he failed once before, or opening the envelope and realizing that he would not be able to answer the question, might have sufficed. From Klácel's letter we learn that the breakdown occurred right at the beginning of the written test, when Mendel had not yet even seen any of the examiners and had therefore had no opportunity to come into a scientific disagreement with any one of them. Also, for all we know, the first written test could have been in physics rather than botany. The argument about the purported cover-up does not hold water either. The second note mentioned by Iltis and referring to the destruction of the examination documents was presumably entered not too long after the aborted examination, when Tschermak was not yet even born and later backdating of the note so that it would fit into the chronological order with the rest of the entries would have nearly been impossible. Destruction of failed examination documents would be unusual, but Mendel did not fail at the second attempt, he did not *complete* it. The clerks at the university might not have therefore seen any reason why to keep the material pertaining to it.

A more recent discovery of note in the archive of the Examination Committee[54] complicates our account of the examination but by no means undermines it. The note dated July 3, 1856 documents that on that date Mendel applied for admission to the examination and that the examination was scheduled to take place on August 5, 1856. The entry has been interpreted[54,159] to mean that the actual examination took place not on May 5, as Iltis and all biographers who followed him have claimed, but on August 5, instead. This interpretation is, however, indefensible. First of all, the note does not prove that an examination actually took place on August 5; it only indicates that it was *scheduled* for that date. The problem with the August 5 date is the absence of any record that would place Mendel in Vienna at that time. There is no entry in either the accounting register at the abbey or in Mendel's *Heimatschein* (police record) documenting that Mendel traveled to Vienna again in 1856, after the May visit. Finally, there is a simple explanation for both the note and the lack of evidence for Mendel's additional travel to Vienna in 1856. After recovering from his illness, Mendel decided to try once more and resume the examination. He submitted the application for admission to it in July by mail and was informed to report for the examination on August 5. In the meantime, however, he mulled over everything in his head and decided that he was in no condition to face the Examining Committee again. He then either canceled the appointment or simply did not show up for it. After a while, the university classified the attempt as aborted and ordered the relevant material destroyed, as it might have done in cases of other aborted examinations.

Substitute Teacher at the *Oberrealschule*

The new school was the *k.k. Oberrealschule*, or *Realschule* for schort.[162] The incentive for founding the *Oberrealschule* was, as in the case of the Technical Institute, the need for a school that would provide general education like the *Gymnasium*, but mirroring the advances in science and technology of the time. The school was to have six grades and pupils were to enter it after the completion of the elementary school. They would exit it by passing the final examination, the *mittlere Reife*, which, like the *Matura* at the *Gymnasium*, would open for them the doors for admission to technically or agriculturally oriented universities or specialized schools of higher education. Just how timely the creation of the new school was became apparent when it opened in 1851 and over 300 pupils registered for it. The drive on the new school forced the authorities to launch three parallel classes with more than 100 pupils in each of them. In subsequent years the admission numbers grew even higher and overcrowding became one of the school's most serious problems. Its second problem was the lack of a suitable building. Provisionally the authorities placed the school in an old, rather ran-down building, the so-called Schütz house (Fig. 6.5) in the Trnitá (*Dornrösslein* or Thorn) street on the outskirts of the city. In the meantime, they began with the construction of a new building on Jánskát o Janská street in the city's center. A Viennese architect, who designed the building, was clearly influenced by the eclecticism of the *Ringstrasse* style when he gave the building the appearance of a Florentine palace with bow windows and a tall clock tower at its northeastern corner (Fig. 6.21). When completed in 1859, the robust building might have appeared to the pupils more like a prison than an educational institute. It certainly hardly aspired to become one of Brno's architectural jewels.

The large number of students called for a large teaching staff and so led to the hiring of several substitute teachers. Josef A. Auspitz (1812–1889),[37] the school's director, was a professor of accounting, educated at the universities of Vienna and Budapest. As you may recall, he was previously one of the first professors at Brno's newly founded Technical Institute, where Mendel substituted briefly for Helcelet before enrolling at the University of Vienna. When in early 1854 one of Auspitz's staff members at the *Oberrealschule*, the physics teacher Johann Patek, left for Znojmo to become a school director there and Auspitz needed urgently a replacement, he remembered Mendel and approached him to fill the vacancy. Mendel accepted the offer and Napp apparently had no objections. There was, of course, the problem that Mendel was officially not qualified for the position, but Auspitz took care of it. He wrote a letter to the *Landesschulrat* (Regional Educational Council), in which he stressed the urgency of the situation and explained that Mendel actually *was* qualified for the job. He studied physics and was even an assistant at the Physics Institute in Vienna (which of course was not true) and took private zoology lessons at the Court Museum there. Auspitz then added that he himself had an opportunity to evaluate Mendel's teaching capabilities at the Technical Institute and that he found his lectures insightful, logical, and easily comprehensible. The council agreed to waive the requirement of the teaching certificate in Mendel's case, and so on May 26, 1854, Mendel became once again a substitute teacher, even though the official appointment letter signed by Count Lažanský was not issued until July 14, 1854. Mendel was to occupy this post for 14 years.

At the time of Mendel's appointment, the 1853/1854 school year was already coming to a close. Also, 12 days after the appointment, on June 7 and 8, the St. Thomas Abbey suffered the Apostolic Visitation by Bishop Schaffgotsch and his helpers, as described earlier. It was therefore not until September that Mendel assumed fully his new duties. Auspitz assigned him to teach physics and natural history in the two lowest grades. For the teaching of physics in the higher grades, Auspitz hired another substitute teacher, Alexander Zawadski (1798–1868), who joined the school at about the same time as Mendel.[163] Mendel and Zawadski remained colleagues until 1868, when the former became abbot and the latter died. Like Mendel and Kollar, Zawadski came from Silesia, from the city Bielsko at the foothills of the Beskydy Mountains in the southernmost part of present-day Poland. And like Mendel and Kollar, Zawadski had to struggle through life in his pursuit of higher education in natural sciences. A religion teacher, who discovered his gifts, tutored Zawadski privately to prepare him for the *Gymnasium* at Těšín, from which he then continued his studies at the Philosophical Institute at Olomouc and later at the University of Lvov. In the Ukraine's capital, Zawadski advanced

Fig. 6.21 The new *Oberrealschule* at Brno's Janská Street. Here Mendel taught physics and natural history as a substituting professor from 1859 to 1868. Here he also presented his pea hybridization results at the regular meeting of the Natural Science Society in 1865

through teaching posts at the *Realschule* and the Philosophical Institute to a university appointment at the age of 42. In 1848 he placed himself at the head of the revolutionary movement at the university and paid for it with the loss of his job. Unable to find another university employment, Zawadski accepted the position at Brno's *Oberrealschule*. Auspitz might have had a soft spot for victims of the 1848 uprising, since he himself was also one of them. For his own involvement in the revolution, Auspitz was banished from Vienna to the provinces and degraded from a university to a secondary school position. Later, however, he apparently came back into good graces with the government, judging from his 1869 acquisition of the *Hofrat* title and his election into the Provincial Diet. Zawadski's interests were broad, but botany was their focal point. He was an expert on the flora of eastern Galicia and Bukovina, the region from which he originated.

Alongside Zawadski, two other colleagues of Mendel at the *Realschule* shared his interests in natural sciences: Makowsky and Oborny. While Zawadski was 24 years older than Mendel, Makowsky and Oborny were, respectively, 11 and 18 years younger. Alexander Makowsky (1833–1908), native of Svitavy on the Svitava River in west-central Moravia, came from a well-situated family, which had one brother who was the city's mayor and another who was the head of the police department. After completing his studies at Brno's Technical Institute, he joined the staff of the *Oberrealschule* for four years, from 1860 to 1864, and later made his career at the Technical Institute as it transformed first into the Polytechnic Institute in 1867 and then into the German Technical University in 1873. In that year Makowsky became the university's professor of mineralogy and geology and a few years later even its *Rektor*. He devoted the early phase of his scientific career to botany, culminating in the publication of a book on the flora of Brno and its environs, to which Mendel contributed a meteorological table.[164] Later he focused on geology and published geological maps of Moravia and its different regions.

Adolf Oborny (1840–1924), Mendel's other colleague, was born in the small village of Světlá near Telč in southwestern Moravia, to the family of a gamekeeper on the lands of Prince Lichtenstein. After the completion of his studies at Brno's Technical Institute, Oborny joined the staff of the *Oberrealschule* in 1865 as a substitute teacher. In 1870 he passed the qualifying examination for teaching natural history and mathematics and moved to Znojmo to teach at the *Realschule*, where he remained for 27 years. Two years later, from 1899 until his retirement in 1907, he was the director of the *Oberrealschule* in Lipník nad Bečvou. He then returned to Znojmo, where he died. He was a botanist whose four-volume *Flora of Moravia and Austrian Silesia* published from 1884 to 1888 became a solid foundation for all subsequent Moravian naturalists to build on. He was an expert on the flora of the Znojmo region and on the genus *Hieracium* that also interested Mendel.

With these three colleagues Mendel shared an interest in natural sciences and continued to interact with them past their periods of employment at Brno's *Oberrealschule*. They were all members of the Natural Science Section of the Agricultural Society and saw one another at its regular meetings. Some of them liked to visit Mendel at the abbey, where Mendel, once he began his experiments, had always something interesting to show them.

On the staff of the *Oberrealschule* were also two other friars from the St. Thomas Abbey, Benedikt Fogler and Augustin Krátký, but Mendel communicated with either of them very little. The former taught physics, but his interest in this discipline did not go beyond getting the required information into the heads of his pupils. He was a strict disciplinarian and apparently not much liked by his charges. Krátký taught religion and so had little in common with Mendel. When the school attained its full size, its professorial staff had nearly two dozen members. Mendel must have known all of them, but his interactions with most of them had probably been restricted to school matters only.

Mendel's responsibility was to teach *Naturgeschichte* (natural history) and *Naturlehre* (physics) in the two lowest grades, that is, Ia, Ib, IIa, and IIb, and in some years, when there were three parallel classes, also Ic and IIc.[50] Only in the last year of his employment, which lasted from May 26, 1854 to March 31, 1868, did he also teach physics in the third grade (IIIc). In all these 14 years, the classes were quite large, the number of pupils per class ranging from 62 to 117. During the entire period, with the exception of the first year, he was also the form master (*Klassenlehrer*) in one of the second grade classes (IIa, IIb, or IIc). The school year was divided into two semesters, and in each semester Mendel taught two hours per week of physics and two hours of natural history in each grade (I and II) and each class (Ia, Ib, IIa, IIb). In total he therefore taught 16 hours (20 in those years in which there were three parallel classes) per week, or on average four to five hours per day.

Considering that the school paid him only half the salary of an ordinary professor, mere 500 florins conventional currency per year, it certainly got lot of work for little money out of him. His responsibilities, however, did not end in the classroom. In addition to a fair amount of administrative work that came with the teaching, he was also responsible for the school's natural history *Kabinett.* Collections of stuffed or otherwise preserved animals, shells, nests, minerals, rocks, pressed plants, models of crystals, anatomical models posters, charts, and diagrams, microscopes, instruments and equipment for demonstrations in physics classes, and all kinds of other teaching aids were an indispensable part of instruction for every good teacher. The richness and condition of its natural history collections were therefore good indicators of the school's quality of education in the natural sciences. To be functional, the collections had to be kept in a good order. They had to be catalogued and stored in an orderly manner, while damaged or broken items had to be repaired or replaced, new items had to be acquired, and existing items cared for. All these duties kept Mendel on his toes. Seen in this light, his experimental achievements, which we describe in Vol. 2 Chap. 3, will appear so much more remarkable.

As at Znojmo's *Gymnasium*, so too at Brno's *Oberrealschule*, Mendel quickly became popular with both students and colleagues. Some of Mendel's former students were still alive when he became posthumously famous at the beginning of the twentieth century. Mendel's early biographers contacted some of them, while others came forward on their own with reminiscences of their teacher.[43] Their testimony must be approached with caution because it was deposited nearly half a century after the events and might have been colored by Mendel's sudden renown. Nevertheless, the unanimous praise of Mendel by his pupils for the lucidity of his instructions; his enthusiasm for the subject and love of the profession; his friendliness, gentleness, kindness, and compassion for his pupils.[43] must certainly be taken to mean that Mendel must have been an outstanding teacher. We will use some of this testimony in later chapters when we attempt to characterize his personality.

After a while, Mendel got used to being a substitute teacher for the rest of his life, or at least for as long as his services would be needed and the government would allow him. The government did not seem to mind that his employment violated the decree issued on November 11, 1849, and why should it have done so, when it was saving money on Mendel? In the end he quit teaching on his own will, when he was elected abbot in 1868. Peace might have returned to Mendel's mind when he gave up trying to acquire the teacher certificate and focused his energy, instead, on his experimental work.

What Had Mendel Learned in Vienna?

Mendel was registered at the University of Vienna as an extraordinary, that is, special student. One consequence of this status was that he was not examined in any of the courses he took. We have therefore no formal record of how well he learned the subjects he chose to study. It is the choice of the subjects itself that tells us something about Mendel's interests. Here, however, we must distinguish between courses for which he registered and attended and those for which he registered but did not take, for reasons mentioned earlier. After taking these circumstances into account, two biases in Mendel's choices become apparent. The first bias was his preference for physics to the detriment of natural history. And the second bias was a preference for experimental over descriptive subjects. Apparently, he liked better to learn how to advance science through cleverly thought-out experiments than to receive facts without knowing how they were attained. He wanted to know how to ask the right question and then plan, design, execute, and interpret experiments aimed at answering this question. He was much less interested in description, identification, and classification of natural objects. He therefore favored Unger's lectures, from which he learned how things in nature might work, over Fencl's exhortations of nature's catalogue. In other words, Mendel spent the two years in Vienna preparing himself for experiments that he might have been formulating vaguely in his mind, while he should have been memorizing knowledge that would certify him as a teacher. He enjoyed teaching, but the exposure to Doppler, Ettingshausen, and Unger wetted his appetite for creating knowledge, rather than conveying it to minds often resistant to its reception; hence his hesitancy to apply a second time for admission to be examined for the teacher certificate, which to him was like stepping on the road to calvary.

References and Notes

Abbreviation: FM *Folia Mendeliana Supplementum ad Acta musei Moraviae*. Moravské zemské muzeum Brno; published since 1966

[1]*But are they allowed to do that?* is what the emperor allegedly exclaimed in a horrible Viennese dialect, when informed about the uprising in the streets of Vienna. See[2], p.128

[2]Craig GA (1971) Europe, 1815–1914. Harcourt Brace Jovanovich College Publishers, Fort Worth, TX

[3]Beales D, Dawson E (2002) The Risorgimento and the Unification of Italy, 2nd edn. Longman, London

[4](a) Fortescue W (2005) France and 1848:the End of Monarchy. Routledge, London (b) Denholm A (1972) France in revolution, 1848. John Wiley, Sidney

[5]*Čas oponou trhnul—a změněn svět!* With this opening line the Czech bard Jan Neruda greeted the 1848 revolution in his poem Romance about the spring of 1848

[6](a) Bérenger J (1997) A History of the Habsburg Empire 1700–1918. Transl. by Simpson CA. Longman, London (b) Polišenský J (1980) Aristocrats and the crowd in the revolutionary year 1848. A contribution to the history of revolution and counter-revolution in Austria. Transl. by Snider F. State University of New York Press, Albany, NY

[7]The word "diet" derives from the Latin *dies* meaning "day." It came to be used for "a day-journey (i.e., daily) allowance money" for the deputy's attendance at the assembly of delegates. Hence, the German *Landtag* and *Reichstag*, where *Tag* means "day" and *Reich* "land" and "empire," respectively. The word "parliament" stemmed from Old French *parler*, which meant "to speak."

[8](a) Klíma A (1994) Češi a Němci v revoluci 1848–1849. Nebesa, Praha. (b) Klíma A (1950) Rok 1848 v Čechách. Rovnost, Praha

[9]Hahn HJ (2001) The 1848 revolutions in German-Speaking Europe. Pearson Education, Harlow, England

[10](a) Trapl M (1977) České národni obrození na Moravě v době předbřeznové a v revolučních letech 1848–1849. Blok, Brno (b) Dřímal J, Peš V (1969–1973) Dějiny města Brna. Vols. 1 and 2. Blok, Brno

[11]Count Josip Jelačič (1801–1859) was a Croatian political and military leader. After the outbreak of the revolution in Vienna, the emperor promoted Jelačič to a general and made him commander of all troops in Croatia. Shortly afterward, however, when the general went on to support the Croatian independence movement, the emperor dismissed him from the post, only to reinstate him 3 months later and involve him in the suppression of the Hungarian revolution

[12]Křížkovský also set to music another revolutionary poem, one entitled *The University*, penned by the Viennese student H.J. Franklin. The poem was recited at a mass rally in Brno on March 17, 1848. Křížkovský's song could subsequently be heard sung by the National Guards as they marched through Brno's streets

[13]Letter of A. Rambousek to M. Klácel. Quoted in ref.[165], pp. 15–16

[14]Calabria is the southernmost part of the Italian peninsula. Originally, the name applied to both the heel and the toe of the "Italian boot," but later came to be restricted to the toe only, which is separated from Sicily by the narrow Strait of Messina. In the mid-nineteenth century, the mountainous region of Calabria became a hideout of the revolutionaries led by Giuseppe Garibaldi (1807–1882). The revolutionaries wore a hat, which had a broad base, narrowing upward. The hat then became a symbol of revolution in general

[15]In Brno, though not only there, many Germans believed Napp to be of Czech extraction, which he clearly was not. In Mendel's family in Hynčice, too, grandmother Rosina Mendel told her grandsons that Napp did not like uncle Gregor because the uncle was a German. One of the grandsons, Ferdinand Schindler, then relayed this information to Mendel's biographer, Hugo Iltis (see letters 2 and 6 in[164]). According to the grandmother, this was the reason why Napp did not want her son to pass the examination to become a professor and did not want him to study at the University of Vienna. None of it is true, of course, and the question is who invented these stories, the mother or her son? It is unthinkable that Gregor would have come up with such insinuations, which he knew were not true. It *is* possible, however, that frustrated with the developments in the abbey, he complained to his mother about them and that she interpreted her son's hints her own way

[16]Footnote added by A. Matalová to the article by Czihak and Sladek[26] (p. 34)

[17]The friend was Jan Peter Jordan, professor of Slavic languages at the University of Leipzig

[18]Dvořáková Z (1976) František Matouš Klácel. Melantrich, Praha

[19]Orel V, Verbík A (1984) Mendel's involvement in the plea for freedom of teaching in the revolutionary year of 1848. FM 19:223–233

[20]Orel V (1996) Gregor Mendel. The first geneticist. Oxford University Press, Oxford, Translated by Stephen Finn

[21]Weiling FJ (1998/1999) G. Mendel und die Eingabe der sechs Capitularen des Stiftes St. Thomas an den Österreichischen Reichstag von 8.August 1848. FM 33/34; 5–10

[22]Nivet C (2006) 1848: Gregor Mendel le moine qui voulait etre citoyen. Médecine/Sciences 22:430–433

[23]Weiling F (1993/1994) Die Original-Handschrift der Pisum-Arbeit Mendels. FM 28/29:19–36

[24]In many religions, including the monotheistic ones, the supreme god has many assistants, serving as intermediaries between him and the mortals. The Greeks called them *angelos*, which means "messengers." In the Judeo-Christian mythology, as in some other traditions, some of the angels revolted against God's rule and were subdued and expelled from the heaven, to become emblematic pursuers of evil on earth. John Milton rehearses the Christian story of the fallen angels in *The Paradise Lost*, while Anatole France turns it into a persiflage in *The Revolt of the Angels*

[25]Others have argued that Napp did not know about the petition, on the grounds that if had he known about it, he would not have allowed it to leave the abbey (See[26])

[26]Czihak G, Sladek P (1991/1992) Die Pesönlichkeit des Abtes Cyrill Franz Napp (1792–1862) und die innere Situation des Klosters zu Beginn der Vesuche Gregor Mendels. FM 26/27; 31–66

[27]After the dissolution of the *Reichstag*, the documents concerning its activities, including the petition of the six St. Thomas friars, were transferred from Kroměříž back to Vienna and ultimately ended up in the state archives. Not in the general archive, as one might have expected, but in the military archive. There, in 1955, Erika Weinzierl-Fischer found the petition accidentally while searching for materials pertaining to Mendel. See[28]
[28]Weinzierl-Fischer E (1955) Die Kirchenfrage auf dem Österreichischen Reichstag 1848/49. Mitteilungen des Österreichischen Stätsarchivs 8:160–190
[29]Redlich J (1929) The Emperor Francis Joseph of Austria: a Biography. Archon Books, New York, NY
[30]The word *hussar* derives from Hungarian *huszár*, which meant originally "highwayman." Later it came to be used as a designation for a member of the light cavalry in Hungary and Croatia. Hussars were known for their ruthlessness
[31]Musil's *Der Man ohne Eigenschaften* was originally published serially from 1932 to 1942. Rowohlt Verlag, Hamburg, published the whole work in one volume in 1952. The novel was rendered into English under the title *The Man Without Qualities* by Eithne Wilkins and Ernst Kaiser and published by Putnam's Sons in New York, NY 1953
[32]New Catholic Encyclopedia. 15 volumes, 2nd edn. Thomson—Dale, Detroit 2003
[33]Buchberger M, Höffer J, Rahner K (eds) (1930) Lexikon für Theologie und Kirche, vol 1–11. Herder, Freiburg
[34]Wolny G (1855–63) Kirchliche Topographie von Mähren. Vols. 1 and 2. Brünn
[35]Zabel J (1975) Zweihundert Jahre Bistum Brünn (1777–1977). Schriftenreihe des Sudetendeutschen Priesterwerkes, Vol. 21. Königstein/Taunus
[36]Other forms of spelling the name are *Scha(a)f(f)gotsch(e)*
[37]Wurzbach C von (ed) (1856–1891) Biographisches Lexikon des Kaiserthums Oesterreich. Vol. 1–60. Verlag der k.k. Hof- und Staatsdruckerei, Wien
[38]Huber K (1978) Die Apostolische Visitation des St. Thomas Kloster in Alt-Brünn. Archiv für Kirchengeschichte von Böhmen-Mähren-Schlesien 5:190–236
[39]The correspondence concerning the visitation was carried out in Latin. Czihak and Sladek published its German translation as a set of appendices to their article (see[26])
[40](a) Sajner J (1967) Gregor Johann Mendel und Znajm. Forschung, Praxis, Fortbildung. Organ für die gesamte praktische und theoretische Medizin 18: 677–685. In addition to containing details of Mendel's 1-year sojourn to Znojmo, this article is appended also with transcripts of the chief documents pertaining to it. (b) Wisnar J (1909) Gregor Johann Mendel. Ein Gedenkblatt dem Andenken an den genialen Forscher gewidmet. Znaim 1909. (c) Wisnar J. Gregor Johann Mendel. In Jahresbericht des k. k. Gymnasium in Znaim für das Schuljahr 1908/1909 Verlag des k. k. Gymnasium, Znaim pp.21–35. (d) Siegel L (1909) Znaim, die erste Lehrstätte des grossen Naturforschers Johann Gregor Mendels. Znaimer Wochenblatt 60(5): 1–2
[41]This interpretation, favored by most Mendel's biographers, is contradicted by the testimony of his mother Rosina, as mediated by Ferdinand Schindler. In at least three letters addressed to H. Iltis (Nos. 2, 6, and 10 in[164], dated in Butovice September 6, 1902, July 22, 1909, and April 6, 1912, respectively), Schindler repeated the claim that Napp was *un*favorably disposed toward Mendel, supposedly because he (Napp) was Czech and as such favored the Czech over the German faction in the abbey. The specific accusations included that Napp did not give Mendel enough time to prepare himself for the teachers' examination, he did not allow him to take this examination, and he did not want to allow him to study in Vienna. Ferdinand Schindler was surprised to hear from Iltis that Mendel did take this examination twice and failed. Was Mendel misinforming his family about what was happening in the abbey, perhaps because he was ashamed to admit his failures? Or was all this the delusion of an aging mother who cannot admit to herself that her favorite son was not infallible? Whichever might have been the case, this version of the story is clearly invalidated by the known facts
[42]Mendel's "Autobiographie" (1850) In[168], pp. 74–77. For an English translation, see A. Iltis: Gregor Mendel's autobiography. J Heredity 45:231—234, 1954. Purkyně University at Brno published a bibliophilic edition of the autobiography under the title *Gregorii Mendel*

Autobiographia Iuvenilis. Ad centesimum Quinquagesium J. G. Mendel natalen. Universitas Purkyniana Brunensis 1972

43Iltis H (1924) Gregor Johann Mendel. Leben, Werk und Wirkung. Julius Springer, Berlin, An English translation by E. and C. Paul was published under the title *Life of Mendel* by George Allen & Unwin, London 1932

44Lutterer I, Majtán M, Šrámek R (1982) Zeměpisná jména Československa. Slovník vybraných zeměpisných jmen s výkladem jejich původu a historického vývoje. Mladá fronta, Praha

45Hosák L (1967) Přehled historického místopisu Moravy a Slezska v období feudalismu do roku 1848. Profil, Ostrava

46(a) Vrbka A (1927) Gedenkbuch der Stadt Znaim 1226–1926. A. Bartosch, Nikolsburg. (b) Vrbka A (1922) Das Znaimer Ländchen. Sudetendeutscher Verlag F. Kraus, Reichenberg

47In the hierarchy of the abbey, Václav Šembera (1807–1881) was the third in command. After Napp, the abbot, and Vorthey, the prior, Šembera was the subprior. He was the cousin of Alois Vojtěch Šembera (1807–1882), then a well-known Czech patriot, writer, journalist, linguist, and historian of literature, born in Nové Mýto near Pardubice, Bohemia. Whether the subprior just happened to be on the road when Mendel arrived at Znojmo or was checking on him is unclear. At the abbey the subprior occupied himself with administration, including the training and supervision of the novices, and some pastoral work

48Mendel's letter of October 31, 1849, to Rambousek is published on p. 128 of[163]

49See two letters to Rambousek of July 14, (1852?) and October 14 (1852?) published in ref. 165, pp. 128 and 129

50Richter O (1943) Johann Gregor Mendel wie er wirklich war. Verhandlungen des Naturforschendes Vereines, Brünn 74(2):1–262

51For the text of Schaffgotsch's letter on Mendel's election to the abbot's office, (see Richter[50], pp. 59–61)

52The minister's letter was published in the *Reichs-Gesetz-und Regierungsblatt für das Kaiserthum Österreich, Jahrgang 1849, No. 36*

53The first usage of the word is credited to J. B. Lamarck (*Hydrogeology* 1802) and G. R. Treviranus (*Biologie oder Philosophie der lebenden Natur für Naturforscher und Ärzte,* Vol. 2 1802)

54Gicklhorn R (1973) Gregor Mendels Lehramtsprüfung und Studienzeit in Wien. Biologische Rundschau 7(4):145–159

55The original letter was published by Sajner[40a] (document No. 2). An English translation appeared in[168]. The extract from the letter that we quote is our own translation

56The original German transcription of the documents is provided by Sajner[40a] (see document No.3) and its English translation by Ryan[168]. Both this and the preceding document (see[55]) are written in what Iltis[43] called "abominable German." In our translation of the excerpts, we did not try, at least not intentionally, to preserve the bureaucratic flavor of these letters

57See document No. 12 in Sajner's[40a] article

58Gicklhorn[54] writes that in the *Einreichungs-Protokoll* book the name Gregor Mendel is entered for the first time on April 17, 1850, which would imply that the letter reached the office on that day. In reality, however, the entry date is April 22, while the letter is dated April 17. Since these two dates appear in adjacent columns in the book, she apparently mixed up the dates

59Bauer G (1991) A. von Baumgartner (1793–1865). Diplomarbeit, Wien

60A "Virginia" was a thin, 25 cm-long cigar with a straw inside. To light the cigar, the smoker had to pull out the straw, light it, and with it light the cigar, which then kept burning for several hours

61Svojtka M (2007) Eindrücke aus der Frühzeit der geologischen Erforschung Ostgaliziens (Ukraine): Leben und erdwissenschaftliches Werk von Rudolf Kner (1810–1869). Geologie und Paläontologie Universität Innsbruck. *Geo. Alp* Sonderband 1, pp 145–154

62Adalbert Stifter (1805–1868) was an Austrian novelist, poet, and painter born in Horní Planá, a village in Bohemia, near the Austrian border. Friedrich Nietzsche called *Der Nachsommer* one of two greatest nineteenth-century novels in the German language, the second being Gottfried Keller's *Der Grüne Heinrich* (The Green Henry)

[63]Theodor Georg von Karajan (1810–1873), no relation to the famous German conductor, was the son of a Greek merchant, who settled in Vienna and was elevated to imperial peerage for his services to the Austrian Monarchy in trade and industry. The son occupied the professional post only briefly (from 1850 to 1852) and then became *Kustos* (keeper) of the imperial court library, a position that back then counted more than that of a professor

[64]The documents pertaining to Mendel's first attempt to become a certified professor of natural history and physics are deposited at the Museum on Natural History, the University of Illinois, Urbana-Champaign, Illinois in the United States (see Hoffmeister DF, Henriksen HC (1979) The collection of Mendeliana at the University of Illinois, Urbana-Champaign. FM 14: 281–284). Here is how they got there. Mendel's biographer, Hugo Iltis (1882–1952), professor of biology at Brno's German *Gymnasium,* borrowed the documents from the archive of the University of Vienna in 1913 to use them for a demonstration following his lecture on Mendel at the Conference of the Society of German Natural Scientists and Physicians (for details see[54]). Instrumental to the lending was Professor J. Wiesner at the Vienna University. When Wiesner died shortly afterward, the archive failed to request the return of the documents. When Iltis later (in 1939) immigrated to the United States, he took with him all the documentation he assembled on Mendel, including the examination papers. In the USA, he organized a Mendel Museum at the Mary Washington College at Fredericks, Virginia, where he became professor of biology (see Iltis H (1943) The Mendel Museum at the Mary Washington College. Scientific Monthly 56: 386–387). In 1955, his son sold, after his father's death, the collection of Mendeliana for $6,000 to the University of Illinois, where it is located to this day

[65]Munzar J (1971) G Mendel's erste, bis jetzt unveröffentlichte Abhandlung über Meteorologie. FM 6:185–187

[66]Gicklhorn R (1973) Wurde Gregor Mendel bei der Lehramtsprüfung an der Wiener Universität ungerecht behandelt? Biologische Rundschau 11(2):73–84

[67]Orel V, Czihak G, Wieseneder H (1983) Mendel's examination paper on the geological formation of the earth of 1850. FM 18:227–272

[68]George W (1967) Mendel and the classification of mammals. FM 2:23–28

[69]Of the six sources Mendel mentions, Munzar[65] has identified four: (a) von Baumgartner A, von Ettingshausen A (1842) Die Naturlehre nach ihren gegenwärtigen Zustande mit Rücksicht auf mathematische Begründung. 7. Auflage, Wien (b) Hessler JF (1850) Handbuch der Physik. Wilhelm Braumüller, Wien (c) Hofer J (1850) Populäre Physik zunächst für Real- und Gymnasialschüler und auch zum Selbststudium für Jedermann. Wilhelm Braumüller, Wien (d) Kunzek A (1847) Leichtfasslichen Darstellung der Meteorologie. Wilhem Braumüller, Wien

[70]The sources Mendel lists are these: (a) von Humbolt A. *Der Kosmos*. 5 Bände, Cotta'scher Verlag, Stuttgart 1845–1862. (b) von Burmeister H (1843) Die Geschichte der Schöpfung. Verlag O.Wigand (c) von Leonhard KE (1845) *Die populäre Geologie oder die Naturgeschichte der Erde*. F. Schweizerbart, Stuttgart. (d) von Bromme, Das Mineralreich. This last source remains unidentified. Mendel might have meant the third volume of the three-volume work: Gistel J, Gistel J, Bromme T (1850) Handbuch der Naturgeschichte aller drei Reiche. Hoffmann, Stuttgart

[71]Weiling F (1970) Die Meteorologie als die wahrscheinliche Quelle der statistischen Kenntnisse J.G. Mendels. FM 5:73–85

[72]For the printed version of Baumgartner's evaluation, see[66]. Its free translation is our own. As stated earlier, the original documents pertaining to the examination are deposited at the Urbana-Champaign museum (see[64]), and their facsimiles and printed versions have been published by Gicklhorn (see[54,66]) and Orel et al.[67]

[73]In addition to the documents at the Urbana-Champaign museum, there is another one, whose original is deposited at the Mendelianum in Brno. The document is signed by Kner and dated November 11, 1850. Its printed version has been published by Gicklhorn[66], who writes: "while the piece in Urbana is drawn up by a member of the examination committee or a secretary, the one presented here (i.e., the Brno document} has been prepared as a memo by Kner by his own hand." She explains the late dating of the document by the fact that the oral examination took

place when the summer vacation had begun already and the writing of the protocol was therefore postponed until the beginning of the fall semester. It is, however, most unlikely that the highly technical Urbana document could have been composed by another committee member or a secretary. First because the committee probably did not meet until at the oral examination, and second no committee member other than Kner had the expertise to write it. Our description is a précis of Kner's evaluation, paraphrasing his statements

[74]Merriam–Webster's Collegiate Dictionary (1996) 10th edn. Merriam—Webster, Springfield, MA

[75]The original source of information about these events and their dating is Iltis[43], who apparently had an access to the correspondence pertaining to them. The documents have since been unaccounted for

[76]Baumgartner expressed his unhappiness with the Ministry's decision on the margin of Spallek's letter

[77]We say "presumably" because the letter is no longer available and in the secondary literature there is confusion on this point. Iltis[43] writes: *Am 1. August wurde Mendel eingeladen nach Wien zu kommen und sich im Bureau Sr. Exzellenz des Direktors A. v. Baumgartner im Finanzministerium zu melden.* In the English version of Iltis' book, this sentence is translated thus: "On August 1 Mendel was instructed to go to Vienna and to report himself at Baumgatner's office in the Ministry of Finance." Some authors have interpreted this sentence to mean that Mendel was in Vienna on August 1. More likely, however, "August 1" was the date on which the Ministry of Education mailed a letter, instructing Mendel to be in Vienna on August 12

[78]We know the date of the *Klausurarbeit* in natural history with certainty because it appears on the document with Mendel's answers. We know the date of Mendel's *Klausurarbeit* in physics from the date on Baugartner's evaluation—assuming that the evaluation was written on the same day as the examination. The originals of both these documents are deposited in Brno's Mendelianum and their printed versions have been published

[79]To view a copy of the original document, see reference[54]. Even on the poor-quality reproduction, the crossed out text is still legible. The original document is in the Urbana-Champaign museum.

[80]Kner R (1849) Lehrbuch der Zoologie zum Gebrauch für höhere Lehranstalten. Wien

[81]In the English version of Iltis' biography[43], *Pfotentier*, *Flatterfüsler*, *Krallenfüsler*, and *Huftiere* are erroneously translated as quadrupeds, plantigrades, clawed ungulates, and hoofed ungulates, respectively

[82]Our summary of Kner's criticism is based on Iltis[43], who quotes extensively the original document, which is apparently no longer available. A summary of Kner's evaluation appears in another document, however (see[73])

[83]In the middle of the nineteenth century, quills as writing utensils began to be replaced by metal pens. Since, however, the Austrian Monarchy was not exactly famous for welcoming progress and innovation, we assume that at the University of Vienna Mendel used quills to write his examination papers

[84]The original of Baumgartner's evaluation of Mendel's *Klausurarbeit* in physics is deposited in Brno's Mendelianum and in printed version published in reference[66]

[85]It is not clear which physics book Mendel studied for the *Klausur* and oral examinations. Since in his essay part of the examination he gave the Baumgartner-Ettingshausen 1842 textbook as the source, he should have been acquainted with the mathematical approach. If, on the other hand, Mendel relied mainly on the Hofer book,[69c] intended for laypeople and so mostly devoid of mathematics, he came to Vienna unprepared for Baumgartner's questions

[86]Gicklhorn[54,66] published the facsimile and the printed version of the protocol. The original document is deposited in the Urbana-Champaign museum

[87]Both letters are deposited at Brno's State Archive and their printed version appears in Sajner[40a]

[88]The whereabouts of this report are unknown. We know about its existence only because Iltis[43] saw it and quoted from it

[89]Kner's report is deposited at Brno's Mendelianum. Gicklhorn (see[66]) published its printed version

90 Napp's letter to Baumgarten thanking him for his opinion of Mendel

91 Hellmer K (1899) Geschichte der deutschen Technischen Hochschule in Brünn. In: Festschrift der k.k. Technischen Hochschule in Brünn zur Feier ihres fünfzigjährigen Bestehens und der Vollendung des Erweiterungsbaues in October 1899. Brünn

92 The protocol signed by Josef Auspitz was found in the institute's archive. Richter (see[50], pp. 64–68) quotes from it extensively

93 Also spelled Kolenati, but the correct spelling is Kolenatý, since it is a typical Czech name. The other members of the staff were Schindler (the director), Hrubý, Auspitz, Teirich, Kořistka, Marin, Ringhofer, Quaddrat, Regner, von Bleileben, and Helcelet

94 Weiling F (1986) Das Wiener Universitätsstudium 1851–1853 des Entdeckers der Verebungsregeln Johannn Gregor Mendel. FM 21:9–40

95 (a) May AJ (1968) Vienna in the Age of Franz Josef. University of Oklahoma Press, Norman, OK (b) Hürlimann M (1970) Vienna. The Viking Press, New York, NY. (c) Johnston WM (1980) Vienna Vienna: The Golden Age 1815–1914. Clarckson N. Potter, New York, NY. (d) Lehne I, Johnson L (1995) Vienna. The Past and the Presemt, 2nd edn. Ariadne Press Riverside, CA. (e) Brook S (1994) Vienna (Eyewitness Travel Guides). Dorling Kindersley, London

96 Květ R (2002) Staré stezky v České Republice, 2nd edn. Moravské zemské muzeum, Brno

97 Letter of March 24, 1853. Mendel wrote it on Maundy Thursday in Brno, where he was spending the Easter holidays. For the printed version of the letter, see[165]; the original letter is deposited at the Mendelianum in Brno. Curiously, Mendel misspelled the assassin's name making it look as if it were Czech. In reality, the attacker was a deranged Hungarian by the name János Libenyi

98 It may seem odd that Franz Joseph would allow the construction of a parliament building provocatively alluding to democratic principles. In the 1870s, however, the developments at home and abroad forced him to relax the reins over his empire. The constitution of a quasi-democratic parliament was one of the signs of this relaxation

99 Kink R (1854) Geschichte der kaiserlichen Universität zu Wien, 2 Vols. Carl Gerold, Wien

100 *Consistory* (from Latin *consistere*, to stand together) is a term that has undergone evolution in its meaning. In ancient Rome, *consistorium* referred to the place at which the emperor's council met. Later, it was applied to the council itself. The word was then taken over by the Christian Church as a designation for the assembly of the clergy, especially that of the pope in Rome. Scandinavian, German, and Austrian universities adopted the term in the sense of a "university board", that is, an assembly administering the institution

101 Gliboff S (1998) Evolution, revolution, and reform in Vienna: Franz Unger's ideas on descent and their post-1848 reception. J Hist Biol 31(2):179–209

102 Roth E (1971) A tale of three cities. Cassell, London, pp 41–43

103 Orel V, Marvanová L, Sajner J (eds) (1965) Iconographia Mendeliana pictorial publication issued for the Centenary of the publishing of the discovery of the principles of heredity by Gregor Johann Mendel in Brno. Moravian Museum in Brno, Brno

104 The oldest matriculation book of the University of Vienna is from 1377. Originally, the books were intended to register all members of the university for a legal purpose—as a certification that the individuals listed in the book were under the jurisdiction of the *Rektor* and the *Consistory* court of the university. However, this special legal status was abolished in 1783 and from that time on the matriculation book served merely as a record-keeping document

105 This circumstance caused some vexation to Mendel's early biographers (see[43]). They found indications that Mendel studied at the University of Vienna, but they could not find his name in the matriculation book. The puzzle was resolved when the existence of the extraordinary form of matriculation came into light (see[54]). Unfortunately, the loose sheet with Mendel's registration could, by then, no longer be found

106 Ulrich J (1908) Gregor Joh. Mendel. Biografische Skizze. Rainer Hosch, Neutitschein 1907. Reprint from the Illustrierter Neutitscheiner Volkskalender für das Jahr

107 Mendel's letter to the *Consistorium*, which he brought with him, when he registered at the University of Vienna

[108]Nivet C (2008) Was Gregor Mendel a clandestine student at Vienna University in 1851? 3rd International Conference of the European Society of the History of Science. Book of Abstracts, pp. 134–135

[109]Matalová A (2011) Was Gregor Mendel *élève protégé* in Doppler's Institute of Phyics at Viennese University? FM 47(1):5–12

[110]Reproduced in[103], the original is deposited at Brno's Mendelianum

[111](a) Schuster PM (2005) Moving the Stars. Christian Doppler, his Life, His Works and Principle, and the World After. Translated from German by Lily Wilmes. Living Edition Publishers, Pöllauberg, Austria. (b) Woodruff AE, in[169], vol. 4, pp. 167–168

[112]Not Linz, as some Mendel's biographers[170] claim

[113]In case you have forgotten, Doppler's effect refers to the apparent variation in frequency of sound, light, or other waves depending on the relative motion of the wave source to the wave recipient, or vice versa. Stated simply, when, for example, a sound of a constant pitch, such as the whistle of a locomotive, moves toward a person standing on a station platform, it seems to have a higher pitch than that of the train moving away. The illusion arises because the sound waves coming from the approaching train are more crowded (have a higher frequency and hence pitch) relative to the person on the platform, whereas those of the moving-away train are spread farther apart than the actual waves emitted by the whistle

[114]Dick A (1986) Gründung und Ausstattung des k. k. physikalischen Institutes der Universität Wien FM: 21: 41–48. The address was Hauptstrasse 104. The building no longer stands; during WWII a bomb hit it and damaged it considerably. Later, during the construction of the underground railway, the house was demolished

[115]Weilig F (1994) Johann Gregor Mendel. Der Mensch und Forscher. Forscher in der Kontroverse. Medizinische Genetik 5:35–49, 208–222, 274–288, 379–393, 1993; 6:35–50, 241–252

[116]The word derives from the French *elév*, meaning pupil, student ward, or beginner. Some authors[43,50,168] claim that Doppler's *Eléven* had a special standing among his students in that they were his assistants or demonstrators and that Mendel was one of them. Others (see[94]) argue that the designation *Eléven* applied to all of Doppler's student and that Mendel was *not* Doppler's assistant

[117]Die Statuten des physikalischen Institutes in Wien. *Zeitschrift für die österreichischen Gymnasien* 1: 458–466, Carl Gerold, Wien 1850

[118]George W (1982) Gregor Mendel and Andreas von Ettingshausen. FM 17:213–216

[119]Count, and later Prince, Andrey Kyrillovich Razumovsky (1752–1836) served as Russian Ambassador in Vienna from 1792 to 1809. As the representative of his country, he commissioned, at his own expense, the construction of a magnificent new embassy, which became known as the Palais Razumovsky or Razumovsky Palace. Built in 1806/1807, the palace became a dignified abode for the prince's large collection of ancient and modern art. Its great ballroom became the place of the premiere performance of Beethoven's Fifth Symphony. Beethoven was a frequent guest at the palace, and he expressed his gratitude for the prince's support by dedicating to him a set of string quartets, which to this day are called Razumovsky Quartets. On New Year's Eve 1814, the prince held a ball at which the guest of honor was the new Tsar of Russia, Nicholas I. To accommodate all the guests, Razumovsky extended the ballroom into a temporary structure, which he heated by a flue from the palace. After all the guests had left, fire broke out in the extension and spread into the palace and the rooms with the art treasures. The prince never recovered from this blow. After his death, his wife sold the palace and the new owners later rented its rooms to the city

[120]In Italian, *Belvedere* means a "beautiful view." The word is commonly used for buildings, often with gardens offering beautiful vistas

[121](a) Reichardt RW (1862) Gallerie österreicher Botaniker. V. Eduard Fenzl. Oesterreichische botanische Zeitschrift 12(1): 1–11 (b) Kotschy T (1867) Eine Lebensskizze von Dr, Eduard Fenzl. Wien

[122]The word *Kustos* derives from the Latin *custos*, a "guardian," and in the present context denotes a scientist in charge of a museum, that is, a *curator*. As *Naturalien* were designated "natural objects," here those assembled for the purpose of teaching, education, and study. One meaning of the word *Kabinett* was the part of an aristocratic palace, in which a prince conducted transactions concerning the state (hence titles such as *Kabenettminister* or *Kabinettsekretär*). It, however, also denoted a room or a set of rooms, in which were kept especially valuable objects of art or science, particularly those not displayed in general exposition halls. The *Hof-Naturalien-Kabinett* then developed into the natural history museum. The botanical division of the museum was located in a building on the premises of the botanical garden. (*Hof,* in the present context, meant the imperial court.)

[123](a) Sachs J (1875) Geschichte der Botanik vom 16. Jahrhundert bis 1860. B. Oldenbourg, München. (b) Möbius M (1937) Geschichte der Botanik von den ersten Anfängniss bis zur Gegenwart. Gustav Fischer, Jena

[124](a) Leitgeb H (1870) Franz Unger. Leuschner & Lubensky, Graz. (b) Reyer A (1871) Leben und Wirken des Naturhistorikers Dr. Franz Unger. Leuscher & Lubensky, Graz. (c) Wiesner J (1902) Franz Unger. K. k. zoologisch-botanische Gesellschaft, Wien. (d) Olby R. Unger, Franz. In[169], vol. 13, pp. 542–543

[125]Olby R in[169] (see[124d]) states that Unger was born in "Der Gute Amthof" as if this was the name of Unger's birthplace. In reality, however, he was born on the estate (Gut, in German) called *Amthof*

[126](a) Unger F (1846) Grundzüge der Anatomie und Physiologie der Pflanzen. Carl Gerold, Wien. (b) Endlicher S, Unger F (1843) Grundzüge der Botanik. Carl Gerold, Wien

[127]The Johanneum of Joanneum was an interesting educational experiment, in which teaching and research were carried out at a museum, both natural and technical. It derived its name from Archduke Johann von Habsburg (1782–1859), who inaugurated the project in 1811 in Graz, Styria. Initially, it covered natural sciences, technology, as well as humanities, but with time it underwent a variety of modifications. In recent time, the project disembogued in the separation of the museum part (the current Universal Museum Joanneum) and the teaching and research part (the current University of Graz)

[128]Unger F (1851) Die Urwelt in ihren verschiedenen Bildungsperioden. Fr. Beck, Wien. The first edition is now a collectors' item, but both the lithographs (reduced in size) and the text (translated into English) have been reissued recently in Rudwick, M. J. S. Scenes From Deep Time. Early Pictorial Representation of the Prehistoric World. The University of Chicago Press, Chicago, IL 1992

[129]The holy Bible containing the old and the new testament. Authorized King James Version. Camex International, New York, NY 1989

[130]Olby RC. Franz Unger and *Wiener Kirchezeitung*: an attack on one of Mendel's teachers by the editor of a Catholic newspaper. *FM*: This article, however, is full of errors, including wrong attributions, mixed up dates, and mixing up Unger with Fenzl regarding their deanship

[131]The title presumably alludes to the worship of *Isis* in ancient Rome as the goddess of nature and so to Unger's alleged paganism. *Philister* is used here in the sense of someone with a materialistic outlook (see[169])

[132]Wunderlich R (1982) Der wissenschaftliche Streit über die Enstehung des Embryos der Blütenpflanzen im zweiten Viertel des 19. Jahrhunderts (bis 1856) und Mendels "Versuche über Pflanzen-Hybriden". FM 17:225–242

[133]Gärtner KFV (1849) Versuche und Beobachtungen über die Bastarderzeugung im Pflanzenreich, mit Hinweisung auf die ähnlichen Erscheinungen im Thierreiche. Suttgart

[134]After Mendel's death his copy of Gärtner's book was placed in the Library of the St. Thomas Abbey

[135]Fitzinger L (1861) Vincenz Kollar. Almanach der kaiserlichen Akademie der Wissenschaften 11: 154–169. Wien. The 1799 birthdate in[20] is in error

[136]Kohn M (1947) Joseph Redtenbacher. J Chem Educ 24(8):366–368

[137]Orel V, Kuptsov VI (1983) Preconditions for Mendel's discovery in the body of knowledge in the middle of the nineteenth century. In: Orel V, Matalová A (eds) Gregor Mendel and the Foundation of Genetics. Moravian Museum, Brno, pp 189–227

[138]Strumf FL. *Die Fortschritte der Chemie in ihrer Anwendung auf Gewerbe, Künste und Pharmacie*. T. C. F. Enslin, Berlin 1853. The title page of this book bearing Mendel's signature is deposited at the University of Chicago Library. How it got there is a mystery. Either it was a part of the collection of the Mendelian documents Hugo Iltis took with him to the United States (see[64]) or a visitor to the Saint Thomas Abbey received (stole?) it and took it to the USA[20]. The whereabouts of the multivolume book are unknown

[139]Sturtevant AH (1967) Mendel and the gene theory. In[172], pp. 11–15

[140]Thus, for example, the removal of a hydrogen atom from methane, CH_4, gives the methyl radical—CH_3. Free radicals are highly reactive because they contain unpaired electrons. Hence, the interaction of two methyl radicals gives rise to ethane CH_3CH_3. In Mendel's time, when nothing was known about the existence of electrons and about the nature of bonds resulting from the sharing of electrons, all fragmented molecules were called radicals. In modern usage a radical is any atom or a group of atoms with an unpaired electron

[141]Monaghan FV, Corcos AF (1983) Possible influences of some nineteenth century chemical concepts on Mendel's ideas about heredity. J Heredity 74:297–299

[142]Grunert P (2006) Leben und Werk von Lukas Friedrich Zekeli (1823–1881): Mehr als eine Fussnote in der Geschichte der Erdwisseschaften in Österreich? Berichte der Geologischen Bundesanstalt Wien 69:24–26

[143]Stern C, Sherwood ER (eds) (1966) The origin of genetics. A Mendel source book. W.H. Freeman, San Francisco, CA

[144]The first person to use the denotation "scientist" was apparently William Whewell in 1840. See Ross S (1962) Scientist: The story of the word. Ann Sci 18: 65–85

[145]Mendel G (1853) Über Verwüstung am Gartenrettig durch Raupen. *Verhandlungen des zoologisch-botanischen Vereins in Wien* 3:116–118

[146]Mendel G (1853) Beschreibung des sog. Erbsenkäfers, *Bruchus pisi*. Mitgeteilt von V. Kollar. *Verhandlungen des zoologisch-botanischen Vereines in Wien* 4:27–30

[147]Hall P (1998) Cities in civilization. Culture, innovation, and urban order (Chapter 5: The city as pleasure principle: Vienna 1880–1910). Weidenfeld & Nicolson, London

[148]The chocolate cake was named after its inventor, Franz Sacher, Prince Meternich's pastry chef. It was originally served exclusively at the expensive Hotel Sacher, but by the time Mendel arrived in Vienna, imitations of the cake could be had at different *Konditoreien*

[149]The letter mentioned earlier, is dated "Wien14/7," presumably 1852. Its photocopy is deposited at Brno's Mendelianum. Its printed version is accessible in[165]

[150]The spiritual exercises (*Exircitien*) do not seem to have been connected with the Summer Feast days, because these, in 1852, all took place in early June: Trinity Sunday on June 6; Corpus Christi on Thursday June 10; Sacred Heart of Jesus on Friday, June 18; and Immaculate Heart of Mary on Saturday, June 19. Rather, they might have been meant to be like maneuvers intended to pacify the bishop

[151]A quote from the Apocalypse, the Book of Revelation 7: 9–12 (see[129]), which in the context reads: *O quam gloriosum est regnum, in quo cum Christo gaudet omnes Sancti. Amicti stolis albis, sequuntur Agnum quocumque ierit.* In translation: "O how glorious is the kingdom, where all the Saints rejoice with Christ. Dressed in white robes, they follow the Lamb wheresoever he goes." It pertains to the All Saints Day and the All Souls Day, celebrated on November 1 and November 2, respectively

[152]The translation of the word *hujus* is uncertain. Commonly suggested English equivalent is "this." In the context of the quoted sentence, it presumably means "in the last week of this month (i.e., July)."

[153]Josefa Smekal (1804–1873) was the wife of the sacristan, the man in charge of the sacristy and of the ceremonial equipment, sacristy being the room in the church where sacred vessels and

vestments are kept and where the clergy vests (clothes with a garment) are kept. She was the housekeeper of the abbey

154 This information is based on a letter of Alois Schindler to Hugo Iltis, dated December 22, 1922 (letter No.15 in[166]). There is no historical support for the claims that Mendel conducted the ceremony or that he assisted Kahlig at the Mass (see[173]). He did, however, join in the holy matrimony his nephew Alois Sturm in Dolní Vražné as well as another nephew, Anna Sturm, in the church at Staré Brno (they were the son and the daughter of Mendel's elder sister Veronika). He also promised Alois Schindler to sanctify his conjugal bond, but at the time of Alois' marriage in 1888, Mendel already rested for four years at the Central Cemetery of Staré Brno

155 Kříženecký J (1963) Mendels zweite erfolglose Lehramtsprüfung im Jahre 1856. Südhoffs Archiv Zeitsschrift für Wissenschaftsgeschichte 47:304–310

156 Hagemann R (2008) Mendels starke persönliche Motivation für seine Vererbungsversuche. BIOspectrum 14:770–772

157 Excerpts from the accounting book have been published by Richter; see[50], Table 1 on p. 26.

158 Iltis[43], p.60, states: "In the protocol of the Examining Committee for *Realschule* teachers it is written that Mendel applied for admission to sit the examination and that this took place on May 5, 1856. A later note states that the (written) papers were *skartiert* (destroyed)." It is, however, not clear what Iltis actually saw: Was it really the actual protocol (minutes) of the examination or merely someone's notes about the examination? If the former, why didn't he copy it? After all for a biographer it was a very important document!

159 Sajner J (1976) Johann Gregor Mendel. Leben und Werk. Ein Bildbuch. Augustinus Verlag, Würzburg 1974 (2nd edn)

160 Klácel's letter to Bratránek dated May 8, 1856

161 Much of what we now know about Mendel, we owe to his first major biographer, Hugo Iltis (1882–1952). Although his life overlapped with Mendel's by 2 years only, Iltis was still able to interview some of the persons who knew the founder of genetics well. To one of them, he refers in his book[43] as "Inspector Novotny," without ever giving his first name or any other particulars about him. Now, "Nowotny" is one of the commonest Czech surnames, spelled as "Novotný," and this fact complicates a search for this man's identity. The particular person was apparently Adolph Nowotny, whom Mendel knew from Brno's *Naturforschender Verein*, of which both men were members

162 The name is variously translated as "modern school," "modern technical school," or "technical secondary school," but since there is no exact English equivalent to the school, we leave it untranslated. *Obere* means "higher" and *Schule* means "school" but *real* does not mean the same thing in German and in English. In the present context, German *real* refers to something factual, and *Realien* are school subjects such as geography, physics, and chemistry, which are all based on facts, in contrast to, for example, classical languages or ideas, which were subjects taught, until then, at a *Gymnasium*

163 (a) Frey T (1869) Nekrolog auf Zawadski. Verhandlungen des naturforschenden Vereins in Brünn, 7; 22–25 (b) Orel V (1972) Professor Alexander Zawadski (1798–1868) -Mendel's superior at the Technical Modern School in Brno. FM 7; 13–20 (c) Szybalski W (2010) Professor Alexander Zawadski of Lvov University -Gregor Mendel's mentor and inspirer. Biopolym Cell 26(2): 83–86

164 Makowsky A (1863) Die Flora des Brünner Kreises. Verhandlubgen des Naturforschenden Vereines, Abhandlungen 1:43–210

165 Kříženecký J (1965) Gregor Johann Mendel 1822–1884. Texte und Quellen zu seinem Wirken und Leben. J.Ambrosius Barth, Leipzig

166 Simunek M. Hossfeld, Thümler U, Sekerák J (eds.) The Letters on G J Mendel. Correspondence of William Bateson, Hugo Iltis, and Erich von Tschermak-Seysenegg with Alois and Ferdinand Schindler. Studies in the History of Sciences and Humanities vol. 28. Prague 2011

167 Neumann AA (1930) Acta et epistolae eruditorum monasterii ord. S. Augustini Vetero-Brunae. Vol. 1 (A) 1819–1850. Sumptibus monasterii Vet Brunensis, Brno

[168] Ryan PE. Gregory Mendel. Abbot and discoverer of the laws of heredity I –XXXI. *The Messenger*, March 1938–November 1941. This series of articles was published anonymously in the journal of the Roman Catholic Diocese of Convigton, Kentucky, on the request of Bishop Howard. The bishop sent R.L. de Waegenaere to Brno to collect material for the article and then instructed Father Ryan, his secretary, to write the articles. (This information comes from a letter Professor Edward O. Dodson, University of Ottawa, Canada, wrote to Dr. Vítězslav Orel, Mendelianum, Brno. The letter is deposited at the Mendelianum.)

[169] Gillispie CC (ed) (1981) Dictionary of scientific biography, vol 8. Charles Scribner's Sons, New York, NY

[170] Olby RC (1985) Origin of Mendelism. Constable, London 1966. 2nd edn. University of Chicago Press

[171] Evans IH (1981) Brewer's dictionary of phrase and fable, Centenaryth edn. Harper & Row Publishers, New York, NY

[172] Brink RA (ed) (1967) Heritage from Mendel. In: Proceedings of the Mendel Centennial Symposium Sponsored by the Genetics Society of America 1965. The University of Wisconsin Press, Madison, WI

[173] Heinenen W (1944) Der junge Genius Johann Gregor Mendel. Ein biographisher Roman, 2nd edn. Fels Verlag Dr Wilhelm Spael KG, Essen

Index

J. Klein and N. Klein, *Solitude of a Humble Genius - Gregor Johann Mendel: Volume 1*,
DOI 10.1007/978-3-642-35254-6, © Springer-Verlag Berlin Heidelberg 2013

MIX
Papier aus verantwortungsvollen Quellen
Paper from responsible sources
FSC® C105338

If you have any concerns about our products,
you can contact us on
ProductSafety@springernature.com

In case Publisher is established outside the EU,
the EU authorized representative is:
Springer Nature Customer Service Center GmbH
Europaplatz 3, 69115 Heidelberg, Germany

Printed by Libri Plureos GmbH
in Hamburg, Germany